T0225898

SPACECRAFT TRAJECTORY OPTIMIZATION

This is a long-overdue volume dedicated to space trajectory optimization. Interest in the subject has grown, as space missions of increasing levels of sophistication, complexity, and scientific return – hardly imaginable in the 1960s – have been designed and flown. Although the basic tools of optimization theory remain an accepted canon, there has been a revolution in the manner in which they are applied and in the development of numerical optimization. This volume purposely includes a variety of both analytical and numerical approaches to trajectory optimization. The choice of authors has been guided by the editor's intention to assemble the most expert and active researchers in the various specialties presented.

Bruce A. Conway is a Professor of Aeronautical and Astronautical Engineering at the University of Illinois, Urbana-Champaign. He earned his Ph.D. in aeronautics and astronautics at Stanford University in 1981. Professor Conway's research interests include orbital mechanics, optimal control, and improved methods for the numerical solution of problems in optimization. He is the author of numerous refereed journal articles and (with John Prussing) the textbook *Orbital Mechanics*.

Cambridge Aerospace Series

Editors: Wei Shyy and Michael J. Rycroft

1. J.M. Rolfe and K.J. Staples (eds.): *Flight Simulation*
2. P. Berlin: *The Geostationary Applications Satellite*
3. M.J.T. Smith: *Aircraft Noise*
4. N.X. Vinh: *Flight Mechanics of High-Performance Aircraft*
5. W.A. Mair and D.L. Birdsall: *Aircraft Performance*
6. M.J. Abzug and E.E. Larrabee: *Airplane Stability and Control*
7. M.J. Sidi: *Spacecraft Dynamics and Control*
8. J.D. Anderson: *A History of Aerodynamics*
9. A.M. Cruise, J.A. Bowles, C.V. Goodall, and T.J. Patrick: *Principles of Space Instrument Design*
10. G.A. Khoury and J.D. Gillett (eds.): *Airship Technology*
11. J. Fielding: *Introduction to Aircraft Design*
12. J.G. Leishman: *Principles of Helicopter Aerodynamics, 2nd Edition*
13. J. Katz and A. Plotkin: *Low Speed Aerodynamics, 2nd Edition*
14. M.J. Abzug and E.E. Larrabee: *Airplane Stability and Control: A History of the Technologies that made Aviation Possible, 2nd Edition*
15. D.H. Hodges and G.A. Pierce: *Introduction to Structural Dynamics and Aeroelasticity*
16. W. Fehse: *Automatic Rendezvous and Docking of Spacecraft*
17. R.D. Flack: *Fundamentals of Jet Propulsion with Applications*
18. E.A. Baskharone: *Principles of Turbomachinery in Air-Breathing Engines*
19. D.D. Knight: *Numerical Methods for High-Speed Flows*
20. C. Wagner, T. Hüttl, P. Sagaut (eds.): *Large-Eddy Simulation for Acoustics*
21. D. Joseph, T. Funada, and J. Wang: *Potential Flows of Viscous and Viscoelastic Fluids*
22. W. Shyy, Y. Lian, H. Liu, J. Tang, D. Viieru: *Aerodynamics of Low Reynolds Number Flyers*
23. J.H. Saleh: *Analyses for Durability and System Design Lifetime*
24. B.K. Donaldson: *Analysis of Aircraft Structures, 2/e*
25. C. Segal: *The Scramjet Engine: Processes and Characteristics*
26. J. Doyle: *Guided Explorations of the Mechanics of Solids and Structures*
27. A. Kundu: *Aircraft Design*
28. M. Friswell, J. Penny, S. Garvey, A. Lees: *Fundamentals of Rotor Dynamics*
29. B.A. Conway (ed.): *Spacecraft Trajectory Optimization*
30. R.J. Adrian and J. Westerweel: *Particle Image Velocimetry*
31. S. Ching, Y. Eun, C. Gokcek, P.T. Kabamba, and S.M. Meerkov: *Quasilinear Control Theory: Performance Analysis and Design of Feedback Systems with Nonlinear Actuators and Sensors*

Spacecraft Trajectory Optimization

Edited by

Bruce A. Conway
University of Illinois at Urbana-Champaign

CAMBRIDGE
UNIVERSITY PRESS

CAMBRIDGE
UNIVERSITY PRESS

32 Avenue of the Americas, New York NY 10013-2473, USA

Cambridge University Press is part of the University of Cambridge.

It furthers the University's mission by disseminating knowledge in the pursuit of education, learning and research at the highest international levels of excellence.

www.cambridge.org
Information on this title: www.cambridge.org/9781107653825

© Cambridge University Press 2010

First published 2010
First paperback edition 2014

A catalogue record for this publication is available from the British Library

Library of Congress Cataloguing in Publication data
Spacecraft trajectory optimization / edited by Bruce A. Conway.
    p.   cm. – (Cambridge aerospace series ; 29)
Includes bibliographical references and index.
ISBN 978-0-521-51850-5
1. Trajectory optimization–Mathematical models.   2. Linear programming.
I. Conway, Bruce A.   II. Title.   III. Series.
TL1075.S65 2010
629.4'11–dc22   2010025018

ISBN 978-0-521-51850-5 Hardback
ISBN 978-1-107-65382-5 Paperback

Contents

Preface   *page* xi

1   **The Problem of Spacecraft Trajectory Optimization**   1
     Bruce A. Conway

     1.1   Introduction   1
     1.2   Solution Methods   3
     1.3   The Situation Today with Regard to Solving Optimal
            Control Problems   12
     References   13

2   **Primer Vector Theory and Applications**   16
     John E. Prussing

     2.1   Introduction   16
     2.2   First-Order Necessary Conditions   17
     2.3   Solution to the Primer Vector Equation   23
     2.4   Application of Primer Vector Theory to an Optimal
            Impulsive Trajectory   24
     References   36

3   **Spacecraft Trajectory Optimization Using Direct Transcription
     and Nonlinear Programming**   37
     Bruce A. Conway and Stephen W. Paris

     3.1   Introduction   37
     3.2   Transcription Methods   40
     3.3   Selection of Coordinates   52
     3.4   Modeling Propulsion Systems   60
     3.5   Generating an Initial Guess   62
     3.6   Computational Considerations   65
     3.7   Verifying Optimality   71
     References   76

4   **Elements of a Software System for Spacecraft Trajectory**
     Optimization   79
     Cesar Ocampo

     4.1   Introduction   79
     4.2   Trajectory Model   80
     4.3   Equations of Motion   85
     4.4   Finite Burn Control Models   85
     4.5   Solution Methods   90
     4.6   Trajectory Design and Optimization Examples   93
     4.7   Concluding Remarks   110
     References   110

5   **Low-Thrust Trajectory Optimization Using Orbital Averaging**
     and Control Parameterization   112
     Craig A. Kluever

     5.1   Introduction and Background   112
     5.2   Low-Thrust Trajectory Optimization   113
     5.3   Numerical Results   125
     5.4   Conclusions   136
     Nomenclature   136
     References   138

6   **Analytic Representations of Optimal Low-Thrust Transfer**
     in Circular Orbit   139
     Jean A. Kéchichian

     6.1   Introduction   139
     6.2   The Optimal Unconstrained Transfer   141
     6.3   The Optimal Transfer with Altitude Constraints   145
     6.4   The Split-Sequence Transfers   157
     References   177

7   **Global Optimization and Space Pruning for Spacecraft**
     Trajectory Design   178
     Dario Izzo

     7.1   Introduction   178
     7.2   Notation   179
     7.3   Problem Transcription   179
     7.4   The MGA Problem   181
     7.5   The MGA-1DSM Problem   183
     7.6   Benchmark Problems   186
     7.7   Global Optimization   190
     7.8   Space Pruning   194
     7.9   Concluding Remarks   197

Appendix 7A   198
Appendix 7B   199
References   200

8   **Incremental Techniques for Global Space
     Trajectory Design**   202
     Massimiliano Vasile and Matteo Ceriotti

     8.1   Introduction   202
     8.2   Modeling MGA Trajectories   203
     8.3   The Incremental Approach   209
     8.4   Testing Procedure and Performance Indicators   216
     8.5   Case Studies   221
     8.6   Conclusions   234
     References   235

9   **Optimal Low-Thrust Trajectories Using Stable Manifolds** . . .   238
     Christopher Martin and Bruce A. Conway

     9.1   Introduction   238
     9.2   System Dynamics   240
     9.3   Basics of Trajectory Optimization   247
     9.4   Generation of Periodic Orbit Constructed as an
           Optimization Problem   250
     9.5   Optimal Earth Orbit to Lunar Orbit Transfer:
           Part 1—GTO to Periodic Orbit   253
     9.6   Optimal Earth Orbit to Lunar Orbit Transfer:
           Part 2—Periodic Orbit to Low-Lunar Orbit   256
     9.7   Extension of the Work to Interplanetary Flight   259
     9.8   Conclusions   260
     References   261

10   **Swarming Theory Applied to Space Trajectory Optimization** . 263
     Mauro Pontani and Bruce A. Conway

     10.1 Introduction   263
     10.2 Description of the Method   266
     10.3 Lyapunov Periodic Orbits   269
     10.4 Lunar Periodic Orbits   274
     10.5 Optimal Four-Impulse Orbital Rendezvous   277
     10.6 Optimal Low-Thrust Orbital Transfers   284
     10.7 Concluding Remarks   290
     References   291

Index   295

Preface

It has been a very long time since the publication of any volume dedicated solely to space trajectory optimization. The last such work may be Jean-Pierre Marec's *Optimal Space Trajectories*. That book followed, after 16 years, Derek Lawden's pioneering *Optimal Trajectories for Space Navigation* of 1963. If either of these books can be found now, it is only at a specialized used-book seller, for "astronomical" prices.

In the intervening several decades, interest in the subject has only grown, with space missions of sophistication, complexity, and scientific return hardly possible to imagine in the 1960s having been designed and flown. While the basic tools of optimization theory – such things as the calculus of variations, Pontryagin's principle, Hamilton-Jacobi theory, or Bellman's principle, all of which are useful tools for the mission designer – have not changed in this time, there has been a revolution in the manner in which they are applied and in the development of numerical optimization. The scientists and engineers responsible have thus learned what they know about spacecraft trajectory optimization from their teachers or colleagues, with the assistance, primarily, of journal and conference articles, some of which are now "classics" in the field.

This volume is thus long overdue. Of course one book of ten chapters cannot hope to comprehensively describe this complex subject or summarize the advances of three decades. While it purposely includes a variety of both analytical and numerical approaches to trajectory optimization, it is bound to omit solution methods preferred by some researchers. It is also the case that a solution method espoused by one author and shown to be successful for his examples may prove completely unsatisfactory when applied by a reader to his own problems. Even very experienced practitioners of optimal control theory cannot be certain *a priori* of success with any particular method applied to any particular challenging problem.

The choice of authors has been guided by the editor's intention to assemble the most expert and active researchers in the various specialties presented. The authors were given considerable freedom to choose their subjects, and while this may yield a somewhat eclectic volume, it also yields chapters written with palpable enthusiasm that are relevant to contemporary problems.

Bruce Conway
Urbana, Illinois

1 The Problem of Spacecraft Trajectory Optimization

Bruce A. Conway

Dept. of Aerospace Engineering, University of Illinois at Urbana-Champaign, Urbana, IL

1.1 Introduction

The subject of spacecraft trajectory optimization has a long and interesting history. The problem can be simply stated as the determination of a trajectory for a spacecraft that satisfies specified initial and terminal conditions, that is conducting a required mission, while minimizing some quantity of importance. The most common objective is to minimize the propellant required or equivalently to maximize the fraction of the spacecraft that is not devoted to propellant. Of course, as is common in the optimization of continuous dynamical systems, it is usually necessary to provide some practical upper bound for the final time or the optimizer will trade time for propellant. There are also spacecraft trajectory problems where minimizing flight time is the important thing, or problems, for example those using continuous thrust, where minimizing flight time and minimizing propellant use are synonymous.

Except in very special (integrable) cases, which reduce naturally to parameter optimization problems, the problem is a continuous optimization problem of an especially complicated kind. The complications include the following: (1) the dynamical system is nonlinear; (2) many practical trajectories include discontinuities in the state variables, for example, there may be instantaneous velocity changes (also known as "ΔV's") from use of rocket motors or from planetary flyby (or "gravity assist") maneuvers, there may be instantaneous changes in spacecraft mass from staging or from using the rocket motor, or there may be sudden changes due to coordinate transformations necessary as the spacecraft moves from the gravitational sphere of influence of one body to that of another; (3) the terminal conditions, initial, final or both, may not be known explicitly, for example, for an interplanetary trajectory, the positions of the departure and arrival planets depend on the terminal times, which are often optimization variables; (4) there may be time-dependent forces, for example, the perturbations from other planets during an interplanetary trajectory can only be determined after the positions of the planets are determined using an ephemeris; (5) the basic structure of the optimal trajectory may not be *a priori* specified but is instead subject to optimization. For example, the optimal number of impulses or the optimal number of planetary flybys (or even the planets to use for the flybys) may

1

not be known. The VEEGA trajectory for Galileo [1] is an example; this was not the only feasible trajectory but was determined to be the optimal flyby sequence.

There are many types of spacecraft trajectories. Until 1998 (and the very successful Deep Space 1 mission), spacecraft were propelled only impulsively, using chemical rockets whose burn duration is so brief in comparison to the total flight time that it is reasonable to model it as instantaneous. Between impulses, the spacecraft motion, as a reasonable first approximation, can be considered Keplerian. Interplanetary cases add the possibility of planetary flyby maneuvers, which again, as a first approximation, may be modeled as nearly instantaneous velocity changes, preceded and followed by Keplerian motion. The impulsive transfer case, even including flybys, is thus a parameter optimization problem with the parameters being such quantities as the timing, magnitude, and direction of the impulsive ΔV's and the timing and altitude of gravity assist maneuvers. Of course, for extremely accurate spacecraft trajectory optimization, the resulting approximate trajectories must be reconsidered with the perturbations of other solar system bodies, the effect of solar radiation pressure, and other small but not insignificant effects included.

While the potential benefits of low-thrust electric propulsion have been known for many years, it has only been relatively recently that spacecraft missions have been flown using this technology, for example in the NEAR and Deep Space 1 missions. Electric propulsion produces very small thrust, so that typical spacecraft acceleration is on the order of 10^{-5} g, and thus thrust is used either continuously or nearly so. The continuous thrust optimal control problem is qualitatively different from the impulsive case as there are now no integrable arcs and the control itself, for example the thrust magnitude and direction, have continuous time histories that must be modeled and determined. If the electric power is provided by solar cells, the variation of power available with distance from the sun must also be taken into consideration. A qualitatively similar continuous thrust case is that of solar sail-powered spacecraft, which of course are also subject to variation in effectiveness as they move away from the sun.

While orbit transfer, for example LEO-GEO transfer, and interplanetary trajectories have been the focus of the bulk of research into spacecraft trajectory optimization, there are certainly many other applications of optimal control theory and numerical optimization to astrodynamics. Recent interesting problems include: (1) multi-vehicle navigation and maneuver optimization for cooperative vehicles, for example a fleet of small satellites in a specified formation [2]; (2) multi-vehicle noncooperative maneuver optimization, for example pursuit-evasion problems such as the interception of a maneuvering ICBM warhead by an intercepting spacecraft or missile [3]; (3) so-called "low-energy" transfer using invariant manifolds of the three-body problem, alone [4] or in combination with conventional or low-thrust propulsion [5] and; (4) trajectory optimization for a spacecraft sent to collide with a threatening Earth-approaching asteroid, with the objective of maximizing the subsequent miss distance of the asteroid at its closest approach to Earth [6] [7]. These are only a few of many examples that could be drawn from recent literature and from the programs of the principal conferences in the subject.

Necessary conditions for optimality for every one of these types of spacecraft trajectory optimization problems may be derived using the calculus of variations (COV). Unfortunately the solution of the resulting system of equations and boundary conditions is either difficult or impossible. For certain simplified but still very useful cases of either impulsive-thrust or continuous-thrust orbit transfer, the analytical necessary conditions may described using the "Primer Vector" theory of Lawden [8], as will be described briefly in Section 1.2.1 of this Chapter and then in much greater detail in Chapter 2. Analytical solutions for the optimal trajectory (i.e. solutions satisfying the necessary conditions) can be obtained in special cases, for example for very-low-thrust orbit raising [9], even in the presence of some perturbations [10]. However, the vast majority of researchers and analysts today use numerical optimization. Numerical optimization methods for continuous optimal control problems are generally divided into two types. *Indirect* solutions are those using the analytical necessary conditions from the calculus of variations [11]. This requires the addition of the costate variables (or adjoint variables or Lagrange multipliers) of the problem, equal in number to the state variables, and their governing equations. This instantly doubles the size of the dynamical system (which alone, of course, makes it more difficult to solve). *Direct* solutions, of which there are many types, transcribe the continuous optimal control problem into a parameter optimization problem [12] [13] [14]. Satisfaction of the system equations is accomplished by integrating them stepwise using either implicit or explicit (for example Runge-Kutta) rules; in either case, the effect is to generate nonlinear constraint equations that must be satisfied by the parameters, which are the discrete representations of the state and control time histories. The problem is thus converted into a nonlinear programming problem. There is a comprehensive survey paper by Betts [15] that describes direct and indirect optimization, the relation between these two approaches, and the development of these two approaches.

1.2 Solution Methods

In just the decade since the publication of Betts' survey paper [15], there has been considerable advancement of direct numerical solutions for optimal control problems. There also has been even more development and improvement, in relative terms, of a qualitatively different approach to solving such problems, one using evolutionary algorithms. The best known of these are genetic algorithms (GA) [16]. Another evolutionary algorithm, the Particle Swarm Optimizer (PSD) will be discussed in Chapter 10. The evolutionary algorithms have two principal advantages over other extant methods; they are comparatively simple and thus easy to learn to use, and they are generally more likely, in comparison to conventional optimizers, to locate global minima. In addition, there has been progress in analytical solutions such as those using *primer vector* theory [8] [17], "shape based" trajectories [18] [19], or Hamilton-Jacobi theory.

All of the solutions may be broadly categorized as being either *analytical* or *numerical*, though of course the analytical solutions (with only a few exceptions such as the Hohmann transfer) use numerical methods and the numerical solutions include some methods that explicitly use the analytical necessary conditions for optimality. In the following sections, the analytical and numerical solution methods will be defined and various examples, some historical and some very recent, will be presented for many of the methods that fall within these categories. This is not intended to be a survey and will be unapologetically incomplete, as the subject is a vast one with a large literature. Rather, the intention in this introductory chapter is to describe the problem of spacecraft trajectory optimization, categorize the solution approaches, provide a small amount of history, and describe the "state of the art" so that the work of the various book chapter authors describing their approaches to the problem will be in context.

1.2.1 Analytical Solutions

This is the original approach for space trajectory optimization, the oldest example of which (1925) is due to Hohmann's conjecture [20] regarding the optimal circular orbit to circular orbit transfer. (The proof of the optimality of the Hohmann transfer came much later [21] [22].) Most of the analytical solutions are based on the necessary conditions of the problem that come from the calculus of variations (COV). Suppose that the system equations may be written in form

$$\dot{x} = f(x,\ u,\ t) \tag{1.1}$$

where x represents an n-dimensional state (vector) and u represents the m-dimensional control (vector). The state vector is problem dependent; there are many choices available. Typically, conventional elliptic elements, equinoctial variables, or Delaunay variables are used for problems that are Keplerian or nearly-Keplerian, for example, very low-thrust orbit raising. Another common choice is spherical polar coordinates. Cartesian coordinates are typically used for three-body problems. The control u is typically a control of thrust magnitude and direction or its equivalent, for example the orientation of a solar-sail spacecraft with respect to the Sun. The problem has some initial conditions specified, that is,

$$x_i(0) \text{ given for i} = 1, 2, ..., k \text{ with k } \leq \text{ n} \tag{1.2}$$

and some terminal conditions, or functions of the terminal conditions, specified as the vector

$$\Psi[x(T),\ T] = 0. \tag{1.3}$$

The objective may be written in the Bolza form as

$$J = \phi[x(T),\ T] + \int_0^T L[x,\ u,\ t]\,dt \tag{1.4}$$

where ϕ is a terminal cost function while the integral expresses a cost incurred during the entire trajectory.

The first step in deriving the conditions for an extremum of (1.4) subject to the system (1.1) and the boundary conditions (1.3) is to define a system Hamiltonian

$$H - L + \lambda^T f$$

Then, in terms of H and the other quantities introduced, the necessary conditions become [11]

$$\dot{\lambda} = -\left(\frac{\partial H}{\partial x}\right)^T \text{ with boundary condition } \lambda(T) = \left[\left(\frac{\partial \phi}{\partial x}\right) + v^T \left(\frac{\partial \Psi}{\partial x}\right)\right]^T_{t=T} \quad (1.5)$$

$$\frac{\partial H}{\partial u} = 0. \quad (1.6)$$

The system of equations (1.1)–(1.6) constitutes a two-point-boundary-value problem (TPBVP); some boundary conditions on the states are specified at the initial time and some boundary conditions on the states and adjoints are specified at the terminal time. In addition, if the terminal time is unspecified (that is free to be optimized), as is often the case, an additional scalar equation obtains

$$\left[\frac{\partial \phi}{\partial t} + v^T \left(\frac{\partial \Psi}{\partial t}\right) + \left(\frac{\partial \phi}{\partial x} + v^T \left(\frac{\partial \Psi}{\partial x}\right)\right) f + L\right]_{t=T} = 0. \quad (1.7)$$

For all but the most elementary optimal control problems, the solution of this TPBVP is challenging and numerical solutions are required. Despite this, it is interesting that when this set of necessary conditions is applied to the optimal space trajectory problem, which is by no means elementary, several very useful observations may be made.

The system equations of motion (1.1) may be written in the form

$$\dot{x} = \bar{f} = \begin{bmatrix} \dot{\bar{r}} \\ \dot{\bar{v}} \end{bmatrix} = \begin{bmatrix} \bar{v} \\ \bar{g}(\bar{r}) + \Gamma \hat{u} \end{bmatrix} \quad (1.8)$$

where g(r) is the gravitational acceleration, Γ is the thrust acceleration magnitude, and $\hat{u}$ is a unit vector indicating the thrust direction.

To minimize the velocity change required, one chooses the integrand in the cost function (1.4) to be L = Γ the acceleration provided by the motor; then the integral will represent the ΔV provided by the motor. The Hamiltonian then becomes

$$H = \Gamma + \bar{\lambda}_r^T \bar{v} + \bar{\lambda}_v^T \left[\bar{g}(\bar{r}) + \Gamma \hat{u}\right] = \Gamma \left[1 + \bar{\lambda}_v^T \hat{u}\right] + \bar{\lambda}_r^T \bar{v} + \bar{\lambda}_v^T \bar{g}(\bar{r}). \quad (1.9)$$

Because H is linear in u, equation (1.6) does not obtain. The optimal control is instead chosen according to Pontryagin's Minimum Principle, stating that at any time on the optimal trajectory, the control variables are chosen in order to minimize the Hamiltonian. Thus the first simple observation is that the thrust pointing unit vector is chosen to be parallel to the opposite of the adjoint (to the velocity) vector, i.e. $-\bar{\lambda}_v(t)$. Because of its physical significance to the problem, this (adjoint) vector is referred to as the *primer vector* [8]. A second simple observation is that with this choice of thrust direction, it is then optimal in this case to choose the thrust magnitude Γ at its maximum possible value if the "switching function"

$$\left[1 + \bar{\lambda}_v^T \hat{u}\right] \tag{1.10}$$

is negative and choose $\Gamma = 0$ if the switching function is positive. The adjoint vector $\bar{\lambda}_v(t)$ is governed by the system equations (1.5) with the Hamiltonian (1.9).

In addition, it is straightforward to show that if the Hamiltonian H is not explicitly time dependent, then H is a constant on the optimal trajectory. This result is not necessarily useful for *obtaining* the optimal control but can be of great use in determining, by its use *a posteriori*, the accuracy of the numerical solution of the TPBVP, that is, a good solution will have H the same, to several significant figures, when evaluated at any point on the numerical solution [14] [23].

Finally, while the necessary conditions guarantee only that the trajectory represents an extremum of the cost, by the nature of the space trajectory problem, there is clearly no upper bound to the fuel that could be consumed on a feasible trajectory (other than consuming all the fuel available). So one may be confident that a solution is a local minimum and not a local maximum.

Further results can be obtained from a description of the necessary conditions in terms of the primer vector, and these will be described in Chapter 2. It will suffice to say here that while the primer vector is defined, and has the significance with regard to optimal thrust direction found above, this is of course true only on the optimal trajectory. The improvement of a known, nonoptimal trajectory via primer vector theory was first discussed by Lion and Handelsman [17]. Jezewski and Rozendaal [24] showed under what conditions an optimal N impulse trajectory could be improved by the addition of another impulse, and where and with what direction to apply it.

Solution of the analytical necessary conditions is possible for some special cases. One useful example is the case of very-low-thrust orbit raising. With certain assumptions, it is possible to find approximate solutions of the analytical necessary conditions. Many of these are found in a survey paper of the subject by Petropoulos and Sims [25]. The most common simplifications include: assuming that the thrust direction is always tangential; assuming that the thrust pointing is always in the direction of the velocity vector; or assuming that the orbit is always circular. Surprisingly, exact solutions also exist in certain cases, including this low-thrust orbit raising, even in the presence of nonspherical Earth perturbations [10]. This will be discussed in Chapter 7. The mathematics and analysis become very involved.

The solution of the TPBVP resulting from (or constituting) the necessary conditions becomes quite difficult for other problems, particularly those with path constraints (typically on the state variables or on functions of the state variables) or constraints on total fuel available.

Many methods have been developed to solve the TPBVP numerically. The most obvious and well known is probably shooting (an archetype of shooting applied to spacecraft trajectory optimization may be found in the paper by Breakwell and Redding [26]) but there are other methods including finite-difference methods [27] [28] and collocation [12] [13] [14]. The long-recognized difficulty of the "indirect" approach to determining the optimal trajectory is that the initial costate variables of the TPBVP are unknown and further that the nonlinearity of the problem means that the vector flow is very sensitive to some or all of these initial costate variables. A further difficulty is that the costate variables lack the physical significance of the state variables so that estimating the order of magnitude or even the sign of the initial costates is very difficult. For problems with constrained arcs, another difficulty that arises is discontinuity of controls and costate variables at the junctions of constrained and unconstrained arcs. This also increases the difficulty of solving the associated TPBVP.

Another solution method that satisfies both the necessary and sufficient conditions for optimality is the method of Static/Dynamic control (SDC) of Whiffen [29] [30]. The term static refers to decision variables that are discrete, such as launch dates or planetary flyby dates, while the term dynamic refers to controls that have a continuous variation in time, such as thrust pointing angle time histories. SDC is a general nonlinear optimal control algorithm based on Bellman's principle of optimality [11]. The implementation of SDC in the program Mystic is a very capable low-thrust spacecraft trajectory optimizer.

A recent, qualitatively different approach to the determination of optimal space trajectories is that of Guibout and Scheeres [31]. In this work, the dynamical system of state and costate variables (the vector field) is solved for specified terminal conditions and final time by solving the associated Hamilton-Jacobi (H-J) equation. The solution of the H-J equation is a generating function for a canonical transformation. Once this solution is determined, the initial value of the costate vector may be found; the optimal trajectory and the optimal control may then be found by forward integration of the flow field. Scheeres et al. show an example of an optimal rendezvous in the vicinity of a nominal circular orbit [32].

1.2.2 Numerical Solutions via Discretization

Many recent methods for solving optimal control problems seek to reduce them to parameter optimization problems that can then be solved by a NLP problem solver. One principal way in which such methods are distinguished is with regard to what quantities are parameterized. In one popular method, the collocation method that will be discussed in Chapter 3, it is possible to parameterize the state variables and the costate variables (that is, to solve the TPBVP). It is also possible in collocation to parameterize only the state variables and the control variables, as will be discussed in the next section. A third possibility, yielding the smallest number of parameters for a given problem, is to parameterize only the control variables and some free terminal states, but then the system equations must be numerically integrated (as opposed to the implicit integration that occurs in collocation). This is referred to as "control parameterization" and will be discussed in Chapter 5.

Of course all of the solutions described in the previous section are obtained numerically, that is, they will employ methods such as numerical integration, solving TPBVP problems using "shooting" methods, or solving boundary value problems by converting them into nonlinear programming (NLP) problems. What is meant in this section by "numerical solution" is solutions that do not explicitly employ the analytical necessary conditions of the COV, for example, solutions that do not employ the costate (adjoint) variables of the problem or solutions that satisfy the H-J-B equation or Bellman's principle for discrete systems.

Why would one want to avoid the use of the necessary conditions, particularly when the resulting trajectory has a "guarantee" of being a local extremum (that one loses in a numerical solution) and has other benefits previously discussed, such as information about sensitivity to terminal conditions and guidance toward improving a solution by for example, adding/subtracting thrust arcs? The principal reason is the lack of robustness of the various methods for solving the Euler-Lagrange TPBVP stemming, as previously mentioned, from the nonlinearity of the problem and a lack, in the general case, of a systematic means for determining a sufficiently good approximation to the initial adjoint variables of the problem.

A variety of direct solution methods have been developed. They are best categorized by the way in which they handle the discretization of the equations of motion, which appear as function-space constraints in the original optimal control problem. A more complete survey will be presented in Chapter 3. In the last two decades, however, the most successful approach is arguably one in which the continuous problem is discretized and state and control variables are known only at discrete times. Satisfaction of the equations of motion is achieved by employing an explicit or implicit numerical integration rule that needs to be satisfied at each step; this results in a large NLP problem with a large number of nonlinear constraints. This approach was termed "direct transcription" by Canon et al. [33]. While known to mathematicians in the 1960s and 1970s, it became known in the aerospace community principally through two papers. Dickmanns and Well [34] used the collocation scheme to solve the TPBVP of the indirect method. This approach is significantly more robust than shooting methods because it eliminates the sequential nature of the shooting solution, with its forward numerical integration, in favor of a solution in which simultaneous changes in all of the discrete state and costate parameters are made in order to satisfy algebraic constraints (while minimizing the objective of course).

However, the most useful development for space trajectory optimization was the observation in 1987 by Hargraves and Paris [12] that it was not necessary to use this approach to solve the indirect TPBVP, that in fact the adjoint variables (which had been used to determine the optimal control from Pontryagin's principle) could be removed from the solution provided that discrete control variables were introduced as additional NLP parameters. This significantly improved the robustness of the method; by eliminating the adjoint variables, the problem size is reduced almost by half, and there is no longer a need to provide the NLP problem solver with an estimate of the adjoint variables, something that is always problematic. A fortunate coincidence is that at about the same time (1980s), the NLP technology required to efficiently and robustly solve large problems became available (and has been continuously improved since then) [35] [36]. The astrodynamics community swiftly embraced this method. Many optimal spacecraft trajectories have since been determined using direct methods. The direct method has also been significantly developed in the last two decades. There are now many approaches, differing primarily (for collocation methods) on how the implicit integration rules are constructed [37]. The most common approaches are to use trapezoidal [38] or Hermite-Simpson [12] integration rule constraints, or higher-degree rules from the same Gauss-Lobatto family [13] or a Gauss-pseudospectral method [39]. There also exist commercial software packages implementing direct methods for general optimal control problems, for example DIDO [40] and SOCS [38], and even solvers specifically for space and launch vehicle trajectory optimization, for example OTIS [41] and ALTOS [42] [43].

It would be accurate to say that the great majority of optimal space trajectories are now determined numerically, with methods that do not make explicit use of the analytical necessary conditions of the problem, as will be described briefly below and in detail in Chapter 2. However, that does not mean that the necessary conditions are no longer useful. On the contrary, they provide useful information that many numerical solutions naturally lack. For example, primer vector theory can provide important information on how a solution may be improved, for example by adding thrust arcs or coast arcs or by adding impulses for an impulsive trajectory. The solution of the TPBVP of the necessary conditions also provides information on the sensitivity of the solution to changes in terminal conditions and constraints.

Fortunately, without solving the TPBVP, it is possible to make use of some of these beneficial features of the solution of the necessary conditions, as will be described in Chapter 3. This occurs because of a correspondence between the final adjoint variables of the continuous TPBVP and some Kuhn-Tucker multipliers generated in (some) numerical solutions of the trajectory optimization problem [13] [14]. With these multiplier variables available, it is possible, for example, to compute the value of the system Hamiltonian over the entire trajectory time history. For many problems in which H should be a constant, this can provide a check on the accuracy of the numerical solution. Or, knowing the final adjoints and final states from, for example, a direct solution using collocation and NLP, one can integrate the E-L equations backward to the initial time. If the initial states are recovered, one can then say that the numerical solution satisfies the analytical necessary conditions and thus represents an extremal path.

1.2.3 Evolutionary Algorithms

A qualitatively different approach, recently applied to spacecraft trajectory optimization, is the use of "evolutionary" algorithms (EA). The best known of the EAs is the genetic algorithm (GA). EA's are numerical optimizers that determine, using methods similar to those found in nature, an optimal set of discrete parameters that has been used to characterize the problem solution. The EA's have two principal advantages over all of the direct and indirect solution methods previously described in this chapter: they require no initial "guess" of the solution (in fact they generate a population of initial solutions randomly), and they are more likely than other methods to locate a global minimum in the search space rather than be attracted to a local minimum.

All of the EAs require that the problem solution be capable of being described by a relatively small, in comparison to the vector of parameters of a nonlinear program, set of discrete parameters. This can be accomplished, for spacecraft trajectory optimization problems, in a number of ways:

(1) If the trajectory can naturally be described by a finite set, for example an impulsive thrust trajectory, the parameters will be such things as times, magnitudes, and directions of impulses. Between impulses the trajectories may be determined by solving Lambert's problem. In this case a small number of parameters will suffice to completely describe the solution.

(2) If the trajectory contains non-integrable arcs, for example low-thrust arcs, it is still the case that much of the trajectory can be described with a small number of parameters such as departure and arrival dates and times for the beginning and end of thrust arcs. Quantities that must be described continuously, such as thrust magnitude or pointing time history, can be parameterized using, for example, polynomial equations in time. Then the additional parameters are a small number of polynomial coefficients [44]

(3) Low-thrust arcs can also be described using "shape-based" methods [18] [19]. In this approach, a shape, which is an analytical expression for the trajectory, can be generated from a small number of parameters such that the resulting trajectory will actually be a solution of the system equations of motion. Unfortunately the thrust time history that allows this beneficial result can only be determined *a posteriori*. An EA is then used to choose the parameters defining the shape to satisfy the boundary conditions of the problem and to minimize the cost. The resulting trajectory may not be realizable, as it may require greater thrust than is available. However the trajectory may well be satisfactory as an initial guess for a more accurate method, for example a direct method such as collocation [12] [13] [14].

In the simplest form of the genetic algorithm, the set of parameters describing the solution is written as a string or sequence of numbers. Suppose that this sequence is converted to binary form; it is then similar to a chromosome but consisting only of two possible variables, a 1 or a 0. Every sequence can be "decoded" to yield a trajectory whose cost or objective value can be determined. The first step in the GA is the generation of a "population" of sequences using a random process. The great majority of these randomly generated sequences will have very large costs; many may even be infeasible. The population is then improved using three natural processes: *selection*, *combination*, and *mutation*. Selection removes the worst sequences and may also, via *elitism*, guarantee that the best sequence survives into the next generation unchanged. Following selection, remaining sequences are used as "parents," that is partial sequences from two parents are combined to form new individuals. Finally, mutation changes a randomly chosen bit in a small fraction of the population.

The process is then repeated; the cost of every individual in the new generation is determined. Since the best individual from the previous generation was retained, the objective may improve but cannot worsen. In practice, there is generally rapid improvement in the early generations; if the process locates the global minimum then, of course, improvement will cease. Termination of the algorithm is usually done either after a fixed number of generations or after the objective has reached a plateau. Of course neither of these termination conditions guarantees that a minimum has been found, nor are there necessary conditions for optimality with this method. Additional shortcomings are that there is no way to enforce satisfaction of boundary conditions; normally a "penalty function" approach is taken in which unsatisfied boundary conditions are added to the cost, and that the solution will be less accurate than a typical direct solution (and even less accurate than an indirect solution). Nevertheless, the method has been very useful when applied to optimizing space trajectories, either for finding approximate extremals [44] or when used to provide an initial guess for more accurate methods, for example collocation with NLP.

Betts [15] notes that one significant advantage of the GA in comparison to all other solution methods is how straightforward it is to use. There are many GA routines available (a commonly used one is found in MATLAB) so the user need only provide a subroutine for decoding the sequence to evaluate the cost (which for space trajectory problems can be as simple as a routine that integrates the system equations of motion) provide bounds on the parameters, and then provide values for certain constant parameters that control the evolutionary processes.

There are other EAs that have begun to prove very useful in the determination of optimal space trajectories. One qualitatively different method is particle swarm optimization (PSO). In PSO, some number (say 100) of particles are randomly distributed in a N-dimensional decision parameter space. The objective value is determined for the solution vector corresponding to each particle. Taking an anthropomorphic view, it is then assumed that the particles can communicate so that all know the objective value for all the others. Let $x_i(n)$ denote the position of particle i at the nth time step. At the next iteration, the particles take a step $v_i(n+1)$ in the parameter space so that the new position of particle i becomes

$$x_i(n+1) = x_i(n) + v_i(n+1) \tag{1.11}$$

with (in one form of the PSO)

$$v_{ij}(n+1) = v_{ij}(n) + c_1 r_{1j}(n) \left[y_{ij}(n) - x_{ij}(n) \right] + c_2 r_{2j}(n) \left[\hat{y}_j(n) - x_{ij}(n) \right] \tag{1.12}$$

where $v_{ij}(n)$ is the velocity (step) for component j of particle i at time step n, $x_{ij}(n)$ is the jth component of the position of particle i at the nth time step, $r_{1j}(n)$ and $r_{2j}(n) \subset U(0,1)$ are random values in the range [0, 1] sampled from a uniform distribution. $y_i(n)$ is the "personal best" position, the best position located by the ith particle since the first time step; $\hat{y}_j(n)$ is the "global best" position, the best position located by the any particle of the swarm since the first time step. The step described in equation (1.12) thus has three components. The first is an "inertia" that causes the particle to move in the direction it had previously been moving, the second "nostalgia" component reflects a tendency for the particle to move toward its own most satisfactory position, and the third "social" component draws the particle toward the best position found by any of its colleagues. The c's are constants that weight the importance of the three components and the r's provide stochasticity to the system.

As with the GA, the process can be terminated after a fixed number of iterations or when the "best" solution has not changed for several iterations. This method has proven quite robust, is also very simple to use, and is particularly good in locating global minima when the solution space contains many local minima. A more thorough description of the PSO method and its application to space trajectory problems will be provided in Chapter 10.

There are many other EAs, for example ant colony optimization (ACO) or differential evolution (DE). The interested reader can easily find information on the use of these methods [45].

1.3 The Situation Today with Regard to Solving Optimal Control Problems

One can safely say, for example by considering papers published recently in astrodynamics journals, that solutions using analytical methods, that is analytical solutions of the first-order necessary conditions, are seldom found. This is due, as previously mentioned, to the complexity of the problem when realistic terminal boundary conditions and when bounds on the controls are present. Also, solutions found numerically using indirect methods, for example with shooting methods, are also becoming less common. This is almost certainly due to the success that has been achieved with direct methods, particularly those using collocation via low-degree rules such as trapezoid or Hermite-Simpson [11] [13], via the pseudospectral method [39], or by higher degree G-L implicit integration [13]. (These collocation methods are all derivable from the same source, as will be seen in Chapter 3.) These methods have proven particularly robust and efficient and have been used to solve many types of problems including low-thrust orbit raising [46], Earth-Moon transfer [47] [5] [48], and interplanetary transfers [49].

An early difficulty faced by users of these methods was that, while robust, it was still necessary to supply a reasonable initial guess of the solution parameters, that is a discretized form of the state and control time-histories on the optimal trajectory, to the NLP problem solver. This, of course, is not always a simple matter. For some cases, for example for low-thrust orbit raising, approximate analytical solutions such as a Lawden spiral, as described in Section 1.2.1, are available and make a very satisfactory initial guess. For other problems, such as the optimal low-thrust Earth-Moon transfer, obtaining a satisfactory initial guess is much more difficult. Today, however, the situation is much improved since evolutionary algorithms such as the GA, which can provide a solution to the problem in their own right, can also be used as "pre-processors" to provide an initial guess of the solution from which a method such as direct collocation with NLP can converge to a much more accurate solution. An additional advantage of this approach is that some of the EAs are better suited to locating the global minimum than are the methods using NLP, as the NLP solver will tend to converge to a local minimum in the neighborhood of the initial guess it is given. Thus starting from a guess provided by an EA is more likely to enable the direct solver to find a global minimum. (Of course there is no guarantee in any case.)

John Betts' observation in 1988 [15] that "one may expect many of the best features of seemingly disparate techniques to merge, forming still more powerful methods" was clearly very prescient.

REFERENCES

[1] D'Amario, L. et al. (1989) Galileo 1989 VEEGA Trajectory Design, *Journal of the Astronautical Sciences*, **37**, 281–306.

[2] Wang, P. K. C., and Hadaegh, F. Y. (1998) *Optimal Formation-Reconfiguration for Multiple Spacecraft, AIAA-1998-4226, AIAA Guidance, Navigation, and Control Conference and Exhibit*, Boston, MA, Collection of Technical Papers. Pt. 2 (A98-37001 10–63)

[3] Pontani, M., and Conway, B. A. (2008) Optimal Interception of Evasive Missile Warheads: Numerical Solution of the Differential Game, *Journal of Guidance, Control and Dynamics*, **31**, 1111–1122.

[4] Marsden, J. E., and Ross, S. D. (2005) New Methods in Celestial Mechanics and Mission Design, *Bulletin of the American Mathematical Society*, **43**, 43–73.

[5] Mingotti, G., Topputo, F., and Bernelli-Zazzera, F. (2007) Combined Optimal Low-Thrust and Stable-Manifold Trajectories to the Earth-Moon Halo Orbits, *New Trends in Astrodynamics and Applications III*, E. Belbruno (ed.) 100–110.

[6] Englander, J., and Conway, B. A. (2009) *Optimal Strategies Found Using Genetic Algorithms for Deflecting Hazardous Near-Earth Objects*, IEEE Congress on Evolutionary Computation, Trondheim, Norway.

[7] Conway, B. A. (1997) Optimal Low-Thrust Interception of Earth-Crossing Asteroids, *J. of Guidance, Control, and Dynamics*, **20**, 995–1002.

[8] Lawden, D. F. (1963) *Optimal Trajectories for Space Navigation*, Butterworths, London.

[9] Kechichian, J. A. (1997) Reformulation of Edelbaum's Low-Thrust Transfer Problem Using Optimal Control Theory, *J. of Guidance, Control and Dynamics*, **20**, 988–994.

[10] Kechichian, J. A. (2000) Minimum-Time Constant Acceleration Orbit Transfer With First-Order Oblateness Effect, *Journal of Guidance, Control, and Dynamics*, **23**, 595–603.

[11] Bryson, A. E., and Ho, Y-C. (1975) *Applied Optimal Control*, Revised Printing, Hemisphere Publ., Washington, DC.

[12] Hargraves, C. R., and Paris, S. W. (1987) Direct Trajectory Optimization Using Nonlinear Programming and Collocation, *Journal of Guidance, Control, and Dynamics*, **10**, 338–342.

[13] Herman, A. L., and Conway, B. A. (1996) Direct Optimization Using Collocation Based on High-Order Gauss-Lobatto Quadrature Rules, *Journal of Guidance, Control, and Dynamics*, **19**, 592–599.

[14] Enright, P. J., and Conway, B. A. (1992) Discrete Approximations to Optimal Trajectories Using Direct Transcription and Nonlinear Programming, *Journal of Guidance, Control, and Dynamics*, **15**, 994–1002.

[15] Betts, J. T. (1998) Survey of Numerical Methods for Trajectory Optimization, *Journal of Guidance, Control and Dynamics*, **21**, 193–207.

[16] Goldberg, D. (1989) *Genetic Algorithms in Search, Optimization, and Machine Learning*, Addison-Wesley Publ., Reading, MA.

[17] Lion, P. M., and Handelsman, M. (1968) The Primer Vector on Fixed-Time Impulsive trajectories, *AIAA Journal*, **6**, 127–132.

[18] Petropoulos, A. E., and Longuski, J. M. (2004) Shape-Based Algorithm for Automated Design of Low-Thrust, Gravity-Assist Trajectories, *Journal of Spacecraft and Rockets*, **41**, 787–796.

[19] Wall, B. J., and Conway, B. A. (2009) Shape-Based Approach to Low-Thrust Trajectory Design, *Journal of Guidance, Control and Dynamics*, **32**, 95–101.

[20] Hohmann, W. (1925). *Die Erreichbarkeit der Himmelskörper*. Verlag Oldenbourg, München.

[21] Prussing, J. E. (1991) Simple Proof of the Global Optimality of the Hohmann Transfer, *Journal of Guidance, Control, and Dynamics*, **15**, 1037–1038.

[22] Barrar, R. B., (1963) An Analytic Proof that the Hohmann-Type Transfer is the True Minimum Two-Impulse Transfer, *Astronautica Acta*, **9**, 1–11.

[23] Herman, A. L. (1995) *Improved Collocation Methods Used for Direct Trajectory Optimization*, Ph.D. Thesis, University of Illinois at Urbana-Champaign.

[24] Jezewski, D. J., and Rozendaal, H. L. (1968) An Efficient Method for Calculating Optimal Free-Space N-Impulse Trajectories, *AIAA Journal*, **6**, 2160–2165.

[25] Petropoulos, A. E., and Sims, J. A. (2004) A Review of Some Exact Solutions to the Planar Equations of Motion of a Thrusting Spacecraft, *DSpace at JPL*, http://hdl.handle.net/2014/8673

[26] Breakwell, J. V., and Redding, D. C. (1984) Optimal Low-Thrust Transfers to Synchronous Orbit, *Journal of Guidance, Control, and Dynamics*, **7**, 148–155.

[27] Fox, L. (1957) *The Numerical Solution of Two-Point Boundary Problems*, Oxford Press, New York.

[28] Keller, H.B. (1968) *Numerical Methods for Two-Point Boundary Value Problems*, Blaisdell, New York.

[29] Whiffen, G. (2006) Mystic: Implementation of the Static Dynamic Optimal Control Algorithm for High-Fidelity, Low-Thrust Trajectory Design, http://trs-new.jpl.nasa.gov/dspace/bitstream/2014/40782/1/06-2356.pdf

[30] Whiffen, G. J., and Sims, J. A. (2002) *Application of the SDC Optimal Control Algorithm to Low-Thrust Escape and Capture Including Fourth Body Effects*, Second International Symposium on Low Thrust Trajectories, Toulouse, France.

[31] Guibout, V. M., and Scheeres, D. J. (2006) Solving Two-Point Boundary Value Problems Using Generating Functions: Theory and Applications to Astrodynamics, in *Modern Astrodynamics*, Gurfil, P. (ed.), Elsevier, Amsterdam, 53–105.

[32] Scheeres, D. J., Park, C., and Guibout, V. (2003) *Solving Optimal Control Problems with Generating Functions*, Paper AAS 03-575.

[33] Canon, M. D., Cullum, C. D., and Polak, E. (1970) *Theory of Optimal Control and Mathematical Programming*, McGraw-Hill, New York.

[34] Dickmanns, E. D., and Well, K. H. (1975) Approximate Solution of Optimal Control Problems Using Third-Order Hermite Polynomial Functions, *Proceedings of the 6th Technical Conference on Optimization Techniques*, IFIP-TC7, Springer–Verlag, New York.

[35] Gill, P., Murray, W., and Saunders, M. A. (2005) SNOPT: An SQP Algorithm for Large-Scale Constrained Optimization. *SIAM Review*, **47**, 99–131.

[36] Gill, P. E. et al. (1993) *User's Guide for NZOPT 1.0: A Fortran Package For Nonlinear Programming*, McDonnell Douglas Aerospace, Huntington Beach, CA.

[37] Paris, S. W., Riehl, J. P., and Sjauw, W. K. (2006) *Enhanced Procedures for Direct Trajectory Optimization Using Nonlinear Programming and Implicit Integration*, Paper AIAA-2006-6309, AIAA/AAS Astrodynamics Specialist Conference and Exhibit, Keystone, Colorado.

[38] Betts, J. T., and Huffman, W. P. (1997) *Sparse Optimal Control Software SOCS*, Mathematics and Engineering Analysis Technical Document, MEA-LR-085, Boeing Information and Support Services, The Boeing Company, Seattle.

[39] Ross, I. M., and Fahroo, F. (2003) Legendre Pseudospectral Approximations of Optimal Control Problems, *Lecture Notes in Control and Information Sciences*, **295**, Springer–Verlag, New York, 327–342.

[40] Ross, I. M. (2004) *User's Manual for DIDO: A MATLAB Application Package for Solving Optimal Control Problems*, TOMLAB Optimization, Sweden.

[41] Paris, S. (1992) *OTIS-Optimal Trajectories by Implicit Simulation, User's Manual*, The Boeing Company.

[42] Well, K. H., Markl, A., and Mehlem, K. (1997) *ALTOS-A Trajectory Analysis and Optimization Software for Launch and Reentry Vehicles*, International Astronautical Federation Paper IAF-97-V4.04.

[43] Wiegand, A., Well, K. H., Mehlem, K., Steinkopf, M., and Ortega, G. (1999) *ALTOS-ESA's Trajectory Optimization Tool Applied to Reentry Vehicle Trajectory Design*, International Astronautical Federation Paper IAF-99-A.6.09

[44] Wall, B. J., and Conway, B. A. (2005) Near-Optimal Low-Thrust Earth-Mars Trajectories Found Via a Genetic Algorithm," *Journal of Guidance, Control, and Dynamics*, **28**, 1027–1031.

[45] Engelbrecht, A. P. (2007) *Computational Intelligence*, 2nd ed., John Wiley & Sons, West Sussex, England.

[46] Scheel, W. A., and Conway, B. A. (1994) Optimization of Very-Low-Thrust, Many Revolution Spacecraft Trajectories, *Journal of Guidance, Control, and Dynamics*, **17**, 1185–1192.

[47] Herman, A. L., and Conway, B. A. (1998) Optimal Low-Thrust, Earth-Moon Orbit Transfer, *Journal of Guidance, Control, and Dynamics*, **21**, 141–147.

[48] Betts, J. T., and Erb, S. O. (2003) Optimal Low Thrust Trajectories to the Moon, *SIAM Journal on Applied Dynamical Systems*, **2**, 144–170.

[49] Tang, S., and Conway, B. A. (1995) Optimization of Low-Thrust Interplanetary Trajectories Using Collocation and Nonlinear Programming, *Journal of Guidance, Control, and Dynamics*, **18**, 599–604.

2   Primer Vector Theory and Applications

John E. Prussing

Department of Aerospace Engineering, University of Illinois at Urbana-Champaign, Urbana, IL

2.1 Introduction

In this chapter, the theory and a resulting *indirect* method of trajectory optimization are derived and illustrated. In an indirect method, an optimal trajectory is determined by satisfying a set of necessary conditions (NC), and sufficient conditions (SC) if available. By contrast, a direct method uses the cost itself to determine an optimal solution.

Even when a direct method is used, these conditions are useful to determine whether the solution satisfies the NC for an optimal solution. If it does not, it is not an optimal solution. As an example, the best two-impulse solution obtained by a direct method is not the optimal solution if the NC indicate that three impulses are required. Thus, post-processing a direct solution using the NC (and SC if available) is essential to verify optimality.

Optimal Control [1], a generalization of the calculus of variations, is used to derive a set of necessary conditions for an optimal trajectory. The *primer vector* is a term coined by D. F. Lawden [2] in his pioneering work in optimal trajectories. [This terminology is explained after Equation (2.24).] First-order necessary conditions for both impulsive and continuous-thrust trajectories can be expressed in terms of the primer vector. For impulsive trajectories, the primer vector determines the times and positions of the thrust impulses that minimize the propellant cost. For continuous-thrust trajectories, both the optimal thrust direction and the optimal thrust magnitude as functions of time are determined by the primer vector. As is standard practice, the word "optimal" is loosely used as shorthand for "satisfies the first-order NC."

The most completely developed primer vector theory is for impulsive trajectories. Terminal coasting periods for fixed-time trajectories and the addition of midcourse impulses can sometimes lower the cost. The primer vector indicates when these modifications should be made. Gradients of the cost with respect to terminal impulse times and midcourse impulse times and positions were first derived by Lion and Handelsman [3]. These gradients were then implemented in a nonlinear

---

Figures 2.2 and 2.4–2.8 were generated using the MATLAB computer code written by Suzannah L. Sandrik [13].

programming algorithm to iteratively improve a nonoptimal solution and converge to an optimal trajectory by Jezewski and Rozendaal [4].

2.2 First-Order Necessary Conditions

2.2.1 Optimal Constant-Specific-Impulse Trajectory

For a constant specific impulse (CSI) engine, the thrust is bounded by $0 \leq T \leq T_{max}$ (where T_{max} is a constant), corresponding to bounds on the mass flow rate: $0 \leq b \leq b_{max}$ (where b_{max} is a constant). Note that one can also prescribe bounds on the thrust acceleration (thrust per unit mass) $\Gamma \equiv T/m$ as $0 \leq \Gamma \leq \Gamma_{max}$, where Γ_{max} is achieved by running the engine at T_{max}. However, Γ_{max} is not constant but increases due to the decreasing mass. One must keep track of the changing mass in order to compute Γ for a given thrust level. This is easy to do, especially if the thrust is held constant, for example, at its maximum value. However, if the propellant mass required is a small fraction of the total mass because of being optimized, a constant Γ_{max} approximation can be made.

The cost functional representing minimum propellant consumption for the CSI case is

$$J = \int_{t_o}^{t_f} \Gamma(t)dt. \tag{2.1}$$

The state vector is defined as

$$\mathbf{x}(t) = \begin{bmatrix} \mathbf{r}(t) \\ \mathbf{v}(t) \end{bmatrix} \tag{2.2}$$

where $\mathbf{r}(t)$ is the spacecraft position vector and $\mathbf{v}(t)$ is its velocity vector. The mass m can be kept track of without defining it to be a state variable by noting that

$$m(t) = m_o e^{-F(t)/c} \tag{2.3}$$

where c is the exhaust velocity and

$$F(t) = \int_{t_o}^{t} \Gamma(\xi)d\xi. \tag{2.4}$$

Note that from Equation (2.4), $F(t_f)$ is equal to the cost J. In the constant thrust case, Γ varies according to $\dot{\Gamma} = \frac{1}{c}\Gamma^2$, which is consistent with the mass decreasing linearly with time.

The equation of motion is

$$\dot{\mathbf{x}} = \begin{bmatrix} \dot{\mathbf{r}} \\ \dot{\mathbf{v}} \end{bmatrix} = \begin{bmatrix} \mathbf{v} \\ \mathbf{g}(\mathbf{r}) + \Gamma\mathbf{u} \end{bmatrix} \tag{2.5}$$

with the initial state $\mathbf{x}(t_o)$ specified.

In Equation (2.5), $\mathbf{g}(\mathbf{r})$ is the gravitational acceleration and $\mathbf{u}$ represents a unit vector in the thrust direction. An example gravitational field is the inverse-square field:

$$\mathbf{g}(\mathbf{r}) = -\frac{\mu}{\mathbf{r}^2}\frac{\mathbf{r}}{r} = -\frac{\mu}{r^3}\mathbf{r}. \tag{2.6}$$

The first-order necessary conditions for an optimal CSI trajectory were first derived by Lawden [2] using classical calculus of variations. In the derivation that follows, an optimal control theory formulation is used, but the derivation is similar to that of Lawden. One difference is that the mass is not considered a state variable but is kept track of separately.

In order to minimize the cost in Equation (2.1), one forms the Hamiltonian using Equation (2.5) as

$$H = \Gamma + \lambda_r^T \mathbf{v} + \lambda_v^T [\mathbf{g}(\mathbf{r}) + \Gamma\mathbf{u}]. \tag{2.7}$$

The adjoint equations are then

$$\dot{\lambda}_r^T = -\frac{\partial H}{\partial \mathbf{r}} = -\lambda_v^T \mathbf{G}(\mathbf{r}) \tag{2.8}$$

$$\dot{\lambda}_v^T = -\frac{\partial H}{\partial \mathbf{v}} = -\lambda_r^T \tag{2.9}$$

where

$$\mathbf{G}(\mathbf{r}) \equiv \frac{\partial \mathbf{g}(\mathbf{r})}{\partial \mathbf{r}} \tag{2.10}$$

is the symmetric 3×3 *gravity gradient matrix*.

For terminal constraints of the form

$$\boldsymbol{\psi}[\mathbf{r}(t_f), \mathbf{v}(t_f), t_f] = \mathbf{0}, \tag{2.11}$$

which may describe an orbital intercept, rendezvous, etc., the boundary conditions on Equations (2.8–2.9) are given in terms of

$$\Phi \equiv v^T \boldsymbol{\psi}[\mathbf{r}(t_f), \mathbf{v}(t_f), tf] \tag{2.12}$$

as

$$\lambda_r^T(t_f) = \frac{\partial \Phi}{\partial \mathbf{r}(t_f)} = v^T \frac{\partial \boldsymbol{\psi}}{\partial \mathbf{r}(t_f)} \tag{2.13}$$

$$\lambda_v^T(t_f) = \frac{\partial \Phi}{\partial \mathbf{v}(t_f)} = v^T \frac{\partial \boldsymbol{\psi}}{\partial \mathbf{v}(t_f)}. \tag{2.14}$$

There are two control variables, the thrust direction $\mathbf{u}$ and the thrust acceleration magnitude Γ, that must be chosen to satisfy the minimum principle [1], that is, to minimize the instantaneous value of the Hamiltonian H. By inspection, the Hamiltonian of Equation (2.7) is minimized over the choice of thrust direction by aligning the unit vector $\mathbf{u}(t)$ opposite to the adjoint vector $\boldsymbol{\lambda}_v(t)$. Because of the significance of the vector $-\boldsymbol{\lambda}_v(t)$, Lawden [2] termed it the *primer vector* $\mathbf{p}(t)$:

$$\mathbf{p}(t) \equiv -\boldsymbol{\lambda}_v(t). \tag{2.15}$$

The optimal thrust unit vector is then in the direction of the primer vector, specifically

$$\mathbf{u}(t) = \frac{\mathbf{p}(t)}{p(t)} \tag{2.16}$$

and

$$\boldsymbol{\lambda}_v^T \mathbf{u} = -\lambda_v = -p \tag{2.17}$$

in the Hamiltonian of Equation (2.7).

From Equations (2.9) and (2.15), it is evident that

$$\boldsymbol{\lambda}_r(t) = \dot{\mathbf{p}}(t). \tag{2.18}$$

Equations (2.8), (2.9), (2.15), and (2.18) combine to yield the *primer vector equation*

$$\ddot{\mathbf{p}} = \mathbf{G}(\mathbf{r})\mathbf{p}. \tag{2.19}$$

The boundary conditions on the solution to Equation (2.19) are obtained from Equations (2.13) (2.14)

$$\mathbf{p}(t_f) = -v^T \frac{\partial \boldsymbol{\psi}}{\partial \mathbf{v}(t_f)} \tag{2.20}$$

$$\dot{\mathbf{p}}(t_f) = v^T \frac{\partial \boldsymbol{\psi}}{\partial \mathbf{r}(t_f)}. \tag{2.21}$$

Note that in Equation (2.20), the final value of the primer vector for an optimal intercept is the zero vector, because the terminal constraint $\boldsymbol{\psi}$ does not depend on $\mathbf{v}(t_f)$.

Using Equations (2.15)–(2.18), the Hamiltonian of Equation (2.7) can be rewritten as

$$H = -(p-1)\Gamma + \dot{\mathbf{p}}^T \mathbf{v} - \mathbf{p}^T \mathbf{g}. \tag{2.22}$$

To minimize the Hamiltonian over the choice of the thrust acceleration magnitude Γ, one notes that the Hamiltonian is a linear function of Γ, and thus the minimizing value for $0 \leq \Gamma \leq \Gamma_{max}$ will depend on the algebraic sign of the coefficient of Γ in Equation (2.22). It is convenient to define the *switching function*

$$S(t) \equiv p - 1. \tag{2.23}$$

The choice of the thrust acceleration magnitude Γ that minimizes H is then given by the "bang-bang" control law

$$\Gamma = \begin{cases} \Gamma_{max} & \text{for} \quad S > 0 \ (p > 1) \\ 0 & \text{for} \quad S < 0 \ (p < 1) \end{cases}. \tag{2.24}$$

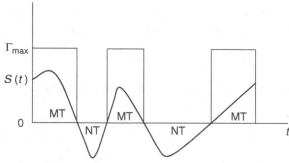

Figure 2.1. Three-burn CSI switching function and thrust profile.

That is, the thrust magnitude switches between its limiting values of 0 (an NT *null-thrust* arc) and T_{max} (an MT *maximum-thrust* arc) each time $S(t)$ passes through 0 [$p(t)$ passes through 1] according to Equation (2.24). Figure 2.1 shows an example switching function for a three-burn trajectory.

The possibility also exists that $S(t) \equiv 0$ [$p(t) \equiv 1$] on an interval of finite duration. From Equation (2.22), it is evident that in this case the thrust acceleration magnitude is not determined by the minimum principle and may take on intermediate values between 0 and Γ_{max}. This IT "intermediate thrust arc" [2] is referred to as a *singular arc* in optimal control [1].

Lawden explained the origin of the term *primer vector* in a personal letter in 1990: "In regard to the term 'primer vector', you are quite correct in your supposition. I served in the artillery during the war [World War II] and became familiar with the initiation of the burning of cordite by means of a primer charge. Thus, $p = 1$ is the signal for the rocket motor to be ignited."

It follows then from Equation (2.3) that if $T = T_{max}$ and the engine is on for a total of Δt time units,

$$\Gamma_{max}(t) = e^{F(t)/c} T_{max}/m_o = T_{max}/(m_o - b_{max}\Delta t). \qquad (2.25)$$

Other necessary conditions are that the variables $\mathbf{p}$ and $\dot{\mathbf{p}}$ must be continuous everywhere. Equation (2.23) then indicates that the switching function $S(t)$ is also continuous everywhere.

Even though the gravitational field is time-invariant, the Hamiltonian in this formulation does not provide a first integral (constant of the motion) on an MT arc, because Γ is an explicit function of time as shown in Equation (2.25). From Equation (2.22)

$$H = -S\Gamma + \dot{\mathbf{p}}^T\mathbf{v} - \mathbf{p}^T\mathbf{g}. \qquad (2.26)$$

Note that the Hamiltonian is continuous everywhere because $S = 0$ at the discontinuities in the thrust acceleration magnitude.

2.2.2 Optimal Impulsive Trajectory

For a high-thrust CSI engine the thrust durations are very small compared with the times between thrusts. Because of this, one can approximate each MT arc as an impulse (Dirac delta function) having unbounded magnitude ($\Gamma_{max} \to \infty$) and zero duration. The primer vector then determines both the optimal times and directions of the thrust impulses with $p \leq 1$ corresponding to $S \leq 0$. The impulses can occur only at those instants at which $S = 0$ ($p = 1$). These impulses are separated by NT arcs along which $S < 0$ ($p < 1$). At the impulse times the primer vector is then a unit vector in the optimal thrust direction.

The necessary conditions (NC) for an optimal impulsive trajectory, first derived by Lawden [2], are shown in Table 2.1.

For a linear system, these NC are also sufficient conditions for an optimal trajectory [5]. Also in [5], an upper bound on the number of impulses required for an optimal solution is given.

Figure 2.2 shows a trajectory (at top) and a primer vector magnitude (at bottom) for an optimal three-impulse solution. (In all of the trajectory plots in this chapter, the direction of orbital motion is counterclockwise.) Canonical units are used. The canonical time unit is the orbital period of the circular orbit that has a radius of one canonical distance unit. The initial orbit is a unit radius circular orbit, shown as the topmost orbit going counterclockwise from the symbol ⊕ at (1,0) to (−1,0). The transfer time is 0.5 original (initial) orbit periods (OOP). The target is in a coplanar circular orbit of radius 2, with an initial lead angle (ila) of 270° and shown by the symbol □ at (0,−2). The spacecraft departs ○ and intercepts □ at approximately (1.8,−0.8) as shown. The + signs at the initial and final points indicate thrust impulses and the + sign on the transfer orbit very near (0,0) indicates the location of the midcourse impulse. The magnitudes of the three ΔVs are shown at the left, with the total ΔV equal to 1.3681 in units of circular orbit speed in the initial orbit.

The examples shown in this chapter are coplanar, but the theory and applications apply to three-dimensional trajectories as well, for example, see Prussing and Chiu [6].

The bottom graph in Figure 2.2 displays the time history of the primer vector magnitude. Note that it satisfies the necessary conditions of Table 2.1 for an optimal transfer.

Table 2.1. *Impulsive necessary conditions*

1. The primer vector and its first derivative are continuous everywhere.
2. The magnitude of the primer vector satisfies $p(t) \leq 1$ with the impulses occurring at those instants at which $p = 1$.
3. At the impulse times the primer vector is a unit vector in the optimal thrust direction.
4. As a consequence of the above conditions, $dp/dt = \dot{p} = \dot{\mathbf{p}}^T \mathbf{p} = 0$ at an intermediate impulse (not at the initial or final time).

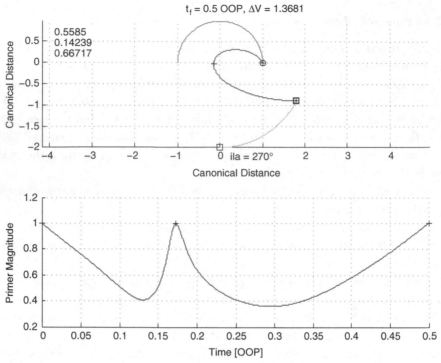

Figure 2.2. Optimal three-impulse trajectory and primer magnitude.

Note also that at a thrust impulse at time t_k

$$\Gamma(t) = \Delta v_k \delta(t - t_k) \tag{2.27}$$

and from Equation (2.4)

$$\Delta v_k = \int_{t_k^-}^{t_k^+} \Gamma(t)dt = F(t_k^+) - F(t_k^-) \tag{2.28}$$

where t_k^+ and t_k^- are times immediately after and before the impulse time, respectively. Equation (2.3) then becomes the familiar solution to the rocket equation:

$$m(t_k^+) = m(t_k^-)e^{-\Delta v_k/c}. \tag{2.29}$$

2.2.3 Optimal Variable-Specific-Impulse Trajectory

A variable-specific-impulse (VSI) engine is also known as a power-limited (PL) engine, because the power source is separate from the engine itself, for example, solar panels, and radioisotope thermoelectric generator. The power delivered to the engine is bounded between 0 and a maximum value $P_{\max}$, with the optimal value being constant and equal to the maximum. The cost functional representing minimum propellant consumption for the VSI case is

$$J = \frac{1}{2} \int_{t_o}^{t_f} \Gamma^2(t)dt. \tag{2.30}$$

Writing Γ^2 as $\boldsymbol{\Gamma}^T\boldsymbol{\Gamma}$, the corresponding Hamiltonian function can be written as

$$H = \frac{1}{2}\boldsymbol{\Gamma}^T\boldsymbol{\Gamma} + \boldsymbol{\lambda}_r^T\mathbf{v} + \boldsymbol{\lambda}_v^T[\mathbf{g}(\mathbf{r}) + \boldsymbol{\Gamma}]. \tag{2.31}$$

For the VSI case, there is no need to consider the thrust acceleration magnitude and direction separately, so the vector $\boldsymbol{\Gamma}$ is used in place of the term $\Gamma\mathbf{u}$ that appears in Equation (2.7).

Because H is a nonlinear function of $\boldsymbol{\Gamma}$, the minimum principle is applied by setting

$$\frac{\partial H}{\partial \boldsymbol{\Gamma}} = \boldsymbol{\Gamma}^T + \boldsymbol{\lambda}_v^T = \mathbf{0}^T \tag{2.32}$$

or

$$\boldsymbol{\Gamma}(t) = -\boldsymbol{\lambda}_v(t) = \mathbf{p}(t) \tag{2.33}$$

using the definition of the primer vector in Equation (2.15). Thus for a VSI engine, the optimal thrust acceleration vector is equal to the primer vector: $\boldsymbol{\Gamma}(t) = \mathbf{p}(t)$.

Because of this, Equation (2.5), written as $\ddot{\mathbf{r}} = \mathbf{g}(\mathbf{r}) + \boldsymbol{\Gamma}$, can be combined with Equation (2.19), as in [7] to yield a fourth-order differential equation in $\mathbf{r}$:

$$\mathbf{r}^{iv} - \dot{\mathbf{G}}\dot{\mathbf{r}} + \mathbf{G}(\mathbf{g} - 2\ddot{\mathbf{r}}) = \mathbf{0}. \tag{2.34}$$

Every solution to Equation (2.34) is an optimal VSI trajectory through the gravity field $\mathbf{g}(\mathbf{r})$. But desired boundary conditions, such as specified position and velocity vectors at the initial and final times, must be satisfied.

Note also that from Equation (2.32)

$$\frac{\partial^2 H}{\partial \boldsymbol{\Gamma}^2} = \frac{\partial}{\partial \boldsymbol{\Gamma}}\left(\frac{\partial H}{\partial \boldsymbol{\Gamma}}\right)^T = \mathbf{I}_3 \tag{2.35}$$

where $\mathbf{I}_3$ is the 3×3 identity matrix. Equation (2.35) shows that the (Hessian) matrix of second partial derivatives is positive definite, verifying that H is minimized.

Because the VSI thrust acceleration of Equation (2.33) is continuous, a recently developed procedure [8] to test whether second-order NC and SC are satisfied can be applied. Equation (2.35) shows that an NC for minimum cost (Hessian matrix positive semidefinite) and part of the SC (Hessian matrix positive definite) are satisfied. The other condition that is both an NC and an SC is the Jacobi no-conjugate-point condition. Reference [8] details the recently developed test for that.

2.3 Solution to the Primer Vector Equation

The primer vector equation, Equation (2.19), can be written in first-order form as the linear system

$$\frac{d}{dt}\begin{bmatrix} \mathbf{p} \\ \dot{\mathbf{p}} \end{bmatrix} = \begin{bmatrix} \mathbf{O}_3 & \mathbf{I}_3 \\ \mathbf{G} & \mathbf{O}_3 \end{bmatrix}\begin{bmatrix} \mathbf{p} \\ \dot{\mathbf{p}} \end{bmatrix} \tag{2.36}$$

where $\mathbf{O}_3$ is the 3×3 zero matrix.

Equation (2.36) is of the form $\dot{\mathbf{y}} = \mathbf{A}(t)\mathbf{y}$, and its solution can be written in terms of a transition matrix $\mathbf{\Phi}(t, t_o)$ as

$$\mathbf{y}(t) = \mathbf{\Phi}(t, t_o)\mathbf{y}(t_o) \tag{2.37}$$

for a specified initial condition $\mathbf{y}(t_o)$.

Glandorf [9] presents a form of the transition matrix for an inverse-square gravitational field. [In that Technical Note, the missing Equation (2.33) is $\mathbf{\Phi}(t, t_o) = P(t)P^{-1}(t_o)$.]

Note that on an NT (no-thrust or coast) arc, the variational (linearized) state equation is, from Equation (2.5),

$$\delta\dot{\mathbf{x}} = \begin{bmatrix} \delta\dot{\mathbf{r}} \\ \delta\dot{\mathbf{v}} \end{bmatrix} = \begin{bmatrix} \mathbf{O}_3 & \mathbf{I}_3 \\ \mathbf{G} & \mathbf{O}_3 \end{bmatrix} \begin{bmatrix} \delta\mathbf{r} \\ \delta\mathbf{v} \end{bmatrix}, \tag{2.38}$$

which is the same as Equation (2.36). So the transition matrix in Equation (2.37) is also the transition matrix for the state variation, that is, the *state transition matrix* [10].

This state transition matrix has the usual properties from linear system theory and is also symplectic [10], which has the useful property that

$$\mathbf{\Phi}^{-1}(t, t_o) = -\mathbf{J}\mathbf{\Phi}^T(t, t_o)\mathbf{J} \tag{2.39}$$

where

$$\mathbf{J} = \begin{bmatrix} \mathbf{O}_3 & \mathbf{I}_3 \\ -\mathbf{I}_3 & \mathbf{O}_3 \end{bmatrix}. \tag{2.40}$$

Note that $\mathbf{J}^2 = -\mathbf{I}_6$, indicating that $\mathbf{J}$ is a matrix analog of the imaginary number i.

Equation (2.39) is useful when the state transition matrix is determined numerically because the inverse matrix $\mathbf{\Phi}^{-1}(t, t_o) = \mathbf{\Phi}(t_o, t)$ can be computed without explicitly inverting a 6×6 matrix.

2.4 Application of Primer Vector Theory to an Optimal Impulsive Trajectory

If the primer vector evaluated along an impulsive trajectory fails to satisfy the necessary conditions of Table 2.1 for an optimal solution, the way in which the NC are violated provides information that can lead to a solution that does satisfy the NC. This process was first derived by Lion and Handelsman [3]. For given boundary conditions and a fixed transfer time, an impulsive trajectory can be modified either by allowing a terminal coast or by adding a midcourse impulse. A terminal coast can be either an initial coast, in which the first impulse occurs after the initial time, or a final coast, in which the final impulse occurs before the final time. In the former case, the spacecraft coasts along the initial orbit after the initial time until the first impulse

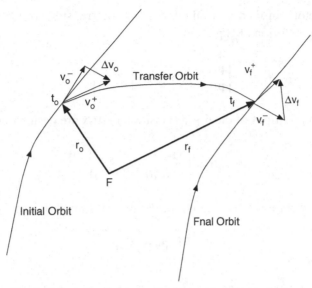

Figure 2.3. A fixed-time impulsive rendezvous trajectory.

occurs. In the latter case, the rendezvous actually occurs before the final time, and the spacecraft coasts along the final orbit until the final time is reached.

To determine when a terminal coast will result in a trajectory that has a lower fuel cost, consider the two-impulse fixed-time rendezvous trajectory shown in Figure 2.3.

In the two-body problem, if the terminal radii $\mathbf{r}_o$ and $\mathbf{r}_f$ are specified along with the transfer time $\tau \equiv t_f - t_o$, the solution to Lambert's Problem [10] [11] provides the terminal velocity vectors $\mathbf{v}_o^+$ (after the initial impulse) and $\mathbf{v}_f^-$ (before the final impulse) on the transfer orbit. Because the velocity vectors are known on the initial orbit ($\mathbf{v}_o^-$ before the first impulse) and on the final orbit ($\mathbf{v}_f^+$ after the final impulse), the required velocity changes can be determined as

$$\Delta \mathbf{v}_o = \mathbf{v}_o^+ - \mathbf{v}_o^- \tag{2.41}$$

and

$$\Delta \mathbf{v}_f = \mathbf{v}_f^+ - \mathbf{v}_f^-. \tag{2.42}$$

Once the vector velocity changes are known, the primer vector can be evaluated along the trajectory to determine if the NC are satisfied. In order to satisfy the NC that on an optimal trajectory the primer vector at an impulse time is a unit vector in the direction of the impulse, one imposes the following boundary conditions on the primer vector

$$\mathbf{p}(t_o) \equiv \mathbf{p}_o = \frac{\Delta \mathbf{v}_o}{\Delta v_o} \tag{2.43}$$

$$\mathbf{p}(t_f) \equiv \mathbf{p}_f = \frac{\Delta \mathbf{v}_f}{\Delta v_f}. \tag{2.44}$$

The primer vector can then be evaluated along the transfer orbit using the 6×6 transition matrix solution of Equation (2.37)

$$\begin{bmatrix} \mathbf{p}(t) \\ \dot{\mathbf{p}}(t) \end{bmatrix} = \mathbf{\Phi}(t, t_o) \begin{bmatrix} \mathbf{p}(t_o) \\ \dot{\mathbf{p}}(t_o) \end{bmatrix} \qquad (2.45)$$

where the 3×3 partitions of the 6×6 transition matrix are designated as

$$\mathbf{\Phi}(t, t_o) \equiv \begin{bmatrix} \mathbf{M}(t, t_o) & \mathbf{N}(t, t_o) \\ \mathbf{S}(t, t_o) & \mathbf{T}(t, t_o) \end{bmatrix}. \qquad (2.46)$$

Equation (2.45) can then be evaluated for the fixed terminal times t_o and t_f to yield

$$\mathbf{p}_f = \mathbf{M}_{fo}\mathbf{p}_o + \mathbf{N}_{fo}\dot{\mathbf{p}}_o \qquad (2.47)$$

and

$$\dot{\mathbf{p}}_f = \mathbf{S}_{fo}\mathbf{p}_o + \mathbf{T}_{fo}\dot{\mathbf{p}}_o \qquad (2.48)$$

where the abbreviated notation is used that $\mathbf{p}_f \equiv \mathbf{p}(t_f), \mathbf{M}_{fo} \equiv \mathbf{M}(t_f, t_o)$, and so on. Equation (2.47) can be solved for the initial primer vector rate

$$\dot{\mathbf{p}}_o = \mathbf{N}_{fo}^{-1}[\mathbf{p}_f - \mathbf{M}_{fo}\mathbf{p}_o] \qquad (2.49)$$

where the inverse matrix $\mathbf{N}_{fo}^{-1}$ exists except for isolated values of $\tau = t_f - t_o$. With both the primer vector and the primer vector rate known at the initial time, the primer vector along the transfer orbit for $t_o \leq t \leq t_f$ can be calculated as using Equations (2.43–2.46, 2.49) as

$$\mathbf{p}(t) = \mathbf{N}_{to}\mathbf{N}_{fo}^{-1}\frac{\Delta \mathbf{v}_f}{\Delta v_f} + [\mathbf{M}_{to} - \mathbf{N}_{to}\mathbf{N}_{fo}^{-1}\mathbf{M}_{fo}]\frac{\Delta \mathbf{v}_o}{\Delta v_o}. \qquad (2.50)$$

2.4.1 Criterion for a Terminal Coast

One of the options available to modify a two-impulse solution that does not satisfy the NC for an optimal transfer is to include a terminal coast period in the form of either an initial coast, a final coast, or both. To do this, one allows the possibility that the initial impulse occurs at time $t_o + dt_o$ due to a coast in the initial orbit of duration $dt_o > 0$ and that the final impulse occurs at a time $t_f + dt_f$. In the case of a final coast, $dt_f < 0$ in order that the final impulse occur prior to the nominal final time, allowing a coast in the final orbit until the nominal final time. A negative value of dt_o or a positive value of dt_f also has a physical interpretation as will be seen.

To determine whether a terminal coast will lower the cost of the trajectory, an expression for the difference in cost between the perturbed trajectory (with the terminal coasts) and the nominal trajectory (without the coasts) must be derived. The discussion that follows summarizes and interprets results by Lion and Handelsman [3]. The cost on the nominal trajectory is simply

$$J = \Delta v_o + \Delta v_f \qquad (2.51)$$

for the two-impulse solution. In order to determine the differential change in the cost due to the differential coast periods the concept of a noncontemporaneous, or "skew" variation is needed. This variation combines two effects: the variation due to being on a perturbed trajectory and the variation due to a difference in the time of the impulse. The variable d will be used to denote a noncontemporaneous variation in contrast to the variable δ that represents a contemporaneous variation, as in Equation (2.38). The rule for relating the two types of variations is given by

$$d\mathbf{x}(t_o) = \delta\mathbf{x}(t_o) + \dot{\mathbf{x}}_o^* dt_o \qquad (2.52)$$

where $\dot{\mathbf{x}}_o^*$ is the derivative on the nominal (unperturbed) trajectory at the nominal final time and the variation in the initial state has been used as an example.

Next, the noncontemporaneous variation in the cost must be determined. Because the coast periods result in changes in the vector velocity changes, the variation in the cost can be expressed, from Equation (2.51) as

$$dJ = \frac{\partial \Delta v_o}{\partial \Delta \mathbf{v}_o} d\Delta \mathbf{v}_o + \frac{\partial \Delta v_f}{\partial \Delta \mathbf{v}_f} d\Delta \mathbf{v}_f. \qquad (2.53)$$

Using the fact that for any vector $\mathbf{a}$ having magnitude a

$$\frac{\partial a}{\partial \mathbf{a}} = \frac{\mathbf{a}^T}{a} \qquad (2.54)$$

the variation in the cost in Equation (2.53) can be expressed as

$$dJ = \frac{\Delta \mathbf{v}_o^T}{\Delta v_o} d\Delta \mathbf{v}_o + \frac{\Delta \mathbf{v}_f^T}{\Delta v_f} d\Delta \mathbf{v}_f. \qquad (2.55)$$

Finally, Equation (2.55) can be rewritten in terms of the initial and final primer vector using the conditions of Equations (2.43–2.44) as

$$dJ = \mathbf{p}_o^T d\Delta \mathbf{v}_o + \mathbf{p}_f^T d\Delta \mathbf{v}_f. \qquad (2.56)$$

The analysis in [3] leads to the result that

$$dJ = -\dot{\mathbf{p}}_o^T \Delta \mathbf{v}_o dt_o - \dot{\mathbf{p}}_f^T \Delta \mathbf{v}_f dt_f \qquad (2.57)$$

The final form of the expression for the variation in cost is obtained by expressing the vector velocity changes in terms of the primer vector using Equations (2.43–2.44) as

$$dJ = -\Delta v_o \dot{\mathbf{p}}_o^T \mathbf{p}_o dt_o - \Delta v_f \dot{\mathbf{p}}_f^T \mathbf{p}_f dt_f. \qquad (2.58)$$

In Equation (2.58), one can identify the gradients of the cost with respect to the terminal impulse times t_o and t_f as

$$\frac{\partial J}{\partial t_o} = -\Delta v_o \dot{\mathbf{p}}_o^T \mathbf{p}_o \tag{2.59}$$

and

$$\frac{\partial J}{\partial t_f} = -\Delta v_f \dot{\mathbf{p}}_f^T \mathbf{p}_f. \tag{2.60}$$

One notes that the dot products in Equations (2.59–2.60) are simply the slopes of the primer magnitude time history at the terminal times, due to the fact that $p^2 = \mathbf{p}^T \mathbf{p}$ and, after differentiation with respect to time, $2p\dot{p} = 2\dot{\mathbf{p}}^T \mathbf{p}$. Because $p = 1$ at the impulse times,

$$\dot{\mathbf{p}}^T \mathbf{p} = \dot{p}. \tag{2.61}$$

The criteria for adding an initial or final coast can now be summarized by examining the algebraic signs of the gradients in Equations (2.59–2.60):

If $\dot{p}_o > 0$, an initial coast (represented by $dt_o > 0$) will lower the cost. Similarly, if $\dot{p}_f < 0$, a final coast (represented by $dt_f < 0$) will lower the cost.

It is worth noting that, conversely, if $\dot{p}_o \leq 0$, an initial coast will not lower the cost. This is consistent with the NC for an optimal solution and represents an alternate proof of the NC that $p \leq 1$ on an optimal solution. Similarly, if $\dot{p}_f \geq 0$, a final coast will not lower the cost. However, one can interpret these results even further. If $\dot{p}_o < 0$, a value of $dt_o < 0$ yields $dJ < 0$, indicating that an *earlier* initial impulse time would lower the cost. This is the opposite of an initial coast and simply means that the cost can be lowered by increasing the transfer time by starting the transfer earlier. Similarly, a value of $\dot{p}_f > 0$ implies that a $dt_f > 0$ will yield $dJ < 0$. In this case, the cost can be lowered by increasing the transfer time by increasing the final time. From these observations, one can conclude that for a time-open optimal solution, such as the Hohmann transfer, the slopes of the primer magnitude time history must be zero at the terminal times, indicating that no improvement in the cost can be made by slightly increasing or decreasing the times of the terminal impulses. Figure 2.4 shows the primer time history for a Hohmann transfer rendezvous trajectory. An initial coast of 0.889 OOP is required to obtain the correct phase angle of the target body for the given ila and there is no final coast.

Figure 2.5 shows an example of a primer history that violates the NC in a manner indicating that an initial coast or final coast or both will lower the cost. The final radius is 1.6, the ila is 90°, and the transfer time is 0.9 OOP.

In this case, the choice is made to add an initial coast, and the gradient of Equation (2.59) is used in a nonlinear programming (NLP) algorithm to iterate on the time of the first impulse. This is a one-dimensional search in which small changes in the time of the first impulse are made using the gradient of Equation (2.59) until the gradient is driven to zero. On each iteration, new values for the terminal velocity

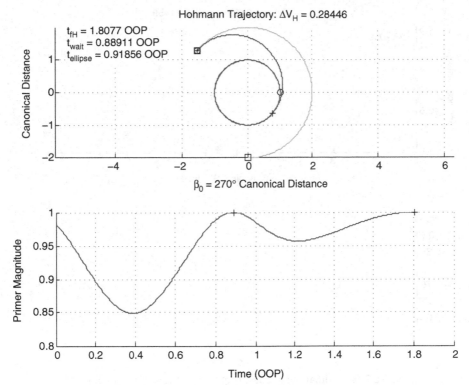

Figure 2.4. Hohmann transfer orbit and primer magnitude.

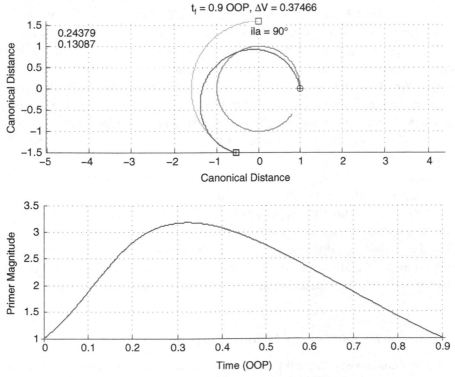

Figure 2.5. Primer magnitude indicating initial/final coast.

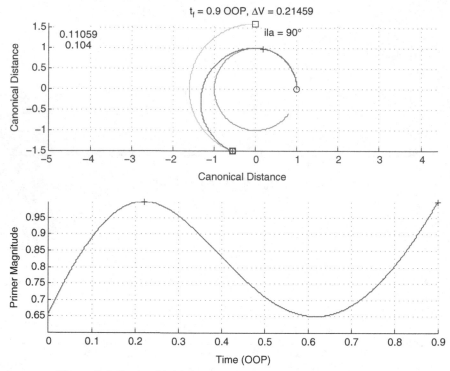

Figure 2.6. Optimal initial coast trajectory and primer magnitude.

changes are calculated by re-solving Lambert's Problem and a new primer vector solution is obtained. Note that once the iteration begins, the time of the first impulse is no longer t_o, but a later value denoted by t_1. In a similar way, if the final impulse time becomes an iteration variable, it is denoted by t_n where the last impulse is considered to be the nth impulse. For a two-impulse trajectory, $n = 2$, but as will be seen shortly, optimal solutions can require more than two impulses. When the times of the first and last impulse become iteration variables, in all the formulas in the preceding analysis, the subscript o is replaced by 1 everywhere and f is replaced by n.

Figure 2.6 shows the converged result of an iteration on the time of the initial impulse.

Note that the necessary condition $p \leq 1$ is satisfied and the gradient of the cost with respect to t_1, the time of the first impulse (at approximately $t_1 = 0.22$), is zero because $\dot{p}_1 = 0$, making the gradient of Equation (2.59) equal to zero. This simply means that a small change in t_1 will cause no change in the cost, that is, the cost has achieved a stationary value and satisfies the first-order necessary conditions. Comparing Figures 2.5 and 2.6, one notes that the cost has decreased significantly from 0.37466 to 0.21459, and that an initial coast is required but no final coast is required.

2.4.2 Criterion for Addition of a Midcourse Impulse

Besides terminal coasts, the addition of one or more midcourse impulses is another potential way of lowering the cost of an impulsive trajectory. The addition of an impulse is more complicated than including terminal coasts because, in the general case, four new parameters are introduced: three components of the position of the impulse and the time of the impulse. One must first derive a criterion that indicates that the addition of an impulse will lower the cost and then determine where in space and when in time the impulse should occur. The where and when will be done in two steps. The first step is to determine initial values of position and time of the added impulse that will lower the cost. The second step is to iterate on the values of position and time using gradients that will be developed, until a minimum of the cost is achieved. Note that this procedure is more complicated than for terminal coasts, because the starting value of the coast time for the iteration was simply taken to be zero, that is, no coast.

When considering the addition of a midcourse impulse, let us assume $dt_o = dt_f = 0$, that is, there are no terminal coasts. Because we are doing a first-order perturbation analysis, superposition applies and we can combine the previous results for terminal coasts easily with our new results for a midcourse impulse. Also, we will discuss the case of adding a third impulse to a two-impulse trajectory, but the same theory applies to the case of adding a midcourse impulse to any two-impulse segment of an n-impulse trajectory. The cost on the nominal, two-impulse trajectory is given by Equation (2.50)

$$J = \Delta\mathbf{v}_o + \Delta\mathbf{v}_f.$$

The variation in the cost due to adding an impulse is given by adding the midcourse velocity change magnitude $\Delta\mathbf{v}_m$ to Equation (2.56)

$$dJ = \mathbf{p}_o^T d\Delta\mathbf{v}_o + \Delta\mathbf{v}_m + \mathbf{p}_f^T d\Delta\mathbf{v}_f. \tag{2.62}$$

The analysis in [3] results in

$$dJ = \Delta\mathbf{v}_m \left(1 - \mathbf{p}_m^T \frac{\Delta\mathbf{v}_m}{\Delta\mathbf{v}_m}\right). \tag{2.63}$$

In Equation (2.63), the expression for dJ involves a dot product between the primer vector and a unit vector. If the numerical value of this dot product is greater than one, $dJ < 0$ and the perturbed trajectory has a lower cost than the nominal trajectory. In order for the value of the dot product to be greater than one, it is necessary that $p_m > 1$. Here again we have an alternative derivation of the necessary condition that $p \leq 1$ on an optimal trajectory. We also have the criterion that tells us when the addition of a midcourse impulse will lower the cost.

> If the value of $p(t)$ exceeds unity along the trajectory, the addition of a midcourse impulse at a time for which $p > 1$ will lower the cost.

Figure 2.7 shows a primer magnitude time history that indicates the need for a midcourse impulse (but not for a terminal coast). The final radius is 2, the ila is 270°, and the transfer time is relatively small, equal to 0.5 OOP.

The first step is to determine initial values for the position and time of the midcourse impulse. From Equation (2.63) it is evident that for a given $\mathbf{p}_m$, the largest

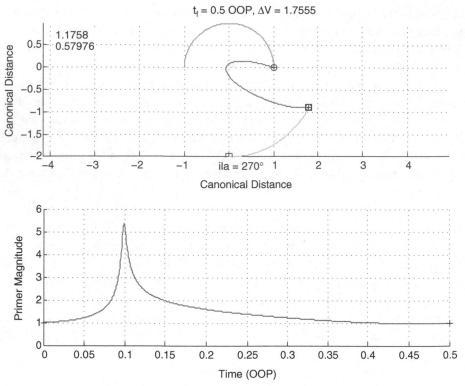

Figure 2.7. Primer magnitude indicating a need for a midcourse impulse.

decrease in the cost is obtained by maximizing the value of the dot product, that is, by choosing a position for the impulse that causes $\Delta\mathbf{v}_m$ to be parallel to the vector $\mathbf{p}_m$ and by choosing the time t_m to be the time at which the primer magnitude has a maximum value. Choosing the position of the impulse so that the velocity change is in the direction of the primer vector sounds familiar because it is one of the necessary conditions derived previously, but how to determine this position is not at all obvious, and we will have to derive an expression for this. Choosing the time t_m to be the time of maximum primer magnitude does not guarantee that the decrease in cost is maximized, because the value of $\Delta\mathbf{v}_m$ in the expression for dJ depends on the value of t_m. However, all we are doing is obtaining an initial position and time of the midcourse impulse to begin an iteration process. As long as our initial choice represents a decrease in the cost, we will opt for the simple device of choosing the time of maximum primer magnitude as our initial estimate of t_m. In Figure 2.7, t_m is 0.1.

Having determined an initial value for t_m, the initial position of the impulse, namely the value $\delta\mathbf{r}_m$ to be added to $\mathbf{r}_m$, must also be determined. Obviously $\delta\mathbf{r}_m$ must be nonzero, otherwise the midcourse impulse would have zero magnitude. The property that must be satisfied in determining $\delta\mathbf{r}_m$ is that $\Delta\mathbf{v}_m$ be parallel to $\mathbf{p}_m$. The analysis of [3] results in

$$\Delta\mathbf{v}_m = \mathbf{A}\delta\mathbf{r}_m \tag{2.64}$$

where the matrix $\mathbf{A}$ is defined as

$$\mathbf{A} \equiv -(\mathbf{M}_{fm}^T \mathbf{N}_{fm}^{-T} + \mathbf{T}_{mo} \mathbf{N}_{mo}^{-1}). \tag{2.65}$$

Next, in order to have $\Delta \mathbf{v}_m$ parallel to $\mathbf{p}_m$, it is necessary that $\Delta \mathbf{v}_m = \varepsilon \mathbf{p}_m$ with scalar $\varepsilon > 0$. Combining this fact with Equation (2.64) yields

$$\mathbf{A} \delta \mathbf{r}_m = \Delta \mathbf{v}_m = \varepsilon \mathbf{p}_m \tag{2.66}$$

which yields the solution for $\delta \mathbf{r}_m$ as

$$\delta \mathbf{r}_m = \varepsilon \mathbf{A}^{-1} \mathbf{p}_m \tag{2.67}$$

assuming $\mathbf{A}$ is invertible.

The question then arises how to select a value for the scalar ε. Obviously too large a value will violate the linearity assumptions of the perturbation analysis. This is not addressed in [3], but one can maintain a small change by specifying

$$\frac{\delta r_m}{r_m} = \beta \tag{2.68}$$

where β is a specified small positive number such as 0.05. Equation (2.67) then yields a value for ε

$$\frac{\varepsilon \left| \mathbf{A}^{-1} \mathbf{p}_m \right|}{r_m} = \beta \Rightarrow \varepsilon = \frac{\beta r_m}{\left| \mathbf{A}^{-1} \mathbf{p}_m \right|}. \tag{2.69}$$

If the resulting $dJ \geq 0$, then decrease ε and repeat Equation (2.67). One should never accept a midcourse impulse position that does not decrease the cost, because a sufficiently small ε will always provide a lower cost.

The initial values of midcourse impulse position and time are now determined. One adds the $\delta \mathbf{r}_m$ of Equation (2.67) to the value of $\mathbf{r}_m$ on the nominal trajectory at the time t_m at which the primer magnitude achieves its maximum value (greater than one).

The primer history after the addition of the initial midcourse impulse is shown in Figure 2.8. Note that $p_m = 1$ but $\dot{\mathbf{p}}_m$ is discontinuous and the primer magnitude exceeds unity, both of which violate the NC. However, the addition of the midcourse impulse has decreased the cost slightly, from 1.7555 to 1.7549.

2.4.3 Iteration on a Midcourse Impulse Position and Time

To determine how to efficiently iterate on the components of position of the mid-course impulse and its time, one needs to derive expressions for the gradients of the cost with respect to these variables. To do this, one must compare the three-impulse trajectory (or three-impulse segment of an n-impulse trajectory) that resulted from the addition of the midcourse impulse with a perturbed three-impulse trajectory.

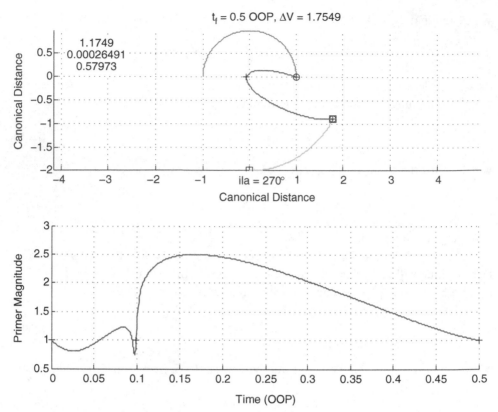

Figure 2.8. Initial (nonoptimal) three-impulse primer magnitude.

Note that, unlike a terminal coast, the values of $d\mathbf{r}_m$ and dt_m are *independent*. (By contrast, on an initial coast $d\mathbf{r}_o = \mathbf{v}_o^- dt_o$ and on a final coast $d\mathbf{r}_f = \mathbf{v}_f^+ dt_f$.)

The cost on the nominal three-impulse trajectory is

$$J = \Delta v_o + \Delta v_m + \Delta v_f \tag{2.70}$$

and the variation in the cost due to perturbing the midcourse time and position is

$$dJ = \frac{\partial \Delta v_o}{\partial \Delta \mathbf{v}_o} d\Delta \mathbf{v}_o + \frac{\partial \Delta v_m}{\partial \Delta \mathbf{v}_m} d\Delta \mathbf{v}_m + \frac{\partial \Delta v_f}{\partial \Delta \mathbf{v}_f} d\Delta \mathbf{v}_f \tag{2.71}$$

which, analogous to Equation (2.56), can be written as

$$dJ = \mathbf{p}_o^T d\Delta \mathbf{v}_o + \mathbf{p}_m^T d\Delta \mathbf{v}_m + \mathbf{p}_f^T d\Delta \mathbf{v}_f. \tag{2.72}$$

The analysis of [3] leads to the result that

$$dJ = \left(\dot{\mathbf{p}}_m^+ - \dot{\mathbf{p}}_m^- \right)^T d\mathbf{r}_m - \left(\dot{\mathbf{p}}_m^{T+} \mathbf{v}_m^+ - \dot{\mathbf{p}}_m^{T-} \mathbf{v}_m^- \right) dt_m. \tag{2.73}$$

In Equation (2.73), a discontinuity in $\dot{\mathbf{p}}_m$ has been allowed because there is no guarantee that it will be continuous at the inserted midcourse impulse, as demonstrated in Figure 2.8.

Equation (2.73) can be written more simply in terms of the Hamiltonian function Equation (2.22) for $p_m = 1$: $H_m = \dot{\mathbf{p}}_m^T \mathbf{v}_m - \mathbf{p}_m^T \mathbf{g}_m$ (for which the second term $\mathbf{p}_m^T \mathbf{g}_m$ is continuous because $\mathbf{p}_m = \Delta \mathbf{v}_m / \Delta v_m$) and $\mathbf{g}_m(\mathbf{r}_m)$ are continuous).

$$dJ = (\dot{\mathbf{p}}_m^+ - \dot{\mathbf{p}}_m^-)^T d\mathbf{r}_m - (H_m^+ - H_m^-)dt_m. \tag{2.74}$$

Equation (2.74) provides the gradients of the cost with respect to the independent variations in the position and time of the midcourse impulse for use in a nonlinear programming algorithm:

$$\frac{\partial J}{\partial \mathbf{r}_m} = (\dot{\mathbf{p}}_m^+ - \dot{\mathbf{p}}_m^-) \tag{2.75}$$

and

$$\frac{\partial J}{\partial t_m} = -(H_m^+ - H_m^-). \tag{2.76}$$

As a solution satisfying the NC is approached, the gradients tend to zero, in which case both the primer rate vector $\dot{\mathbf{p}}_m$ and the Hamiltonian function H_m become continuous at the midcourse impulse.

Note that when the NC are satisfied, the gradient with respect to t_m in Equation (2.76) being zero implies that

$$H_m^+ - H_m^- = 0 = \dot{\mathbf{p}}_m^T(\mathbf{v}_m^+ - \mathbf{v}_m^-) = \dot{\mathbf{p}}_m^T \Delta \mathbf{v}_m = \Delta v_m \dot{\mathbf{p}}_m^T \mathbf{p}_m = 0 \tag{2.77}$$

which, in turn, implies that $\dot{p}_m = 0$, indicating that the primer magnitude attains a local maximum value of unity. This is consistent with the NC that $p \leq 1$ and that $\dot{\mathbf{p}}$ be continuous.

Figure 2.2 shows the converged, optimal three-impulse trajectory that results from improving the primer histories shown in Figures 2.7 and 2.8. Note that the final cost of 1.3681 is significantly less that the value of 1.7555 prior to adding the midcourse impulse. Also, the time of the midcourse impulse changed during the iteration from its initial value of 0.1 to a final value of approximately 0.17.

The absolute minimum cost solution for the final radius and ila value of Figure 2.2 is, of course, the Hohmann transfer shown in Figure 2.4. Its cost is significantly less at 0.28446, but the transfer time is nearly three times as long at 1.8077 OOP. Of this, 0.889 OOP is an initial coast to achieve the correct target phase angle for the Hohmann transfer. Depending on the specific application, the total time required may be unacceptably long.

(As a side note, a simple proof of the global optimality of the Hohmann transfer using ordinary calculus rather than primer vector theory is given in [12].)

REFERENCES

[1] Bryson, A. E., and Ho, Y-C. (1975) *Applied Optimal Control,* Hemisphere Publishing Co., Washington DC.

[2] Lawden, D. F. (1963) *Optimal Trajectories for Space Navigation,* Butterworths, London.

[3] Lion, P. M., and Handelsman, M. (1968) Primer Vector on Fixed-Time Impulsive Trajectories. *AIAA Journal,* **6**, No. 1, 127–132.

[4] Jezewski, D. J., and Rozendaal, H. L. (1968) An Efficient Method for Calculating Optimal Free-Space n-impulse Trajectories. *AIAA Journal,* **6**, No. 11, 2160–2165.

[5] Prussing, J. E. (1995) Optimal Impulsive Linear Systems: Sufficient Conditions and Maximum Number of Impulses, *The Journal of the Astronautical Sciences,* **43**, No. 2, 195–206.

[6] Prussing, J. E., and Chiu, J-H. (1986) Optimal Multiple-Impulse Time-Fixed Rendezvous between Circular Orbits, *Journal of Guidance, Control, and Dynamics,* **9**, No. 1, 17–22. also *Errata,* **9**, No. 2, 255.

[7] Prussing, J. E. (1993) Equation for Optimal Power-Limited Spacecraft Trajectories, *Journal of Guidance, Control, and Dynamics,* **16**, No. 2, 391–393.

[8] Prussing, J. E., and Sandrik, S. L. (2005) Second-Order Necessary Conditions and Sufficient Conditions Applied to Continuous-Thrust Trajectories, *Journal of Guidance, Control, and Dynamics,* **28**, No. 4, 812–816.

[9] Glandorf, D. R. (1969) Lagrange Multipliers and the State Transition Matrix for Coasting Arcs, *AIAA Journal,* **7**, Vol. 2, 363–365.

[10] Battin, R. H. (1999) *An Introduction to the Mathematics and Methods of Astrodynamics,* Revised Edition, AIAA Education Series, New York.

[11] Prussing, J. E., and Conway, B. A. (1993) *Orbital Mechanics,* Oxford University Press, New York.

[12] Prussing, J. E. (1992) Simple Proof of the Global Optimality of the Hohmann Transfer, *Journal of Guidance, Control, and Dynamics,* **15**, No. 4, 1037–1038.

[13] Sandrik, S. (2006) Primer-Optimized Results and Trends for Circular Phasing and Other Circle-to-Circle Impulsive Coplanar Rendezvous. Ph.D. Thesis, University of Illinois at Urbana-Champaign.

3 Spacecraft Trajectory Optimization Using Direct Transcription and Nonlinear Programming

Bruce A. Conway
Dept. of Aerospace Engineering, University of Illinois at Urbana-Champaign, Urbana, IL

Stephen W. Paris
Boeing Research & Technology, Seattle, WA

3.1 Introduction

A spacecraft in flight is a dynamical system. As dynamical systems go, it is comparatively straightforward; the equations of motion are continuous and deterministic, for the unforced case they are essentially integrable, and perturbations, such as the attractions of bodies other than the central body, are usually small. The difficulties arise when the complete problem, corresponding to a real space mission, is considered. For example, a complete interplanetary flight, beginning in Earth orbit and ending with insertion into Mars orbit, has complicated, time-dependent boundary conditions, straightforward equations of motion but requires coordinate transformations when the spacecraft transitions from planet-centered to heliocentric flight (and vice versa), and likely discrete changes in system states as the rocket motor is fired and the spacecraft suddenly changes velocity and mass. If low-thrust electric propulsion is used, the system is further complicated as there no longer exist integrable arcs and the decision variables, which previously were discrete quantities such as the times, magnitudes and directions of rocket-provided impulses, now also include continuous time histories, that is, of the low-thrust throttling and of the thrust pointing direction. In addition, it may be optimizing to use the low-thrust motor for finite spans of time and "coast" otherwise, with the optimal number of these thrust arcs and coast arcs *a priori* unknown.

Since the cost of placing a spacecraft in orbit, which is usually the first step in any trajectory, is so enormous, it is particularly important to optimize space trajectories so that a given mission can be accomplished with the lightest possible spacecraft and within the capabilities of existing (or affordable) launch vehicles. Determining the necessary conditions (the Euler-Lagrange equations, possibly supplemented by Pontryagin's principle) for the optimization of a continuous, deterministic dynamical system of the type corresponding to a spacecraft in flight is also a straightforward problem, although the result, especially for the common case in which staging or impulsive ΔV's are used, is a sophisticated two-point-boundary-value-problem (TPBVP) with interior point constraints.

As described in Chapter 1, the solution of this TPBVP, except for certain special cases, is very difficult. The principal difficulties are attributable to the costate or Lagrange multiplier variables of the problem, equal in number to the system state variables. The use of the analytical necessary conditions of the problem immediately doubles its size; in addition, the costate variables lack the physical significance of the state variables, may differ by several orders of magnitude from the state variables and may have discontinuities at the junctions of constrained and unconstrained arcs in the solution. As also described in Chapter 1, the common approach to the solution of a TPBVP, some form of "shooting," is thus very problematic.

Thus, beginning in the 1960s, new approaches to the solution of the TPBVP resulting from optimization of the dynamic system were sought. Many solutions were developed; the common principles were to convert the continuous problem into a parameter optimization problem and to eliminate the "shooting" approach in favor of a solution method in which all of the free parameters are adjusted contemporaneously. Methods of this type include solutions using finite differences [1] [2] and the collocation method [3]. Collocation is perhaps the best known and most implemented direct transcription method. The state and control histories are represented by discrete values on a mesh. In one basic form of collocation [4], the state history between mesh points (for each state variable) is then represented using Hermite cubic polynomials. The control history may be represented using discrete control parameters or, for example, with a cubic polynomial (in time). The equations of motion are enforced at the mesh points, that is, the slopes of the Hermite cubic polynomials representing the states are in agreement with the system differential equations evaluated at the mesh points. The equations of motion are further enforced using a collocation scheme; a collocation point is chosen at the center of each mesh segment, that is, the Hermite cubic state polynomial is constrained to satisfy the system differential equations between mesh points as well as at mesh points. This yields nonlinear constraint equations involving the state and control variables. If these constraints are satisfied, the system is said to be "implicitly" integrated. Additional constraints, linear and nonlinear, equality and inequality, may be present involving (typically) initial and terminal values of the states and magnitudes of the controls. A nonlinear programming (NLP) problem solver is used to enforce the constraints while simultaneously minimizing the problem objective.

In its original form, for example in the work of Dickmanns and Well [5], the collocation method employed the analytical necessary conditions and thus required a system of governing equations including the costate variables. It was thus, as described in Chapter 1, an "indirect" optimization. It had previously been shown, however, that by discretizing the problem, specifically by representing the states by patched Chebyshev polynomials [5] [6], and by constraining the discrete parameters to satisfy algebraic equations that caused the system to be integrated implicitly, the problem could be converted (or "transcribed") into an NLP problem. That is, the continuous dynamic system could be optimized without the use of the necessary conditions or the costate variables in, as described in Chapter 1, a "direct" optimization. Progress in direct optimization was then rapid. Hargraves and Paris [4] simplified the direct optimization process considerably. They recognized that the collocation method, in the indirect form as developed by Dickmanns and Well, could be converted into a direct optimizer if the NLP problem solver were to be used to minimize the objective directly, in part by adding the controls to the system as parameters to be determined rather than to solve a discrete form of the TPBVP resulting from the analytical necessary conditions, in which the controls are found indirectly through the Maximum Principle.

The collocation method developed by Hargraves and Paris became rapidly adopted. A fortunate circumstance was that at precisely the time when a need arose for efficient NLP problem solvers, capable of solving the large, sparse NLP problems resulting from the application of the collocation method to sophisticated problems, new solvers were developed [8]. The method was further developed and improved; to include other more accurate implicit integration rules [9] [10], to refine the discretization time steps to accommodate the dynamics, for greater accuracy [9] [10], and to enable the use of multiple time scales for problems with both slow and fast dynamics (such as six degrees of freedom vehicle motion) [11] [12].

Other collocation methods have also been developed, the most popular alternate method being a Gauss pseudospectral scheme [13]. However, all of the direct collocation methods that assume the states are represented by basis functions (in time) can be shown to be derivable from the same analysis [14]. Other direct transcription methods have been developed that employ implicit integration but not collocation, that is, those that do not assume that the state variable histories are described by polynomials. One such method is based on explicit Runge-Kutta (RK) integration and parallel shooting and has been shown to be particularly beneficial for very-low-thrust orbit transfers where the states change only very slowly but the controls change rapidly within every revolution [15].

Direct transcription schemes have a number of advantages over other numerical optimization methods. Since there are no costate variables, the problem size is reduced by a factor of two, and the problematic estimation of initial costates is avoided. Also, to a degree, one does not have to specify *a priori* the precise structure of the problem. For example, for a spacecraft trajectory problem, if a solution structure of multiple coast/thrust arcs is assumed, unneeded arcs will collapse to zero length. Similarly, if an impulsive thrust solution is assumed, unneeded impulses will be given zero magnitude. Direct transcription schemes also generally show much better robustness, that is, show ability to converge to an optimal trajectory from poor initial guesses, in comparison to other numerical optimization methods.

This chapter will begin by introducing the direct transcription process in several of these implementations and it will describe how a potential user might choose the most advantageous method for a given problem. Then the method, which has up to then been applied to an arbitrary dynamical system, will be specialized to the case of spacecraft trajectory optimization. This requires a discussion of choice of coordinates, for example polar or Cartesian coordinates, conventional elliptic elements, or Delaunay variables, for a given problem. It also requires discussion of control modeling and parameterization, grid refinement, and transformations of coordinates required for interplanetary flight when a spacecraft leaves the sphere of influence (SOI) of one body and enters another. Another important topic is the successful generation of an initial guess of the solution, that is, a guess of the optimal trajectory that is sufficiently accurate that the NLP solver can improve on it and converge to the local minimum of the objective.

3.2 Transcription Methods

3.2.1 A Basic Collocation Method (Using Hermite Polynomials)

This classic approach transforms an optimal control problem into a nonlinear programming problem. To do this, the state functions $x(t)$ and control functions $u(t)$ are represented by piecewise polynomials and collocation is used to satisfy the differential equations of motion. The first step with this approach is to subdivide the problem into a sequence of smaller trajectory arcs, which we will call phases. The length of each phase is defined as $T_{pi} = E_{i+1} - E_i$. For each phase, the interval $[E_i, E_{i+1}]$ is further subdivided into N segments, as shown in Figure 3.1.

Let the ratio of the length of the jth segment to T_{pi} be denoted as τ_{ji}. Thus, the length of the jth segment in the ith stage is $\tau_{ji} T_{pi}$. Using Hermite interpolation, cubic polynomials are defined for each state on each segment using values of the states at the nodes (the boundaries at the segments) and the state time derivatives, as defined by the equations, at the nodes. The values of the states are then selected (by nonlinear programming) to force the interpolated derivatives to agree with the differential equations at the center of the segment. This procedure is illustrated in Figure 3.2.

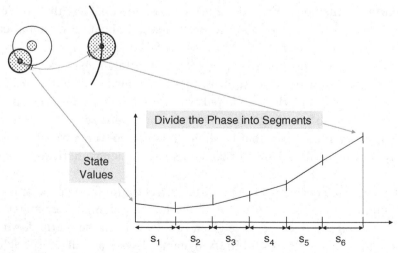

Figure 3.1. Trajectory optimization problem broken into phases.

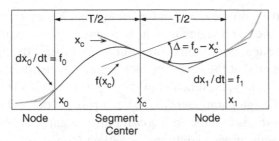

Figure 3.2. Illustration of implicit integration.

The basic procedure can be derived for any polynomial order. The classic cubic derivation is as follows. Let the states x be represented on each segment by cubics of the form

$$x = C_0 + C_1 S + C_2 S^2 + C_3 S^3 \tag{3.1}$$

where to simplify the discussion, the segment length S is transformed such that $S \in [0, 1]$. Let $x(0) = x_0$, $x(l) = x_1$, $dx/ds(0) = x_0'$, $dx/ds(l) = x_1'$. Differentiating Equation (3.1) and evaluating at $S = 0$ and $S = 1$ yields

$$\begin{bmatrix} 1 & 0 & 0 & 0 \\ 0 & 1 & 0 & 0 \\ 1 & 1 & 1 & 1 \\ 0 & 1 & 2 & 3 \end{bmatrix} \begin{bmatrix} C_0 \\ C_1 \\ C_2 \\ C_3 \end{bmatrix} = \begin{bmatrix} x_0 \\ x_0' \\ x_1 \\ x_1' \end{bmatrix}. \tag{3.2}$$

Inverting the 4×4 matrix yields

$$\begin{bmatrix} C_0 \\ C_1 \\ C_2 \\ C_3 \end{bmatrix} = \begin{bmatrix} 1 & 0 & 0 & 0 \\ 0 & 1 & 0 & 0 \\ -3 & -2 & 3 & -1 \\ 2 & 1 & -2 & 1 \end{bmatrix} \begin{bmatrix} x_0 \\ x_0' \\ x_1 \\ x_1' \end{bmatrix}. \tag{3.3}$$

Now using Equation (3.3), evaluating Equation (3.1) at $S = 1/2$, and transforming to segment length $T = g\tau_{ji} T_{pi}$, we see that the interpolated value of x at the center of the segment is

$$x_c = (x_0 + x_1)/2 + T(f_0 + f_1)/8 \tag{3.4}$$

where f_i, is the time rate of change as defined by the equations of motion evaluated at x_i.

(Note $x_0' = Tf[x_0, u(t_0), t_0, \omega]$.) In the same way, the slope at the center

$$x_c' = -3(x_0 - x_1)/2T - (f_0 - f_1)/8$$

is obtained. Evaluating the equations of motion at x_c gives f_c. We can then define the defect (the error in representing the dynamics) at the center of the segment as

$$\Delta = f_c - x$$
$$= f_c + 3(x_0 - x_1)/2T + (f_0 + f_1)/4 \tag{3.5}$$

x_0 and x_1 are varied to enforce $\Delta = 0$. If the cubic polynomial is capable of representing the solution on the given segment, then selecting x_0 and x_1 to drive Δ to zero will produce an accurate approximation to the solution of the state dynamics. The defects for each state evaluated at the center of each segment constitute a set of nonlinear algebraic equations that are a function of the states and controls at each node, the events E, and the design parameters ω. The controls are defined at $S = [0, 1/2, 1]$. The boundary conditions and the path constraints evaluated at both the nodes and centers of the segments provide additional equations (constraints) to be satisfied. All of the independent variables may be collected into a single vector P defined by

$$P^T = [Z^T, E^T, \omega^T] \tag{3.6}$$

Where

$$Z^T = [x_0^T, u_0^T, x_1^T, u_1^T, \ldots \ldots x_{N+1}^T, u_{N+1}^T). \tag{3.7}$$

Collecting all of the nonlinear equations into a single vector equation yields

$$C^T = (\Delta^T, B_N^T, H_N^T) \tag{3.8}$$

with

$\Delta^T = (\Delta_{00}, \Delta_{01}, \ldots, \Delta_{ij} \ldots)$ where $\Delta_{ij} = $ defect for ith state at jth node
$B_N = $ collection of all boundary conditions
$H_N = $ collection of all path constraints

The performance index J is just a function of P. The trajectory optimization problem stated above can be expressed as: minimize $\Phi(P)$, subject to

$$l \le \left\{ \begin{array}{c} P \\ AP \\ C(P) \end{array} \right\} \le u. \tag{3.9}$$

AP is composed of all the linear relationships from Equation (3.8), l and u are the upper and lower bounds for the parameter vector. For equality constraints $l = u$. The upper and lower bounds for the great majority of the nonlinear constraints are usually set to zero, because this forces the solver to choose values for the parameters that satisfy the EOMs. (There may be a small number of additional nonlinear constraints, for example boundary conditions).

System (3.6)–(3.9) constitutes a nonlinear programming problem.

The extension of the preceding analysis to higher order integrators is straightforward. Again, a phase is subdivided into segments. The segments do not need to be of equal length. The rule of thumb, analogous to multistep explicit integrators, is that more segments should be placed where there are rapid changes in the states, controls or constraints.

Within a segment, the values of the states and controls are defined at discrete time points called nodes. For convenience, we define the placement of the nodes relative to the normalized segment lengths. An example set of nodes within a segment is shown in Figure 3.3.

The placement of nodes within a segment can be arbitrary; however, experience has shown benefits to placing the nodes at the Legendre-Gauss-Lobatto (LGL) points. This is accomplished by mapping the segment onto the interval -1 to 1 and using the LGL points in "standard" form. The odd points are called cardinal nodes (τ_1, τ_3, τ_5), while the even points (τ_2, τ_4) are labeled interior nodes. Typically, the number of cardinal nodes per segment is constant within a phase.

The state values specified at the cardinal nodes will be used as parameters for the resulting nonlinear programming problem, while the state values at the interior nodes will be obtained by interpolation. The equations of motion are evaluated to provide time derivative values at the cardinal nodes. Making use of both states values and their time derivatives at the cardinal nodes, Hermite interpolation is used to construct a polynomial representation of the states. This polynomial representation is then evaluated at the interior nodes to compute interpolated values of the states, which in turn are used to compute the state time derivatives at the interior nodes by evaluating the equations of motion. The polynomial representation is also differentiated and evaluated at the interior nodes to compute yet another set of time derivatives. The differences between the slopes as defined by the state representation and the equations of motion at the interior nodes are called the defects. The control functions are represented by values at both the cardinal and interior nodes. This procedure is illustrated in Figure 3.4.

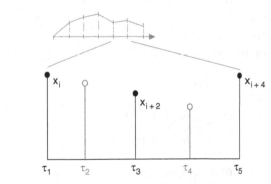

Figure 3.3. Cardinal node segment structure.

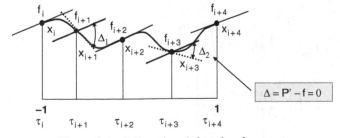

Figure 3.4. Collocation defect development.

This procedure can also be summarized as:

Using a Hermite interpolating polynomial, for each state, we can write

$$x(\tau) = \{u(\tau)\}^T\{x\} + \{v(\tau)\}^T\{f\} = \sum_{i=1}^{(n+1)/2} u_i(\tau)x(\tau_i) + \sum_{i=1}^{(n+1)/2} v_i(\tau)f(\tau_i). \quad (3.10)$$

The state values at the interior nodes can then be computed as

$$\{x_*\} = [E]\{x\} + \frac{T}{2}[F]\{f\}. \quad (3.11)$$

The matrices $[E]$ and $[F]$ are determined from the Hermite polynomial; T is segment length in "dimensional" time. This also allows

$$x'(\tau) = \{u'(\tau)\}^T\{x\} + \{v'(\tau)\}^T\{f\} = \sum_{i=1}^{(n+1)/2} u'_i(\tau)x(\tau_i) + \sum_{i=1}^{(n+1)/2} v'_i(\tau)f(\tau_i) \quad (3.12)$$

Applying our definition of defect, the vector $\{\Delta\}$ of defects at the interior nodes is

$$\{\Delta\} = [A]\{x\} + \frac{T}{2}[B]\{f\} + \frac{T}{2}[C]\{f_*\}. \quad (3.13)$$

with

$\{x\}$ states at the cardinal nodes

$\{f\}$ slopes from the equations of motion at the cardinal nodes

$\{f_*\}$ slopes from the equations of motion at the interior nodes, x_*

T segment length (needed to convert from "τau" space to "dimensional" time

$[A]$ Matrix of constants

$[B]$ Matrix of constants

$[C]$ Identity Matrix multiplied by -1

The elements of $[A]$ and $[B]$ are just

$$a_{i,j} = u'_j(\tau_i)$$

$$b_{i,j} = v'_j(\tau_i)$$

which are easily built by applying the product rule of differentiation to ACM algorithm 211 [16] (used to compute $u\&v$)

$$u_j(\tau) = \left[\prod_{\substack{i=1 \\ i \neq j}}^{(n+1)/2} \left(\frac{(\tau - \tau_i)}{(\tau_j - \tau_i)} \right)^2 \right] \cdot \left(1 - (\tau - \tau_j) \cdot 2 \cdot \sum_{\substack{i=1 \\ i \neq j}}^{(n+1)/2} \frac{1}{(\tau_j - \tau_i)} \right)$$

$$v_j(\tau) = \left[\prod_{\substack{i=1 \\ i \neq j}}^{(n+1)/2} \left(\frac{(\tau - \tau_i)}{(\tau_j - \tau_i)} \right)^2 \right] \cdot (\tau - \tau_j). \quad (3.14)$$

Notice that the elements of $[A]$, $[B]$, $[C]$, $[D]$, $[E]$, and $[F]$ only depend on the values of τ (the node locations), which are fixed. This means that the matrices need only to be computed once and stored. Combining the segments results in the matrices taking on a banded structure, as shown below for a 2 segment, 3 cardinal node case

$$
\begin{Bmatrix} \Delta_1 \\ \Delta_2 \\ \Delta_3 \\ \Delta_4 \end{Bmatrix} = \begin{bmatrix} -1.6654 & 1.4963 & 0.1701 & 0 & 0 \\ -0.1701 & 1.4963 & 1.6654 & 0 & 0 \\ 0 & 0 & -1.6654 & 1.4963 & 0.1701 \\ 0 & 0 & -0.1701 & 1.4963 & 1.6654 \end{bmatrix} \begin{Bmatrix} x_1 \\ x_3 \\ x_5 \\ x_7 \\ x_9 \end{Bmatrix}
$$

$$
+ \frac{T}{2} \begin{bmatrix} -0.1386 & -0.6530 & -0.0450 & 0 & 0 \\ -0.0450 & -0.6530 & -0.1386 & 0 & 0 \\ 0 & 0 & -0.1386 & -0.6530 & -0.0450 \\ 0 & 0 & -0.0450 & -0.6530 & -0.1386 \end{bmatrix} \begin{Bmatrix} f_1 \\ f_3 \\ f_5 \\ f_7 \\ f_9 \end{Bmatrix}
$$

$$
+ \frac{T}{2} \begin{bmatrix} -1 & 0 & 0 & 0 \\ 0 & -1 & 0 & 0 \\ 0 & 0 & -1 & 0 \\ 0 & 0 & 0 & -1 \end{bmatrix} \begin{Bmatrix} f_2 \\ f_4 \\ f_6 \\ f_8 \end{Bmatrix} \tag{3.15}
$$

Again we combine the state values at the cardinal nodes, along with the event times and design parameters, in a single array Z. A constraint array is assembled from the boundary conditions, defects, and path constraint. This results in a tractable nonlinear programming problem.

3.2.2 Pseudospectral Methods

Pseudospectral techniques have gained popularity within the last several years [17] [18] [19] [20]. They are very similar to the method of collocation. The major differences are with the manner in which the representative polynomial is constructed and the computation of the defects. When the pseudospectral method is employed, a polynomial representation of the states is formed and a number of constraints are enforced so that the slopes of the polynomials agree with the differential equations at a finite number of points. Values of the states are established at the cardinal and interior nodes. Making use of the state values, Lagrange interpolation is used to construct a polynomial interpolation of the states. This polynomial is differentiated and evaluated at all the nodes to compute interpolated time derivative values of the states. The equations of motion are evaluated to provide the values of the physical time derivatives. The difference between the two sets of time derivatives forms the defects. The procedure is shown in Figure 3.5.

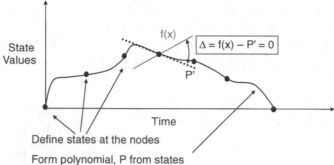

Define states at the nodes

Form polynomial, P from states

At each node require

　　Slope of the polynomial, P' = state time derivative, f(x)(rhs)

Figure 3.5. Illustrating the pseudospectral procedure.

To summarize, knowing the values of x at discrete time points τ_i, then

$$p(\tau) = \{w(\tau)\}^T \{x\} = \sum_{i=1}^{n} w_i(\tau) x(\tau_i). \qquad (3.16)$$

This is simply the Lagrange form of the interpolating polynomial where

$$w_i(\tau) = \frac{\prod_{j=1, j \neq i}^{n} (\tau - \tau_j)}{\prod_{j=1, j \neq i}^{n} (\tau_i - \tau_j)}. \qquad (3.17)$$

The derivative is simply

$$p'(\tau) = \{w'(\tau)\}^T \{x\} = \sum_{i=1}^{n} w'_i(\tau) x(\tau_i). \qquad (3.18)$$

Since we want the form

$$\{p'\} = [D] \{x\} \qquad (3.19)$$

([D] is known as the derivative matrix). The defects, for a first order system, can be written as

$$\{\Delta\} = \{p'\} - \{f\} = [D]\{x\} - \{f\} = 0. \qquad (3.20)$$

The elements of [D] are just $d_{i,j} = w'_j(\tau_i)$, which are easily computed by using the product rule of differentiation applied to $w_j(\tau_i)$.

Using this formulation for the pseudo-spectral method, the values of [D] are dependent only on the chosen node spacing. This allows for other pseudospectral variants to be generated by simply changing the nodal distribution.

The methods described in this section generally use a limited number of cardinal nodes per segment (<5) and use many segments. Historically the pseudospectral methods use many nodes per segment (>20) with a limited number of segments (1 or 2).

3.2.3 A Direct Method Not Using Collocation: R-K Parallel-Shooting

The collocation method has proved to be useful and robust for many problems, however, there are some problems that are best transcribed with a different approach. The method of direct transcription with Runge-Kutta (RK) integration and parallel-shooting is a direct method that also converts the optimization problem into a NLP problem [15]. For an optimization problem formed by a single arc, this method discretizes time into a sequence of stages that can be described by the partition $[t_0, t_1, \ldots, t_N]$, with $t_0 = 0, t_N = t_f$, where $t_0 < t_1 < \cdots < t_N$ and letting $h_i = t_i - t_{i-1}$ for $i = 1, \ldots, N$. The mesh points t_i are called nodes, and the $[t_{i-1}, t_i]$ intervals are termed segments. The state variables are approximated by parameter values at the nodes. Considering a single integration step per segment, the control variables are approximated at the nodes t_i, and at the center points, $t_{i-1} + h/2$, by $u_i = u(t_i)$ for $i = 0, \ldots, N$ and $v_i = u(t_{i-1} + h/2)$ for $i = 1, \ldots, N$. It is possible, however, to use multiple integration steps within each segment. This allows a decrease in the size of the time interval without significantly increasing the size of the NLP problem.

Let p be the number of integration steps in each segment. As before, the states and control variables are provided at the nodes. The controls within each segment are provided by $v_{ij} = u(t_{i-1} + jh/2p)$ for $j = 1, 2, \ldots, 2p - 1$ and for $i = 1, 2, \ldots, N$, as shown in Figure 3.6 for $p = 3$.

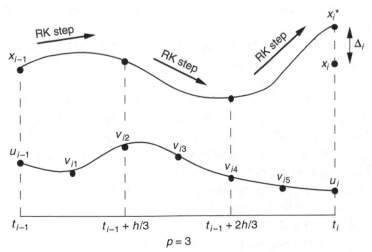

Figure 3.6. Structure for 3-step RK parallel-shooting scheme.

During the first RK step of this case, the governing equations are integrated forward from t_{i-1} to $t_{i-1} + h/p$ using the controls u_{i-1}, v_{i1}, and v_{i2}, and the fourth-order Runge-Kutta process [15]

$$y_{i1}^1 = x_{i-1} + \frac{1}{2p} hf(x_{i-1}, u_{i-1}) \tag{3.21}$$

$$y_{i1}^2 = x_{i-1} + \frac{1}{2p} hf(y_{i1}^1, v_{i1}) \tag{3.22}$$

$$y_{i1}^3 = x_{i-1} + \frac{1}{p} hf(y_{i1}^2, v_{i1}) \tag{3.23}$$

$$y_{i1}^4 = x_{i-1} + \frac{1}{6p} h \left[f(x_{i-1}, u_{i-1}) + 2f(y_{i1}^1, v_{i1}) + 2f(y_{i1}^2, v_{i1}) + f(y_{i1}^3, v_{i2}) \right]. \tag{3.24}$$

The second step makes use of the result of the first step, that is, it uses as initial states the values provided by the forward step Equation (3.24) and performs the integration from $t_{i-1} + h/p$ to $t_{i-1} + 2h/p$ using the controls v_{i2}, v_{i3}, and v_{i4}. Finally, the third step utilizes the result of the second one and executes the integration from $t_{i-1} + 2h/p$ to t_i using the controls v_{i4}, v_{i5}, and u_i, yielding an approximation of the state, x_i^*, at the node t_i as shown in Figure 3.6. The constraint equation

$$\Delta_i = y_{i3}^4 - x_i = 0; \quad \text{for} \quad i = 1, \ldots, N \tag{3.25}$$

is then applied at the right side node. If Equation (3.25) is satisfied, the system differential equations have then been integrated using the RK method.

Although $p - 1$ estimates of the state vector are computed inside each segment, they do not appear explicitly as parameters in the NLP problem, generating significant savings in the number of variables employed. System control variable parameters are specified much more frequently than system state variables. This is especially beneficial for problems, such as low-thrust trajectory optimization, where the control changes rapidly while the states, for example the orbit elements, change only slowly.

The NLP parameters can then be arranged as a single vector P that collects all the independent variables and the problem again becomes a NLP problem of form (3.6)–(3.9). Once the NLP problem is clearly defined, it can be solved by using dense or sparse solvers such as NPSOL and SNOPT [21]. SNOPT is prefered because it can take advantage of the sparsity present in the constraint Jacobian.

The advantage of the RK parallel-shooting method in comparison to collocation, for certain problems, is seen in the cartoon of Figure 3.6. For the three-step form of the algorithm shown in the figure, note that the control variable is specified at seven points but the state is an NLP parameter at only the left node and the right node of the segment. In comparison to H-S collocation or even the G-L higher-degree collocation described in Section 3.2, there are many more opportunities to insert controls in the mesh, that is, at points where the states need not also be defined. This becomes advantageous for problems in which the state variables change on a much slower timescale than the control variables. An obvious example is very-low-thrust orbit raising [22]. In this trajectory the states, if modeled for example using elliptic elements or equinoctial variables, change only slowly (excepting of course the longitude) but the optimal thrust pointing angle changes with a period corresponding to the orbit period. Thus a discretization similar to that shown in the cartoon, with a segment width corresponding approximately to one revolution, would be a satisfactory and economical (in terms of number of NLP parameters) one for that problem.

3.2.4 Comparison of Direct Transcription Methods

When applying direct transcription methods, there are several important things the user should be aware of. First, the discretization in time must be sufficient to adequately capture the problem dynamics. Second, care needs to be taken to verify that the optimal solution is actually found. As will be shown, a solution to the NLP problem is not necessarily the optimal trajectory. To illustrate these points we will examine a simple low-thrust transfer.

The objective of this problem is to maximize the final radius for a spacecraft with constant thrust to journey between two circular orbits (nominally Earth & Mars) in fixed time. The initial position and velocity are fixed. The final location is free. The final velocity vector is constrained to reflect circular motion at the final position. This problem has been solved by a variety of methods [3]. The vehicle characteristics are taken from Bryson & Ho [22]. Thrust = 0.85 lbs, propellant flow = 13.9 lbs/day, $w(0) = 10,000$ lbs, $r(0) = 1.0$ AU and $t_f = 193.0$ days. The thrust vector has a constant magnitude and a variable direction described by the control ϕ. The system is converted to canonical form. The dynamics are

$$
\begin{aligned}
\dot{u} &= \frac{v^2}{r} - \frac{\mu}{r^2} + \frac{T \sin \phi}{m} \\
\dot{v} &= \frac{uv}{r} + \frac{T \cos \phi}{m} \\
\dot{r} &= u \\
\dot{\theta} &= \frac{v}{r} \\
\dot{m} &= constant
\end{aligned}
\tag{3.26}
$$

where

u radial velocity
v tangential velocity
r radius
θ heliocentric longitude
m mass

The initial estimates of the trajectory and thrust vector history are shown in Figure 3.7. One can see that this initial guess is deliberately poor; it does not get even close to the terminal orbit.

Table 3.1. *Collocation results*

Method	Total Number of Nodes	Number of Nodes Per Segment	Number of Segments	CPU Time	Final Radius (performance Index)	Position Error	Explicit Integration Step Size
Collocation	13	2	6	0.17	1.52539406	−7.92E-03	.03
	13	3	3	0.29	1.52268557	−1.63E-05	.03
	13	4	2	1.05	1.52522892	−1.25E-05	.03
	9	5	1	0.28	1.51559982	5.00E-03	.03
	11	6	1	0.51	1.52035844	−1.25E-03	.03
	13	7	1	1.03	1.52247855	−2.43E-04	.03
	25	2	12	0.36	1.52525097	−1.62E-03	.03
	25	3	6	0.50	1.52521362	−2.16E-04	.03
	25	4	4	0.98	1.52522380	−3.32E-05	.03
	25	5	3	0.66	1.52504836	−1.34E-03	.03
	21	6	2	0.92	1.52523630	−4.59E-05	.03
	25	7	2	1.01	1.52524626	−6.13E-05	.03
	49	2	24	0.85	1.52523062	−1.44E-04	.03
	49	3	12	1.22	1.52524185	−3.60E-05	.03
	49	4	8	1.45	1.52524666	−1.34E-03	.03
	49	5	6	1.96	1.52524714	−2.86E-05	.03
	51	6	5	1.59	1.52525042	−3.84E-04	.03

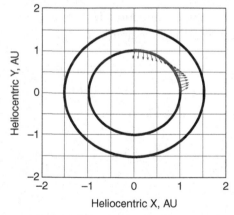

Figure 3.7. Initial Earth-Mars trajectory estimate.

Trajectories were generated using collocation, the pseudospectral method, and multiple shooting. The results when using collocation are shown in Table 3.1. The number of segments and nodes per segment were chosen to give a total number of nodes (both cardinal and interior) to be equal to 13, 25, and 49. Due to the integer relations of the segments and nodes per segment, it is not always possible to generate a specific number of total nodes, so the closest pairing were used. Table 3.1 reports the CPU times, optimal final radius value found, and a position error. This position error is the difference of the final radius as defined by the direct transcription method and the result of explicitly integrating the controls to the final time. A fixed-step,

Table 3.2. *Pseudospectral results*

Method	Total Number of Nodes	Number of Nodes Per Segment	Number of Segments	CPU Time	Final Radius (performance Index)	Position Error	Explicit Integration Step Size
Pseudo-	13	13	1	1.23	1.52379869	−1.48E-04	.03
Spectral	25	25	1	2.67	1.52524543	−3.67E-06	.03
	25	13	2	3.95	1.52515095	−6.24E-06	.03
	49	49	1	44.35	1.52198836	−3.01E-03	.03
	49	25	2	25.79	1.52510706	−2.27E-03	.03
	51	17	3	16.64	1.52524408	1.06E-05	.03

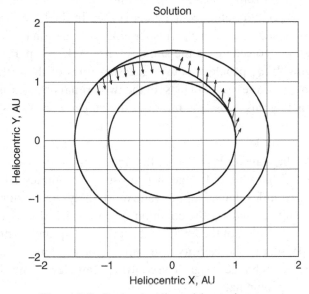

Figure 3.8. Optimal radius raising trajectory.

fourth-order Runge-Kutta method was used for all explicit integration (both for the check case and the multiple shooting scheme). The position error term is the sum of the differences between the position and velocity at the final time generated by the implicit integration and the same quantities obtained by explicitly integrating the optimal controls.

The collocation results show that a good solution can be achieved with a minimal number of nodes, however, care must be taken to ensure that sufficient nodes are used.

The results from using the pseudospectral method mirror those from collocation. Adding nodes in general improves the solution. One noticeable difference from the collocation results is in the amount of time needed to produce a solution. This increase is mainly caused by the less sparse nature of the pseudospectral approach.

If possible, for all of the methods, once an optimal solution is believed to have been found, a second solution using more nodes is recommended.

3.3 Selection of Coordinates

3.3.1 Motivation for Choice of Coordinate System

The choice of a coordinate system on which to express the trajectory optimization problem is fundamental and strongly influences the solution process. The main choice is whether to use coordinates, for example Cartesian or polar coordinates, in which case the system equations become

$$\dot{\bar{x}} = \bar{f} = \begin{bmatrix} \dot{\bar{r}} \\ \dot{\bar{v}} \end{bmatrix} = \begin{bmatrix} \bar{v} \\ \bar{g}(\bar{r}) + \Gamma \hat{u} \end{bmatrix} \tag{3.27}$$

or to use some form of orbital elements, for example conventional elliptic elements, Delaunay variables [24], or equinoctial elements [24]. In the latter case, there is no simple form of the system equations corresponding to Equation (3.27).

Using Cartesian coordinates is the simplest choice. It is a particularly natural choice for the three-body-problem as the Cartesian form of the system equations, with the system origin at the center of mass of the two primary bodies, is the most well known. Unfortunately this is also the most disadvantageous choice for describing a trajectory to be determined using direct transcription and NLP. This is due to the fact that the resulting numerical optimization problem is an NLP problem and the NLP solvers are most robust and efficient when the underlying state variable parameters change slowly and within a limited range. In Cartesian coordinates, all of the position and velocity state variables change rapidly (and most likely change sign periodically).

Another difficulty is that for the common case in which the term $\Gamma \hat{u}$ describes the magnitude and direction of the low-thrust propulsion applied, which direction is usually roughly parallel to the velocity, the unit vector will have measure numbers that rapidly change and change in sign. The control parameters in the NLP parameter vector are thus also problematic.

The situation is significantly improved if polar or cylindrical coordinates are used. The radius will always be positive and change only slowly. Angular position coordinates also change either slowly or rapidly (with the mean motion of the orbit) and predictably. For the usual case in which there is no retrograde motion, the angular velocities are also generally of one sign and do not change rapidly. The thrust pointing direction control, being primarily tangential, usually (for example for the case of low-thrust orbit raising) is described by small angles that change sign periodically, with a frequency related to the orbit frequency (or not at all). The system control parameters are thus in a form that improves or maintains the robustness of the solution using NLP.

The most advantageous choice is some form of orbital elements. The most commonly used are conventional elliptic elements, but if there is the possibility that the orbit will become circular or equatorial, a nonsingular set of elements such as equinoctial elements is a good choice [24]. Even when there is no possibility of a singularity, equinoctial elements are a very good choice to use for describing the orbit. When using orbital elements, it is generally the case that five of the six elements change only slowly (either due to perturbations or to the application of low-thrust propulsion) while the sixth element, which might be true anomaly, mean anomaly, or mean longitude, changes rapidly but with a mean value equal to (or approximately equal to) the mean motion. Of course all of the elements will be positive and can normally be assumed to lie within a quite restricted range, all of which improves the robustness of the numerical solution. When or if the position and velocity of the spacecraft are needed, for example, to satisfy a constraint representing interception or rendezvous with a target, these are easily obtained with analytical expressions relating the elements to Cartesian coordinates and velocities [25]. Another circumstance in which a transformation is required is if an impulsive velocity change of specified magnitude and direction is made. The elements after the impulse can be obtained by evaluating the position and velocity immediately prior to the impulse, then adding the impulse to the current velocity to obtain the post-impulse velocity, and then transforming from the post-impulse position and velocity to the new elements, also using a straightforward procedure [25].

When using orbital elements, the system variational equations are most commonly written in a form in which the thrust term $\Gamma \hat{u}$ is expressed using components in a satellite-fixed radial, tangential, and normal (to the instantaneous orbit plane) basis. Thus, as described above, for cases often encountered, the thrust pointing is nominally tangential (or anti-tangential) so that for the orbit element formulation also, the system control parameters change predictably and within a modest range and improve or maintain the robustness of the solution using NLP.

The equinoctial elements are simply related to the conventional elliptic elements; the semimajor axis a is a common element, the mean longitude λ and the parameters P_1, P_2, Q_1, and Q_2 replace the classical orbit elements e, i, Ω, ω, and f. The parameters P_1 and P_2 are defined by

$$P_1 = e \sin \bar{\omega} \tag{3.28}$$

and

$$P_2 = e \cos \bar{\omega} \tag{3.29}$$

where $\bar{\omega}$ is the "longitude of pericenter" defined by $\bar{\omega} = \Omega + \omega$. The parameters Q_1 and Q_2 are defined by

$$Q_1 = \tan\left(\frac{1}{2}i\right) \sin \Omega \tag{3.30}$$

and

$$Q_2 = \tan\left(\frac{1}{2}i\right) \cos \Omega \tag{3.31}$$

The mean longitude ℓ is defined by $\ell = \bar{\omega} + M = \bar{\omega} + n\Delta t$ where M is the mean anomaly defined by $M = E - e \sin E$ (which is Kepler's equation) and E is the eccentric anomaly. The parameter n is the mean orbital motion and Δt is the time elapsed since periapse passage. Now, the eccentric longitude K is defined by $K = \bar{\omega} + E$. Substituting for $M = \ell - \bar{\omega}$ and $E = K - \bar{\omega}$, Kepler's equation becomes

$$\ell = K + P_1 \cos K - P_2 \sin K \tag{3.32}$$

which is the augmented form of Kepler's equation. The true longitude L is defined as $L = \bar{\omega} + f$, where f is the true anomaly, and L is related to the eccentric longitude K through

$$\sin L = \frac{a}{r}\left[\left(1 - \frac{a}{a+b}P_2^2\right)\sin K + \frac{a}{a+b}P_1 P_2 \cos K - P_1\right] \tag{3.33}$$

and

$$\cos L = \frac{a}{r}\left[\left(1 - \frac{a}{a+b}P_1^2\right)\cos K + \frac{a}{a+b}P_1 P_2 \sin K - P_2\right]$$

where

$$\frac{a}{r} = 1/(1 - P_1 \sin K - P_2 \cos K) \tag{3.34}$$

and

$$\frac{a}{(a+b)} = 1/\left(1 + \sqrt{1 - P_1^2 - P_2^2}\right). \tag{3.35}$$

The classical orbit elements e, i, Ω, and ω may be recovered from the relationships

$$e^2 = P_1^2 + P_2^2, \tan^2\left(\frac{1}{2}i\right) = Q_1^2 + Q_2^2, \tan\bar{\omega} = \frac{P_1}{P_2}, \text{ and } \tan\Omega = \frac{Q_1}{Q_2}. \tag{3.36}$$

The variational equations for the equinoctial elements may be found in Battin [24] and are given below. The time rates of change of the semimajor axis a, mean longitude l, and the parameters P_1, P_2, Q_1 and Q_2 are

$$\frac{da}{dt} = 2\frac{a^2}{h}\left[(P_2 \sin L - P_1 \cos L)A_R + \frac{p}{r}A_T\right] \tag{3.37}$$

$$\frac{dl}{dt} = n - \frac{r}{h}\left\{\left[\frac{a}{a+b}\left(\frac{p}{r}\right)(P_1 \sin L + P_2 \cos L) + 2\frac{b}{a}\right]A_R\right.$$

$$\left. + \frac{a}{a+b}\left(1 + \frac{p}{r}\right)(P_1 \cos L - P_2 \sin L)A_T + (Q_1 \cos L - Q_2 \sin L)A_N\right\} \tag{3.38}$$

$$\frac{dP_1}{dt} = \frac{r}{h}\left\{-\frac{p}{r}\cos L\, A_R + \left[P_1 + \left(1 + \frac{p}{r}\right)\sin L\right]A_T\right.$$

$$\left. - P_2(Q_1 \cos L - Q_2 \sin L)A_N\right\} \tag{3.39}$$

$$\frac{dP_2}{dt} = \frac{r}{h} \left\{ \frac{p}{r} \sin L A_R + \left[P_2 + \left(1 + \frac{p}{r}\right) \cos L \right] A_T \right.$$
$$\left. + P_1 (Q_1 \cos L - Q_2 \sin L) A_N \right\} \tag{3.40}$$

$$\frac{dQ_1}{dt} = \frac{r}{2h} \left(1 + Q_1^2 + Q_2^2\right) \sin L \, A_N \tag{3.41}$$

and

$$\frac{dQ_2}{dt} = \frac{r}{2h} \left(1 + Q_1^2 + Q_2^2\right) \cos L \, A_N \tag{3.42}$$

where $b = a\sqrt{1 - P_1^2 - P_2^2}$ which is the semiminor axis and $h = nab$ which is the specific angular momentum. The useful ratios p/r and r/h are defined by $p/r = 1 + P_1 \sin L + P_2 \cos L$ and

$$r/h = \frac{h}{\mu(1 + P_1 \sin L + P_2 \cos L)} \tag{3.43}$$

respectively. The mean orbital motion is given by $n = \sqrt{\mu/a^3}$. The variables A_R, A_T, and A_N represent the perturbing accelerations; the components of the real forces acting on the spacecraft, apart from the central body attractions in the radial, transverse, and normal directions relative to the instantaneous orbit plane.

If the perturbing accelerations A_R, A_T, and A_N are all zero, then the differential equations (3.37–3.42) describe the trajectory of a spacecraft traveling in an orbit whose elements a, P_1, P_2, Q_1, and Q_2 are fixed but with mean longitude λ advancing at a constant rate. This will of course be the case for a spacecraft coasting, in one gravitational field, between the application of impulses. That might correspond to an interesting orbit transfer problem but, since the coasting arcs are integrable, it reduces to a parameter optimization problem that does not require the direct transcription that is the subject of this chapter. A more interesting and common case, which does require the tools developed in this chapter, is a finite-thrust transfer of a spacecraft that departs one body's gravitational field and enters that of another body (perhaps while en route to a third body as in an interplanetary transfer). The perturbations are best divided into two parts, one due to operation of the low-thrust propulsion system and the other collecting all the other disturbing accelerations, for example, oblateness and other higher-degree terms in the potential of the central body, "third-body" attractions, or solar radiation pressure, that is

$$A_R = \Gamma \cos \alpha \sin \beta + \Upsilon_R$$
$$A_T = \Gamma \cos \alpha \cos \beta + \Upsilon_T \tag{3.44}$$
$$A_N = \Gamma \sin \alpha + \Upsilon_N$$

where Γ is the spacecraft's instantaneous thrust acceleration magnitude, α is the thrust acceleration pointing angle measured out of the instantaneous orbit plane, and β is the in-plane thrust acceleration pointing angle. The angle α is positive if the thrust has a component in the direction of the orbit angular momentum vector. The angle β is zero if the thrust is directed normal to the radius vector and is positive if the thrust has a radially-outward component. The thrust pointing angles α and β are thus control variables for this problem. The thrust magnitude may also be a control, over a continuous range if the engine is throttleable or turned on or off if it is not.

3.3.2 Coordinate Transformations

When the spacecraft leaves the sphere of influence of one body and enters that of another, for example, for an Earth-to-Moon trajectory, the equations of motion, (3.27) or (3.37–3.42), become extraordinarily sensitive to changes in some or all of the state variables, which, as described in previous paragraphs, is very disadvantageous for a numerical solution using direct transcription and NLP. The obvious solution is to switch from one set of coordinates, centered on the body the spacecraft is departing, to a set centered on the body to which the spacecraft is arriving, with the switch occurring at the boundary of this body's sphere of influence.

The system equations (3.37–3.42) will not change because of the transformation, but the equations need to be re-initialized with new values for a, ℓ, P_1, P_2, Q_1, and Q_2 appropriate to the change from one central body to another. The most straightforward way to do this is to note that the six equinoctial elements completely specify the position and velocity (vectors) of the spacecraft with respect to the central body (and vice versa). Therefore, the transformation follows this simple process (at the point or time at which the space craft (s/c) leaves the SOI of body 1 for that of body 2):

1) Determine $\mathbf{r}$ and $\mathbf{v}$ of the spacecraft with respect to (w.r.t) central body 1 using the instantaneous values of a, ℓ, P_1, P_2, Q_1, and Q_2.
2) Determine the position and velocity of the s/c w.r.t central body 3. This will require knowledge of the position and velocity of body 3.
3) The position and velocity found in step (2) uniquely determine the values of the equinoctial elements appropriate for the case in which body 2 is the central body (and body 1 now the perturbing body).

This process is illustrated in the following example of an Earth–Moon low-thrust transfer. At any time

$$r_{ES} = r_{EM} + r_{MS} \tag{3.45}$$

$$v_{ES} = v_{EM} + v_{MS} \tag{3.46}$$

where r_{MS} and r_{ES} are the positions of the spacecraft relative to the Moon and Earth, respectively, r_{EM} is the position of the Moon relative to the Earth, v_{ES} is the velocity vector of the spacecraft relative to an Earth-centered inertial (ECI) coordinate system, v_{EM} is the velocity vector of the Moon relative to the ECI system, and v_{MS} is the velocity vector of the spacecraft relative to a Moon-centered nonrotating (MCNR) coordinate system.

Vectors r_{ES} and r_{MS} may be found from the following equation

$$r = [\mathbf{R}] \begin{bmatrix} r\cos L \\ r\sin L \\ 0 \end{bmatrix} \qquad (3.47)$$

where $r = a(1 - P_1\sin K - P_2\cos K)$ and

$$[\mathbf{R}] = \frac{1}{1 + Q_1^2 + Q_2^2} \begin{bmatrix} 1 - Q_1^2 + Q_2^2 & 2Q_1Q_2 & 2Q_1 \\ 2Q_1Q_2 & 1 + Q_1^2 - Q_2^2 & -2Q_2 \\ -2Q_1 & 2Q_2 & 1 - Q_1^2 - Q_2^2 \end{bmatrix} \qquad (3.48)$$

but with each position vector found as a function of the *local* equinoctial elements.

Similarly, the velocity vectors v_{ES} and v_{MS} of the spacecraft may be determined using

$$v = [\mathbf{R}] \left(\frac{u}{h}\right) \begin{bmatrix} -P_1 - \sin L \\ P_2 + \cos L \\ 0 \end{bmatrix}. \qquad (3.49)$$

but again with each velocity vector found as a function of the *local* equinoctial elements. The position of the Moon relative to the Earth, r_{EM}, may be found as

$$r_{EM} = \begin{bmatrix} a_M(\cos E_M - e_M) \\ a_M\sqrt{1 - e_M^2}\sin e_M \\ 0 \end{bmatrix} \qquad (3.50)$$

in the MCNR basis. It can be expressed in the ECI reference frame as

$$r_{EM} = [\mathbf{S}] \begin{bmatrix} a_M(\cos E_M - e_M) \\ a_M\sqrt{1 - e_M^2}\sin e_M \\ 0 \end{bmatrix} \qquad (3.51)$$

where

$$[S] = \begin{bmatrix} c\Omega_M c\omega_M - c\Omega_M s\omega_M sf_M & -c\Omega_M s\omega_M - s\Omega_M c\omega_M cf_M & s\Omega_M sf_M \\ s\Omega_M c\omega_M + c\Omega_M s\omega_M cf_M & -s\Omega_M s\omega_M + c\Omega_M c\omega_M cf_M & -c\Omega_M sf_M \\ s\omega_M sf_M & c\omega_M sf_M & cf_M \end{bmatrix} \quad (3.52)$$

is the transformation matrix from the MCNR basis to the ECI basis, as described in references [9] and [37]. Note that c denotes cos( ) and s denotes sin( ). The orbit elements Ω_M (longitude of the ascending node), ω_M (argument of perigee), and f_M (true anomaly) for the Moon at any instant in time may be obtained from an ephemeris.

The velocity of the Moon, v_{EM}, may be determined using

$$v_{EM} = [S] \begin{bmatrix} \dfrac{-\mu_E}{h_M} \sqrt{1 - e_M^2} \, \dfrac{\sin E_M}{1 - e_M \cos E_M} \\[2mm] \dfrac{\mu_E}{h_M}(1 - e_M^2) \dfrac{\cos E_M}{1 - e_M \cos E_M} \\[2mm] 0 \end{bmatrix} \tag{3.53}$$

where h_M is the specific angular momentum of the moon.

When Equations (3.47) and (3.49) are used to determine r_{MS} and v_{MS}, these vectors will be expressed naturally on the MCNR basis. So that all of the vectors in Equations (21) and (22) are expressed on the same (ECI) basis, r_{MS} and v_{MS} should be premultiplied by the [S] transformation matrix.

Now Equations (3.45, 3.46) may be written as

$$r_{ES} - r_{EM} = r_{MS} \tag{3.54}$$

$$v_{ES} - v_{EM} = v_{MS} \tag{3.55}$$

Then Equations (3.54) and (3.55) yield six nonlinear constraint equations involving the two sets of equinoctial variables; in these equations, only the variables appropriate to the Earth (body 1) as the central body appear on the LHS, and only the variables appropriate to the Moon (body 2) as the central body appear on the RHS. Enforcing these constraints when the problem is solved accomplishes the transformation of variables from the Earth-departure phase to the Moon-arrival phase. Fortunately the direct transcription with NLP formulation makes this transformation straightforward; these constraints need only be added to the many-times-more-numerous "defect" constraints accomplishing the implicit integration.

The implementation of this transformation in the direct transcription requires the use of a "knot," that is, a segment of zero width in which the state variables (here the equinoctial elements) are discontinuous.

There are some additional considerations with regard to the formulation of the system governing equations on each side of the knot. First note that defining $\Upsilon = \{\Upsilon_R, \Upsilon_T, \Upsilon_N\}^T$ as the "third-body" component of the disturbing acceleration on the motion of the spacecraft

$$\Upsilon_M = -\mu_M \left[\frac{r_{MS}}{r_{MS}^3} + \frac{r_{EM}}{r_{EM}^3} \right] \tag{3.56}$$

will be the perturbing acceleration of the moon during the Earth-departure phase and

$$\Upsilon_E = -\mu_E \left[\frac{r_{ES}}{r_{ES}^3} - \frac{r_{EM}}{r_{EM}^3} \right] \tag{3.57}$$

will be the perturbing acceleration of the Earth during the Moon-arrival phase. With r_{ES} and r_{EM} found, as described above, on the ECI basis, Υ_M found as the right-hand side of (3.37) will also be expressed on the ECI basis. The gravitational perturbation Υ_M must then be resolved into components in the spacecraft-fixed radial, tangential, and normal directions for use in equations (3.44). This is accomplished by the transformation

$$\Upsilon_M \text{ (local)} = \mathbf{T}^T \mathbf{R}^T \Upsilon_M \text{ (ECI)} \tag{3.58}$$

where
$$\mathbf{T} = \begin{bmatrix} \cos L & -\sin L & 0 \\ \sin L & -\cos L & 0 \\ 0 & 0 & 0 \end{bmatrix}. \tag{3.59}$$

Similarly, when r_{MS} is found as described above, on the MCNR basis, and with r_{EM} determined on the MCNR basis through

$$r_{EM} = \begin{bmatrix} a_M(\cos E_M - e_M) \\ a_M\sqrt{1 - e_M^2}\sin E_M \\ 0 \end{bmatrix} \tag{3.60}$$

then Υ_E found as the right-hand side of (23) will also be expressed on the MCNR basis. As described in the previous paragraph, the gravitational perturbation Υ_E must then be resolved into components in the spacecraft-fixed radial, transverse, and normal directions for use in equations (3.44). This is accomplished again by the transformation (3.58–3.59) with the only difference being that the true longitude L is now determined using (3.33) *using equinoctial variables appropriate to the Moon-centered orbit.*

An additional consideration with regard to the system equations is that it is necessary to convert the length, mass, and time units from one local set to the other across the knot representing the coordinate transformation. The most common situation, and perhaps the best in terms of scaling to improve the robustness of the numerical solution, is to use canonical units in which the gravitational parameter $\mu = 1$. However, this does not remove the need for the conversion of elements such as the semimajor axis or accelerations such as the thrust acceleration. Also note that when the transformation is accomplished as described in this section, the choice of the location or the time at which the knot occurs does not affect the accuracy (legitimacy) of the equations of motion; the problem is always a three-body problem. It is probably reasonable to perform the transformation at the SOI of the body of arrival; the precise location may affect the robustness of the numerical solution but in principle should not change the optimal trajectory.

3.3.3 Interplanetary Trajectories

For an interplanetary trajectory, it will normally be the case that two coordinate transformations will be required. The first will be when the s/c leaves the SOI of the departure planet and enters heliocentric flight. The second, of course, is when the s/c transitions from heliocentric flight to arrival within the SOI of the target planet. The example in the previous section for an Earth–Moon trajectory applies directly to these two transformations with an obvious change in the subscripts identifying the bodies.

Another situation often encountered in interplanetary trajectories is the use of planetary flybys or "gravity-assist maneuvers." If these are feasible for the interplanetary flight under consideration they are often fuel-minimizing (though perhaps not time-minimizing). While the method of the previous section could be used to describe the encounter with a planet used for a flyby this is seldom done because as a (good) first approximation the hyperbolic flyby can be described analytically [25]. In particular, this analysis provides the absolute (that is, heliocentric) post-flyby velocity as a function of the pre-flyby velocity and the flyby altitude. With a knowledge of the post-flyby heliocentric position and velocity, it is straightforward to determine the corresponding new orbit elements [25]. The flyby can thus appear in the direct transcription as a knot with the states discontinuous across the boundary. An approximation that is thus made is that the flyby occurs instantaneously, but this is not unreasonable as a first approximation because the flyby certainly does occur rapidly in comparison to the many years required for the complete interplanetary trajectory, for example, of spacecraft such as Galileo or Cassini.

3.4 Modeling Propulsion Systems

3.4.1 Impulsive Thrust Case

This case is quite straightforward. The mass remaining (m_f) following an impulsive ΔV is related to the mass immediately prior to the impulse (m_i) and the exhaust velocity (c) of the rocket according to the Tsiolkovsky equation

$$\Delta V = c \, \ln \frac{m_i}{m_f}. \tag{3.61}$$

The exhaust velocity is related to the specific impulse (I_{sp}) of the rocket motor through

$$c = I_{sp}g \tag{3.62}$$

where by convention g is the value of the acceleration due to gravity at the Earth's surface.

It is thus a simple matter to account for propulsive mass, which is often the objective function to be minimized for space trajectory problems. If the vehicle has stages, the mass of the discarded stage needs to be deducted from the total mass prior to the application of the next impulse or prior to the next use of an electric low-thrust motor, if a combination of impulsive and continuous-thrust maneuvers is used for the trajectory. Note that if there are multiple impulses applied, and the trajectory uses only impulsive ΔV's, then minimizing the sum of all the ΔV's will minimize the propellant consumed, or maximize the final mass.

3.4.2 Continuous Thrust Case

There are several options for formulating an objective function when continuous thrust is used. First it is important to know if the engine is to provide a constant thrust or is throttleable. In the former case, the rate of mass ejection is constant so that minimizing the total time of operation is equivalent to minimizing the total propellant consumed.

For the (perhaps piecewise continuous) constant thrust case, since the mass of the vehicle is decreasing, the acceleration provided by the thrust will monotonically increase. This can be modeled by adding the equation

$$\frac{d\Gamma}{dt} = \frac{1}{c}\Gamma^2 \tag{3.63}$$

to the other system equations. Of course if, as mentioned above, a combination of impulsive and continuous-thrust maneuvers is used for the trajectory, then the value of Γ will need to be re-initialized following each impulsive ΔV, to account for the instantaneous change in spacecraft mass. Similarly, if the spacecraft ever ejects some part during the flight, for example, a fairing or a solid rocket motor shell, the value of Γ will need to be re-initialized. Minimizing propulsive mass consumed is then equivalent to minimizing the final value of the acceleration Γ.

If the engine is throttleable, it is necessary to add the mass variation equation to the system equations

$$\dot{m} = -T/c \tag{3.64}$$

where thrust magnitude T is chosen by the optimizer within the feasible range. Then instantaneous thrust acceleration is given by $\Gamma = T/m$. Mass now becomes a state variable so that the objective can be the value of m at the final time (m_f).

3.4.3 Power-Limited Case

The thrust T may be written using (3.64) as $T = |\dot{m}|\,c$. The power required for a given thrust T is

$$P = \frac{1}{2}|\dot{m}|\,c^2 = \frac{1}{2}cT \le P_{\max}. \tag{3.65}$$

Then, as shown in [26]

$$\frac{d}{dt}\left(\frac{1}{m}\right) = -\frac{\dot{m}}{m^2} = \frac{|\dot{m}|}{m^2} = \frac{\Gamma^2}{2P} \quad \text{since } \dot{m} \le 0. \tag{3.66}$$

Integrating yields

$$\frac{1}{m_f} - \frac{1}{m_0} = \frac{1}{2}\int_{t_0}^{t_f} \frac{\Gamma^2(t)}{P(t)}\,dt. \tag{3.67}$$

Thus maximizing final mass m_f is accomplished by minimizing the integral in Equation (3.67). The power used should thus be the maximum available at any time t. If the available power is constant, for example, at a value P_{max}, the objective function to be minimized is then

$$J = \frac{1}{2} \int_{t_0}^{t_f} \Gamma^2(t)\, dt. \tag{3.68}$$

If the low-thrust electric motor is powered by solar energy, there will be a change in thrust magnitude and effective I_{sp} as the spacecraft's distance from the sun increases, since the solar panels then receive less sunlight. It is reasonable to assume that available power decreases as

$$P = \frac{P_{max}}{r^2} \tag{3.69}$$

where P_{max} is the maximum rated power and r is the distance from the sun (in AU). The NSTAR engine, used for several interplanetary missions, has a specific impulse of 3120 sec and a thrust of 93.4 mN when supplied with 3.52 kW of power (nominal available at 1 AU). Polk et al., in their study characterizing NSTAR's performance [27], include a table of thrust and specific impulse versus available power. Englander [28] has modeled the table results. A power law fit is used to model T_{max}, and a quadratic fit is used to model I_{sp}, in canonical units, where the distance unit is 1 AU and 2π time units correspond to the period of an orbit at 1 AU (that is 1 year)

$$T_{max} = 15.5852 r^{-2.03} \tag{3.70}$$

$$Isp = -2.9862 r^2 + 6.1813 r + 7.0073. \tag{3.71}$$

Exhaust velocity c is then given by

$$c = \frac{Isp}{g_0} = -0.3044 r^2 + 0.6305 r + 0.7143. \tag{3.72}$$

Similar modeling can be done for whatever engine is chosen for the spacecraft.

Another type of low-thrust propulsion is that using a solar sail. When a solar sail is used, its effectiveness also decreases with distance from the sun, but in this case because the total insolation is decreasing. It is reasonable to assume that available thrust decreases with the inverse square of the radius, just as power decreases in Equation (3.65).

3.5 Generating an Initial Guess

Solution via the method of direct transcription with NLP requires that the NLP problem solver be given an initial guess of the vector of NLP parameters. This vector contains the discrete time history of the state and control parameters, which normally number in the hundreds or few thousands, and a small number of additional parameters, for example, times of certain events and possibly the final time. While modern NLP solvers are typically quite robust, it has nonetheless been our experience that a "reasonable" initial guess needs to be provided, especially for large problems. Of course "reasonable" is not a very precise term. In our experience, initial guesses, that is, approximate candidate optimal trajectories, are *more* "reasonable" to the extent that they:

(1) satisfy, at least at the 0th iteration, the system EOM, so that initially all of the nonlinear "defects" are very small;
(2) satisfy any specified initial and terminal constraints;
(3) satisfy the boundary conditions given to the NLP problem solver for the upper and lower bounds for all of the parameters.

Creating an initial guess that does all three of these things would be very difficult in most cases; fortunately that is seldom necessary. There are several approaches for the generation of a satisfactory initial guess.

3.5.1 Using a Known Optimal Control Strategy

Many problems one may encounter are similar in some way to problems whose solution has already been obtained. Of course the precise circumstances will be different but the problems may be similar enough that the optimal control strategy can be approximately determined.

There are a number of obvious examples. One very useful example is low-thrust orbit transfer, sometimes called orbit raising. Analytic solutions are available for a number of cases. Edelbaum [29] derived solutions for transfer from one circular orbit to another, for non-coplanar cases, with continuous, constant acceleration. However, he assumed a constant yaw profile. Wiesel and Alfano [30] generalized this result to allow the yaw angle [α in Equation (3.44)] to change each revolution. Kechichian, in a series of papers [31] [32], has generated analytic solutions that correspond very well to exact solutions for cases even including the effect of Earth shadow on electrically propelled spacecraft and Earth oblateness.

None of these solutions will yield precisely an optimal trajectory because of the assumptions or constraints on thrust acceleration and pointing angle that are made, but they would nonetheless likely be very good suboptimal approximations from which the NLP problem solver could converge to an optimal trajectory without such constraints.

A related but less accurate approach is to assume a control strategy qualitatively similar to that found in these analytic solutions. That is, one finds for many cases of low-thrust orbit transfer or orbit raising, especially for cases where the eccentricity remains small, that the optimal thrust direction is approximately along the velocity vector. There is a simple reason for this; it maximizes the rate of change of the kinetic energy. Thus one way to generate an approximate optimal trajectory is to do a forward numerical integration of the system equations assuming that the thrust is aligned with the velocity (or tangentially) and continue the integration until the desired orbit is crossed. Of course the terminal boundary conditions will not be satisfied but the initial guess may still be "reasonable" by the definition above, since it will satisfy two of the three conditions given.

3.5.2 Shape-Based Methods for Generating Suboptimal Trajectories

Another approach is to use "shape-based" methods for generating guesses that satisfy the EOM and the boundary condition [33]. The basic concept is that the trajectory is described by a geometric shape, for example a logarithmic spiral, a Cartesian oval or an exponential sinusoid. The shape can be made to satisfy both the equations of motion (in one gravity field) and most or all of the boundary conditions. The thrust magnitude and direction are then determined *a posteriori*; this may be problematic because the required thrust magnitude may not be feasible for the actual spacecraft propulsion system. It may also be the case that the actual spacecraft may have only constant thrust capability but modulated thrust is required for the spacecraft to travel the path of the shape.

Petropoulos and Longuski [34] derived a method capable of satisfying all of the orbit transfer boundary conditions for an orbital interception with a shape that is an exponential sinusoid. Petropoulos and Sims [35] have written a survey of such methods. Wall and Conway [33] extended the method to the case of satisfying rendezvous boundary conditions. They also made the shape "more" optimal by using a genetic algorithm to optimize some of the free parameters of the problem, such as departure and arrival dates and number of spacecraft revolutions about the central body. Their paper contains a number of examples (circle-circle transfer, low-thrust escape, low-thrust interception, low-thrust asteroid rendezvous) showing how the shape-based approximation is found, followed by its use as the initial guess for a direct solution using nonlinear programming. In all of the examples the shape-based suboptimal trajectory provides a very satisfactory initial guess for the more accurate direct solution.

3.5.3 Using Evolutionary Methods to Generate an Initial Guess

Using evolutionary algorithms (EA), the best known of which is probably the genetic algorithm (GA), is another approach to finding optimal spacecraft trajectories. Because of the limitations of GAs, these trajectories will necessarily be more inaccurate than trajectories found using direct transcription with NLP or even indirect methods, for example, methods based on shooting. However, for the purpose of providing an initial guess for a direct solution, the suboptimal trajectory found using GA may be completely satisfactory. Chilan [36] has developed a method of using GA to generate initial guesses for the spacecraft trajectory that are always feasible.

An additional benefit of using a GA or some other EA is that an EA is initialized with a population of candidate solutions that is generated randomly. The EA will also explore the parameter space (in part) with randomly generated directions. In combination, this yields an ability to find global minima that may be much improved over methods that begin with a prejudiced initial guess and then use gradient information to improve their objective. This can yield an improved initial guess in comparison to any of the methods described in Sections 3.5.1 and 3.5.2; that is, the NLP problem solver used in the direct method has a tendency to converge to a solution in the neighborhood of the initial guess. This can be a problem when there are many local minima, so an initial guess generator that has already searched the space and (perhaps) found the neighborhood of the global minimum (even if that guess has the inherent inaccuracy of an EA) is potentially very beneficial.

3.6 Computational Considerations

Direction transcription methods are not without issues with "real world" problems. In particular, problems need to be scaled and the width and location of the segments need to be managed to achieve accurate solutions.

3.6.1 Equation Scaling

Our experience with using nonlinear programming for trajectory optimization has shown that realistic trajectory problems must use some form of equation scaling. Most academic problems (Brachistochrone, Van der Pohl, among others) work well without scaling because all variables in the problem have similar magnitude. Trajectory equations expressed in traditional English or metric units have variables with widely varying magnitudes (for example, controls of order 1 and positions of order 10^8). One approach to scaling would be to use a system of units that produce variables with similar magnitudes (the canonical variables described in [37] are one such set of units). Another approach is to keep the force computation in familiar units and scale only the optimization calculations. The computations required are illustrated in Figure 3.9.

The symbol S denotes the scaling transformation, Z represents the unscaled variables, and X represents the scaled variables.

The dimensional states, z_{ijk}, are related to the nonlinear programming states by

$$z_{ijk} = x_m S_{ijk} \text{ with}$$

z   dimensional state
i   state index
j   node index

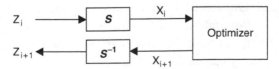

Figure 3.9. Illustration of equation scaling.

k   phase number
x   nonlinear programming variable
S   state scale factor

Typically the user is responsible for inputting values of S_{ijk}. The variables associated with time do not use the offset, but use

$$t_{ik} = x_m * t_{\text{scale}}$$

$$t_{pk} = x_m * t_{\text{scale}}$$

with

t_i initial phase time
t_{scale} time scale factor
t_p phase time length

Additionally the defects, Δ_{ijk}, are scaled as

$$\Delta_{ijk} = \Delta_{ijk}^{d} / d_{\text{scale}ij}$$

with

Δ defect (scaled)
Δ^d dimensional defect (include both implicit integration defects and general
            defects)
d_{scale} defect scale factor (user input)

In general, the user can set t_{scale} to be a bit larger than t_i and t_p.

When the defects divided by the corresponding d_{scale} element are less than a given tolerance (usually 1.E-06), the defect is considered to be zero. Remember that the defects are the state integration errors over a segment. In picking values for d_{scale}, one is defining an acceptable integration error.

We have generated an autoscaling technique that is roughly based upon the above advice and computational experience.

For time, we use

$$t_{\text{scale}} = 10^{INT(log_{10}(max(INT(ti))))}$$

Recognizing that not all state values are of equal importance, (is a 30-meter altitude error the same as a 30-degree flight path error?), we have generated a procedure to relate dissimilar states. Our solution is to employ what we have termed the *method of equivalent position*. We scale the state variables so that the change of a scaled variable maps to a change in position (equivalent position scaling). This procedure has provided improved convergence and accuracy.

For example, when Cartesian coordinates are used, the scaling, in the phase of interest, is constructed from

$$UVW_{\text{scale}} = |V_{\max} - V_{\min}|$$
$$XYZ_{\text{scale}} = UVW_{\text{scale}} * T_p$$

where

$V_{\max}$ maximum velocity component during the phase
$V_{\min}$ minimum velocity component during the phase
T_p phase length (dimensional time).

Then the scaling factors are

$$S_{ijk} = UVW_{\text{scale}} \text{ for the Cartesian velocities}$$
$$S_{ijk} = XYZ_{\text{scale}} \text{ for the Cartesian positions.}$$

A similar procedure can be used for other coordinate frames.

3.6.2 Grid Refinement

To achieve accurate implicit integrations, segments need to be placed in the regions where rapid changes in the states or controls occur. Unfortunately, this knowledge is normally not available until after a solution is found. The procedures for refining an existing grid structure to form a new computational grid are heavily based on methods used for adaptive step size control for explicit numerical integration.

The fundamental procedure used for grid refinement is:

(1) generate a trajectory;
(2) compute the implicit integration errors;
(3) evaluate the errors;
(4) build a new grid.

This process is repeated until the integration errors are driven less than a tolerance or the desired number of grid have been computed. There are many techniques for computing the implicit integration errors. A rather simple but effective procedure used in OTIS [38] is shown in Figure 3.10.

Explicit integration is viewed as the "truth." Explicit integration is started using the states as defined at the beginning of a segment. Those states are propagated to the end of the segment using the currently selected explicit integrator, equations of motion, and vehicle models. The controls are defined using a polynomial representation based on their node values. At the end of a segment, errors are computed for

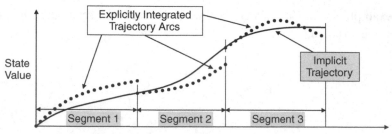

Figure 3.10. Implicit integration error evaluation.

each state as

$$e = z_{\text{explicit}} - z_{\text{implicit}}$$

z_{explicit} are the values of the states as a result of the explicit integration;
z_{implicit} are the implicit integration state values.

As with scaling, a potential problem with this approach is that not all state errors are of equal importance, so we use the scaled variables to compute relative errors.

For Cartesian coordinates

$$e_{peu} = e_u^* t_p^* (segl/2)/S_x \quad \text{converted } x\text{-dot error}$$

$$e_{pev} = e_v^* t_p^* (segl/2)/S_x \quad \text{converted } y\text{-dot error}$$

$$e_{pew} = e_w^* t_p^* (segl/2)/S_x \quad \text{converted } z\text{-dot error}$$

$$e_{pex} = e_x/S_x \quad \text{converted } x \text{error}$$

$$e_{pey} = e_y/S_x \quad \text{converted } y \text{error}$$

$$e_{pez} = e_z/S_x \quad \text{converted } z \text{error}$$

With

e_u	x-dot error
e_v	y-dot error
e_w	z-dot error
e_x	x error
e_y	y error
e_z	z error
t_p	phase time (dimensional)
S_x	positional scale factor
$segl$	normalized segment length

Again, similar procedures are used with other coordinate frames.

Recall that with the form of implicit integration being used, the states are represented as a polynomial of order n. This allows us to conclude that the interpolation error for the jth segment is

$$\varepsilon_j \propto (segl_j^{(n+1)}) \tag{3.73}$$

so in theory

$$\frac{\varepsilon_j}{\varepsilon_{\text{desired}}} = \left(\frac{segl_j}{segln_j}\right)^{(n+1)} \tag{3.74}$$

where segln is the new segment length required to produce the desired error, $\varepsilon_{\text{desired}}$. Solving for segln

$$segln_j = segl_j \left(\frac{\varepsilon_{\text{desired}}}{\varepsilon_j}\right)^{1/(n+1)}. \tag{3.75}$$

Recall that the segl's are normalized. (A phase, in nondimensional time, runs from −1 to +1.) Thus

$$\sum_{j=1}^{\text{NP\_SEG}} segl_j = 2 \tag{3.76}$$

which allows us to conclude, that if

$$\sum_{j=1}^{\text{NP\_SEG}} segln_j \geq 2$$

then there are sufficient segments (more of the existing segments can be expanded) to allow the desired error tolerance to be met.

Directly using Equation (3.75) to define the new segment lengths, segln, can cause problems as shown in Figure 3.11.

Experience has shown that grid refinement works better if the amount a segment is changed is limited. By using limit functions, local segment normalization, inverse interpolation, and the above equation for scgln, we are able to limit the expansion and contraction of the segments. This is illustrated in Figure 3.12.

$$\sum_{j=1}^{\text{NP\_SEG}} segln_j < 2$$

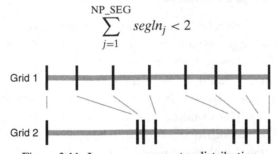

Figure 3.11. Improper segment redistribution.

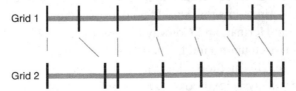

Figure 3.12. Effective grid refinement.

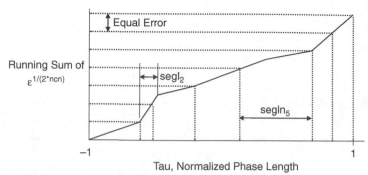

Figure 3.13. New segment distribution procedure.

When the majority of segments should be contracted to met the error tolerance (more segments are needed). Some methods simply halved the segments indicated for contraction. In the worst case, this would double the number of segments and nodes. With higher-order defects, this approach would rapidly overwhelm the available array storage. A different approach must be used. The new number of segments, nsegn, is estimated by

$$nsegn = nseg \left(2 \Big/ \sum_{j=1}^{NP\_SEG} segln_j \right) + 1. \tag{3.77}$$

The new segment spacing is obtained by equally distributing $\varepsilon^{1/(n+1)}$. This is graphically illustrated in Figure 3.13.

In practice the above procedures work well. The early grid refinements tend to gradually add segments, with subsequent grid refinements transitioning to the pure redistribution mode.

3.6.3 Other Grid Refinement Strategies

Betts [39] describes a mesh refinement strategy for use with direct transcription via collocation. In this strategy, two approaches are used in combination to reduce the error below some user-chosen tolerance. One tool is to increase the order of the discretization selectively. Based on the estimated error in a given segment, Betts' method may first improve the result by switching from a low-order-of-accuracy implicit integration rule (for example, the trapezoid rule) to a better rule such as the Hermite-Simpson rule of Equation (3.5). If that is insufficient, additional grid points (nodes) are added within segments that contained the largest error at the most recent solution. Betts recommends placing an upper bound (in fact, five) on the number of new grid points that should be placed within the original interval.

Herman [9] and Herman and Conway [40] propose a grid refinement strategy using an algorithm based upon a method used for step size control in some numerical integration software. This method is known as adaptive quadrature and is described in many numerical analysis textbooks, for example, in that of Conte and de Boor [41]. This automatic node placement (ANP) algorithm examines the accuracy of a solution by comparing subintervals of a time history of two solutions found using two different node distributions. The automatic node placement (ANP) process begins with an arbitrary, initial node distribution. Then an optimal solution is found using this distribution (ANP iteration #1). Next each of the subintervals in the initial distribution is split in half with additional nodes placed at the previous midpoints, and a second solution is found (ANP iteration #2) using this node distribution.

Let A(n,b,i) represent an approximate integral at the nth iteration over subinterval i using the Simpson system constraint (3.5) with integration subintervals of equal length b. Then $A(k + 1, \Delta t_i, i)$ is the optimal state solution over subinterval i on distribution k+1 using two subintervals of length Δt_i (corresponding to ANP iteration) and $A(k, 2\Delta t_i, i)$ is the optimal state solution over subinterval i on distribution k using one subinterval of length $2\Delta t_i$. These two solutions are compared using the relationship

$$E(i) \approx \frac{A(k + 1, \Delta t_i, i) - A(k, 2\Delta t_i, i)}{(2^{p-1} - 1)} \tag{3.78}$$

If $|E(i)|$ is larger than a user-specified threshold, then the subdivision is retained. The process continues until a node distribution is found that provides a solution that satisfies the relationship given for each system equation on every subinterval.

3.6.4 Solving Problems with Variables That Change with Significantly Different Timescales

Another computationally difficult case is the case in which a large proportion of the states change significantly but on different timescales. An example is the combined translational and rotational motion of a space vehicle. The translational motion of the vehicle's center of mass experiences change much more slowly than the coupled rotational motion (about the center of mass). Ordinarily, when using a direct transcription method, the grid on which the problem is to be solved will need to use a segment width appropriate to capture the most rapid (in this case the rotational) motion. Specifying the slowly changing translational states at these grid points is clearly wasteful and yields a much larger, more computational time consuming and less tractable NLP problem than is desirable. Desai and Conway [42] [43] show how it is possible to use two different discretizations for one problem, that is, a fine grid for the rapidly changing states and a coarser grid for the slowly changing states. The very-accurate fifth-degree Gauss-Lobatto transcription [10] is employed. For a challenging example, a reentry vehicle similar to the Space Shuttle, a dramatic reduction in NLP problem size is achieved without loss of solution accuracy.

3.7 Verifying Optimality

3.7.1 Optimality of Assumed Control Switching Structures and the Discrete Switch Function

When a solution is obtained using a direct transcription method, the analytical necessary conditions for optimality are of course not satisfied; there is no certainty that the solution to which the NLP problem solver has converged represents a local minimum of the objective function.

One situation in which this is likely to occur is if the optimal control switching structure is not known. As described previously in this Chapter and in Chapter 1, for the continuous thrust case, it often occurs that the optimal control consists of a sequence of maximum-thrust (MT) arcs and coasting (zero-thrust) arcs. The problem can be solved using direct transcription if a switching structure is assumed *a priori*. If an erroneous structure is assumed, the problem is transcribed, and the corresponding nonlinear programming problem is solved, the result is a suboptimal solution. Thus a method is needed to determine whether a given control structure is indeed optimal.

First consider the case where one of the phases is extraneous, for example, suppose one guessed a three-burn structure, whereas the optimal solution had only two burns. This mistake is indicated when the nonlinear programming routine drives the corresponding MT-arc phase duration to zero. This collapse of an MT arc renders the transcription inefficient, because the variables for the state and control over the zero-duration phase become extraneous, and it may be desirable to re-solve the problem using the corrected structure. The problem of detecting the opposite situation, where the assumed structure does not contain enough phases, is much more difficult. Suppose one guessed a two-burn structure, when in fact the optimal solution included a third burn. The optimal two-burn solution would be obtained, which may be very suboptimal generally, and there would be no immediate indication that a third burn should be added. (In the implementation of their hybrid method, Zondervan et al. [44] encounter similar difficulties and recommend assuming a conservatively large number of burns to avoid this situation.) To solve this problem, Enright and Conway [15] developed a method that uses the fact, determined by Enright [45], that at the solution of the nonlinear programming problem, the (Kuhn-Tucker) multipliers conjugate to the defect constraints in the NLP problem solution provide a discrete approximation to the adjoint variables of the continuous optimal control problem. In conjunction with the primer vector theory presented in Chapter 2, this allows one to determine the optimality of the assumed structure in a very precise fashion.

The key to the method is the observation that if the assumed structure is correct, then the assumed-structure solution must satisfy the general necessary conditions (Section 1.2.1), including the switching condition (1.10). The procedure is as follows:

(1) Assume a particular control switching structure. Transcribe and solve the nonlinear programming problem.
(2) If any phase collapses, eliminate that phase and re-solve the problem, or proceed.
(3) Evaluate the switch function (1.10) over the MT arcs using the discrete approximations to the states and the discrete approximations to the adjoint variables obtained from the multipliers of the nonlinear programming problem. This evaluation of (1.10) is referred to as the "discrete switch function."
(4) If the switch condition (1.10) is violated (for example, if the discrete switch function is negative anywhere) over one or more of the MT arcs, modify the assumed structure (for example, add an MT arc) and re-solve the problem.

This procedure is repeated until the discrete switch function has the appropriate behavior.

3.7.2 Example: Two and Three Thrust-Arc Rendezvous

The next problem considered was a time-fixed circle-to-circle rendezvous. The spacecraft is initially in a circular orbit and is required to rendezvous with a second spacecraft in a coplanar circular orbit of different radius within a specified final time. The impulsive version of this problem has been studied extensively [46] [47], and the optimal number of impulses has been shown to be dependent on the relative phasing of the spacecraft and the final time constraint.

Canonical units normalized to the initial orbit were used. The transcription was performed using the Hermite-Simpson method with cubic controls. Analytical expressions for the problem derivatives were used with the exception of the derivative of the final time constraint, which was handled by finite-differencing. (This constraint involves the calculation of the coast-arc durations from the orbital elements and the initial and final state vectors, and is algebraically intensive.)

It was first assumed that the solution consisted of two MT arcs separated by a coast arc. Initial and final coasts were also allowed and were handled through the initial and final conditions. With 8 segments per phase, there were a total of 18 nodes. The NLP state vector consisted of the states ($n = 5$), controls ($m = 1$) and control derivative ($m = 1$) values at each node, and the MT-arc phase durations (2), resulting in a dimension of 128. There were 8 defects over each of the MT arcs and a single set of matched-integral defects over the coast arc, resulting in 84 nonlinear constraints. The final time constraint and the rendezvous constraint (the spacecraft must be at the same longitude as the target at the final time) provided 2 more nonlinear constraints, resulting in an NLP nonlinear constraint vector of dimension 86. There were also 14 linear constraints for the splines. Initial conditions were fixed, and the other final conditions were simple bounds.

The initial guess had no initial or final coasts. The state over the first burn was set on the initial orbit, and the state over the final burn was set on the final orbit. The thrust angle was initialized to zero. The target radius was 3, the initial thrust acceleration was 0.1, the exhaust velocity was 1.5, the initial lead angle of the target was 4.5 radian, and the final time was 10.0. (All units are canonical with respect to the initial orbit.) The tentatively optimal trajectory is shown in Figure 3.14 and includes an initial coast. The cost was 5.226, with an equivalent Δv of 0.6425.

The discrete switch function was then examined. This is the switch function (1.10) evaluated with the discrete approximations to the adjoints and the states during the MT arcs. If the two-burn structure were optimal, the discrete switch function would remain above zero over the MT arcs (cf. Chapter 2). Figure 3.15 shows the switch function for the first burn and clearly indicates the nonoptimality of the assumed structure.

A three-burn structure (three phases and two sets of matched-integral constraints) was then assumed and the problem was re-solved. This structure required

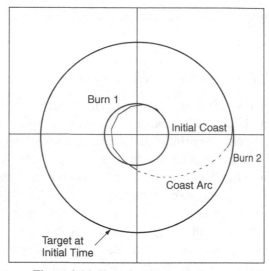

Figure 3.14. Two thrust-arc rendezvous.

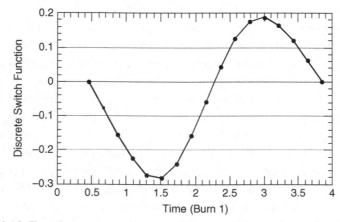

Figure 3.15. Two thrust-arc rendezvous; discrete switching function for first burn.

27 nodes, generating an NLP problem with 192 variables, 129 nonlinear constraints, and 21 linear spline constraints, and requiring 24.6 seconds of CPU time for solution. The optimal trajectory is shown in Figure 3.16. The cost was 4.976, with an equivalent Δv of 0.6045. The impulsive cost, according to Prussing and Chiu [48] is 0.4872 so the gravity loss was 0.1173 or 24%. The first burn is essentially a retro-burn and is driven by the time constraint. Figure 3.17 shows the switch function obtained by a backward integration of the state and adjoint equations for this solution, and verifies the optimality of the assumed burn structure.

3.7.3 Verifying Optimality by Integration of the Euler-Lagrange Equations

As mentioned in Section 7.1, the discrete-adjoint approximation may be exploited to test whether the discrete-approximate solution adequately satisfies the indirect

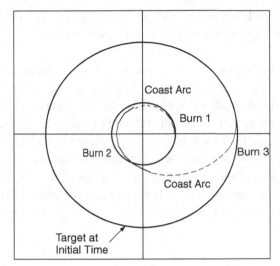

Figure 3.16. Three thrust-arc rendezvous.

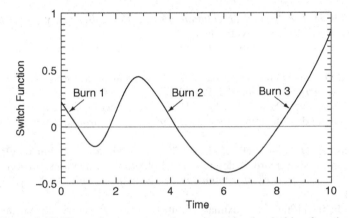

Figure 3.17. Three thrust-arc rendezvous; optimal switching function.

TPBVP. After the correct control switching structure has been confirmed, the adjoints at the final time can be extracted. For the Runge-Kutta parallel-shooting method, these are simply the multipliers corresponding to the last defect (which occurs at the final time). If the Hermite-Simpson collocation method or any of the higher-degree Gauss-Lobatto collocation methods are used, obtaining the adjoints at the final time is problematic because the last defect is located within the last segment, near but not precisely at the final time. If the segment widths are narrow, this small difference may be inconsequential. However, for a precise evaluation, one can use the multipliers corresponding to the terminal constraints (the ν's) and calculate the final adjoints from the terminal conditions (1.5). Then the state equations (3.6) and the adjoint differential equations (1.5) can be numerically integrated backwards from the final time to the initial time using the controls generated by the control optimality conditions (1.6). If the initial state conditions are accurately recovered then the discrete-approximate solution satisfies the optimal control necessary conditions, for example, the Euler-Lagrange TPBVP is solved. This verifies optimality as well as feasibility of the discrete-approximate solution, and provides the ultimate evaluation criterion.

If the system Hamiltonian (1.9) does not explicitly depend on time, which is often the case for spacecraft trajectory problems, then it is known that the Hamiltonian is a constant on the optimal trajectory. Knowledge of the adjoint variables, extracted at discrete points in time from the solution obtained by the NLP problem solver, allows the system Hamiltonian to be evaluated along the trajectory. Of course it should also have a constant value; the degree to which it varies about some nominal value will provide additional information regarding the accuracy of the optimal trajectory that has been obtained. In our experience, it is normally the case that the Hamiltonian will have only a small variation in a good solution, perhaps only in the fourth or fifth decimal place.

REFERENCES

[1] Fox, L. (1975) *The Numerical Solution of Two-Point Boundary Problems*, Oxford University Press, West Sussex, England.

[2] Keller, H. B. (1968) *Numerical Methods for Two-Point Boundary Value Problems*, Blaisdell, New York.

[3] Russell, R. D., and Shampine, L. F. (1972) A Collocation Method for Boundary Value Problems, *Numerical Mathematics*, **19**, 1–28.

[4] Hargraves, C. R., and Paris, S. W. (1987) Direct Trajectory Optimization Using Nonlinear Programming and Collocation, *Journal of Guidance, Control, and Dynamics*, **10**, No. 4, 338–343.

[5] Dickmanns, E. D., and Well, K. H. (1975) Approximate Solution of Optimal Control Problems Using Third-Order Hermite Polynomial Functions, *Proceedings of the 6th Technical Conference on Optimization Techniques*, IFIP-TC7, Springer–Verlag, New York.

[6] Johnson, F. T. (1969) Approximate Finite-Thrust Trajectory Optimization, *AIAA Journal*, **7**, 993–997.

[7] Hargraves, C. R., Johnson, F. T., Paris, S. W., and Rettie, I. (1981) Numerical Computation of Optimal Atmospheric Trajectories, *AIAA Journal of Guidance and Control*, **4**, 406–414

[8] Gill, P. E. et al. (1993) *User's Guide for NZOPT 1.0: A Fortran Package For Nonlinear Programming*, McDonnell Douglas Aerospace, Huntington Beach, CA.

[9] Herman, A. L. (1995) *Improved Collocation Methods Used For Direct Trajectory Optimization*, Ph.D. Thesis, University of Illinois at Urbana-Champaign.

[10] Herman, A. L., and Conway, B. A. (1996) Direct Optimization Using Collocation Based on High-Order Gauss-Lobatto Quadrature Rules, *Journal of Guidance, Control, and Dynamics*, **19**, No. 3, 593–599.

[11] Desai, P. N., and Conway, B. A. (2008) Six-Degree-of-Freedom Trajectory Optimization Utilizing a Two-Timescale Collocation Architecture, *Journal of Guidance, Control, and Dynamics*, **31**, No. 5, 1308–1315.

[12] Desai, P. N., and Conway, B. A. (2008) A Two-Timescale Discretization Scheme for Collocation, *Journal of Guidance, Control and Dynamics*, **31**, No. 5, 1316–1322.

[13] Ross, I. M., and Fahroo, F. (2003) Legendre Pseudospectral Approximations of Optimal Control Problems, *Lecture Notes in Control and Information Sciences*, **295**, Springer–Verlag, New York, 327–343.

[14] Paris, S. W., Riehl, J. P., and Sjauw, W. K. (2006) *Enhanced Procedures for Direct Trajectory Optimization Using Nonlinear Programming and Implicit Integration*, Paper AIAA-2006-6309, AIAA/AAS Astrodynamics Specialist Conference and Exhibit, Keystone, Colorado.

[15] Enright, P. J., and Conway, B. A. (1992) Discrete Approximations to Optimal Trajectories Using Direct Transcription and Nonlinear Programming, *Journal of Guidance, Control, and Dynamics*, **15**, No. 4, 994–1003.

[16] Schubert, G. R. (1963) Algorithm 211, Hermite Interpolation, *Communications of the ACM*, **6**, No. 1.0.

[17] Fahroo, F., and Ross, I. M. (2000) A Spectral Patching Method for Direct Trajectory Optimization, *Journal of the Astronautical Sciences*, **48**, No. 2/3, 269–286.

[18] Boyd, J. P. (2001) *Chebyshev and Fourier Spectral Methods*, 2nd ed., Dover Publications, New York.

[19] Fornberg, B. (1998) *A Practical Guide to Pseudospectral Methods*, Cambridge University Press, Cambridge, UK.

[20] Rae, J. R. (2001) *A Legendre Pseudospectral Method for Rapid Optimization of Launch Vehicle Trajectories*, M.S. Thesis, Massachusetts Institute of Technology.

[21] Gill, P., Murray, W., and Saunders, M. A. (2005) SNOPT: An SQP Algorithm for Large-Scale Constrained Optimization. *SIAM Review*, **47**, 99–131.

[22] Scheel, W. A., and Conway, B. A. (1994) Optimization of Very-Low-Thrust, Many Revolution Spacecraft Trajectories, *Journal of Guidance, Control, and Dynamics*, **17**, No. 6, 1185–1193.

[23] Bryson, A. E., and Ho, Y-C. (1975) *Applied Optimal Control*, Revised Printing, Hemisphere, Publ., Washing, DC.

[24] Battin, R. H. (1987) *An Introduction to the Mathematics and Methods of Astrodynamics*, AIAA Education Series, AIAA Publ., New York, 463–467.

[25] Prussing, J. E., and Conway, B. A. (1993) *Orbital Mechanics*, Oxford University Press, New York.

[26] Prussing, J. E. (1993) Equation for Optimal Power-Limited Spacecraft Trajectories, *Journal of Guidance, Control, and Dynamics*, **16**, No. 2, 391–393.

[27] Polk, J. E. (2001) *Performance of the NSTAR Ion Propulsion System on the Deep Space One Mission*, Paper AIAA 2001-0965, AIAA Aerospace Sciences Meeting, Reno, NV.

[28] Englander, J., and Conway, B. A. (2009) *Optimal Strategies Found Using Genetic Algorithms for Deflecting Hazardous Near-Earth Objects*, IEEE Congress on Evolutionary Computation, Trondheim, Norway.

[29] Edelbaum, T. N. (1961) Propulsion Requirements for Controllable Satellites, *ARS Journal*, **32**, 1079–1089.

[30] Wiesel, W. E., and Alfano, S. (1983) *Optimal Many-Revolution Orbit Transfer*, Paper AAS 83-352, AAS/AIAA Astrodynamics Specialist Conference, Lake Placid, NY.

[31] Kechichian, J. A. (1997) Reformulation of Edelbaum's Low-Thrust Transfer Problem Using Optimal Control Theory, *Journal of Guidance, Control and Dynamics*, **20**, No. 5, 988–994.

[32] Kechichian, J. A. (2000) Minimum-Time Constant Acceleration Orbit Transfer with First-Order Oblateness Effect, *Journal of Guidance, Control, and Dynamics*, **23**, No. 4, 595–603.

[33] Wall, B. J., and Conway, B. A. (2009) Shape-Based Approach to Low-Thrust Trajectory Design, *Journal of Guidance, Control and Dynamics*, **32**, No. 1, 95–101.

[34] Petropoulos, A. E., and Longuski, J. M. (2004) Shape-Based Algorithm for Automated Design of Low-Thrust, Gravity-Assist Trajectories, *Journal of Spacecraft and Rockets*, **41**, No. 5, 787–796.

[35] Petropoulos, A. E., and Sims, J. A. (2004) A Review of Some Exact Solutions to the Planar Equations of Motion of a Thrusting Spacecraft, *DSpace at JPL*, http://hdl.handle.net/2014/8673

[36] Chilan, M. C. (2009) *Automated Design of Multiphase Space Missions Using Hybrid Optimal Control*, Ph. D. Thesis, University of Illinois at Urbana-Champaign.

[37] Herman, A. L., and Conway, B. A. (1998) Optimal Low-Thrust, Earth-Moon Orbit Transfer, *Journal of Guidance, Control, and Dynamics*, **21**, No. 1, 141–147.

[38] Paris, S. W., Hargraves, C. R., et al. (1988) *Optimal Trajectories by Implicit Simulation* (Version 1.2), WRDC-TR-88-3057, Vols. I-IV, the Boeing Company.

[39] Betts, J. T. (2001) *Practical Methods for Optimal Control Using Nonlinear Programming*, Society for Industrial and Applied Mathematics, Philadelphia.

[40] Herman, A. L., and Conway, B. A. (1992) *An Automatic Node Placement Strategy for Optimal Control Problems Discretized Using Third-Degree Hermite Polynomials*, Paper AIAA 92-4511-CP, AIAA/AAS Astrodynamics Conference, Hilton Head Island, SC.

[41] Conte, S. D., and de Boor, C. (1980) *Elementary Numerical Analysis: An Algorithmic Approach*, McGraw-Hill Book Co., New York.

[42] Desai, P. N., and Conway, B. A. (2008) Six-Degree-of-Freedom Trajectory Optimization Utilizing a Two-Timescale Collocation Architecture, *Journal of Guidance, Control, and Dynamics*, **31**, No. 5, 1308–1315.

[43] Desai, P. N., and Conway, B. A. (2008) A Two-Timescale Discretization Scheme for Collocation, *Journal of Guidance, Control and Dynamics*, **31**, No. 5, 1316–1322.

[44] Zondervan, K. P., Wood, L. J., and Caughey, T. K. (1984) Optimal Low-Thrust Three-Burn Orbit Transfer With Plane Changes, *Journal of the Astronautical Sciences*, **32**, 407–427.

[45] Enright, P. J. (1991) *Optimal Finite-Thrust Spacecraft Trajectories Using Direct Transcription and Nonlinear Programming*, Ph. D. Thesis, University of Illinois at Urbana-Campaign.

[46] Lion, P. M., and Handelsman, M. (1968) The Primer Vector on Fixed-Time Impulsive Trajectories, *AIAA Journal*, **6**, No. 1, 127–133.

[47] Jezewski, D. J., and Rozendaal, H. L. (1968) An Efficient Method for Calculating

[48] Optimal Free-Space N-Impulse Trajectories, *AIAA Journal*, **6**, No. 11, 2160–2165.

[49] Prussing, J. E., and Chiu, J-H. (1986) Optimal Multiple Impulse Time-Fixed Transfer between Circular Orbits, *Journal of Guidance, Control, and Dynamics*, **9**, No. 1, 17–22.

4 Elements of a Software System for Spacecraft Trajectory Optimization

Cesar Ocampo
Department of Aerospace Engineering and Engineering Mechanics,
The University of Texas at Austin, Austin, Texas

4.1 Introduction

This chapter presents the main elements associated with a general spacecraft trajectory design and optimization software system. A unified framework is described that facilitates the modeling and optimization of spacecraft trajectories that may operate in complex gravitational force fields, use multiple propulsion systems, and involve multiple spacecraft. The ideas presented are simple and practical and are based in part on the existing wealth of knowledge documented in the open literature and the author's experience in developing software systems of this type.

The goal of any general trajectory design and optimization system is to facilitate the solution to a wide range of problems in a robust and efficient manner. A trade off exists between scope and depth. An attempt is made to strike a balance between the two and describe an approach that has proven to be robust and useful for a broad range of spacecraft trajectory design problems. The ideas and techniques presented here have been implemented in a working operational system known as Copernicus [1, 2]. This system has been used to support the detailed and comprehensive mission design studies associated with NASA's Constellation program [3, 4]. It has also been used to design and optimize the LCROSS [5] mission trajectory which was launched on June 18, 2009.

The system can be used to do any or all of the following:

- Open Loop Simulation: The system can simulate open loop trajectories in any force field, using one or more spacecraft, and including impulsive and/or finite burn maneuvers.
- Nonlinear Root Finding: The system can search for the values of independent variables to satisfy a set of equality constraint functions. This is required for general spacecraft targeting problems such as orbital boundary value problems or determining the initial conditions for periodic orbits in the circular restricted three body model, for example.
- Optimization: The system can extremize the value of a function consistent with general nonlinear constraints. A typical cost function is propellant consumption, but any other computable function can be extremized.

The key component of the system is the trajectory model that is used to construct simple to complex missions. The secondary components include the equations of motion, the propulsion system models, the control parametrization, the selection of the independent variables, the computation of the dependent constraint functions, and the objective function. The third set of components are the fundamental algorithms used to solve the nonlinear equations of motion, the root finding problem, and the optimization problem. This chapter presents an overview of the trajectory model and some of the secondary components of the system.

Two example missions are designed, illustrating some of the capabilities of the system. The first example illustrates the design of a lunar free-return trajectory. This is a simple example where most of the trajectory is ballistic. The second example mission is more complex and involves the placement of one spacecraft into a polar lunar orbit and a second spacecraft on a trajectory that collides with the lunar north pole. In this example one spacecraft performs impulsive maneuvers and the other spacecraft performs low-thrust maneuvers. Both spacecraft are injected into a translunar trajectory by the same booster.

4.2 Trajectory Model

The trajectory model uses a fundamental building block called a segment. A segment is a trajectory arc that can have velocity impulses and/or a finite burn maneuver. The basis for the segment model is derived from the simplest type of trajectory arc that connects two non-intersecting orbits. Depending on how the relationship between different segments is defined, single or multiple spacecraft trajectories that may interact can be modeled. The generality of how segments are related to one another facilitates the modeling of complex missions.

A segment arc connects two node points. The node points are tagged with an epoch that is referenced to a specified, but otherwise arbitrary, reference epoch, t_{epoch}. The reference epoch is chosen to be close to the time frame of the mission being considered. For any segment, the epoch of the initial node is t_0 and the epoch of the final node is t_f. There is no restriction on the values for t_0 and t_f ($t_0 = t_f$, $t_0 < t_f$, $t_0 > t_f$). The segment duration is defined as $\Delta t = t_f - t_0$. The segment terminal nodes are uniquely defined by specifying the values for any of the pairs $(t_0, t_f), (t_0, \Delta t), (\Delta t, t_f)$.

The state vector $\mathbf{x}$ consists of position $\mathbf{r}$, velocity $\mathbf{v}$, and mass m,

$$\mathbf{x} = \begin{pmatrix} \mathbf{r} \\ \mathbf{v} \\ m \end{pmatrix}. \tag{4.1}$$

If t_0 and t_f are distinct, the segment is integrated numerically from t_0 to t_f using the equations of motion. The equations of motion include natural accelerations and can include controlled accelerations due to an engine. Figure 4.1 illustrates the possible segment types showing the possible combinations of coast arcs, finite burn arcs,

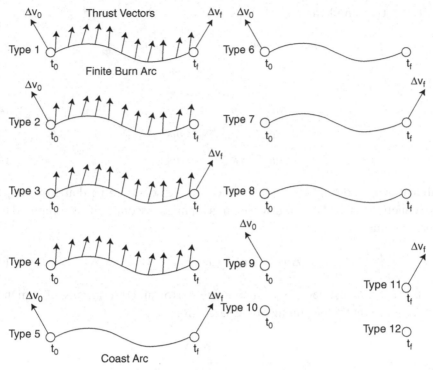

Figure 4.1. Possible segment types.

and impulsive maneuvers. Referring to this figure, it is necessary to be precise in distinguishing between segment types 10 and 12. In segment type 10, t_0 is specified and $\Delta t = 0$; in segment type 12, $\Delta t = 0$ and t_f is specified. Though these are qualitatively similar, the bookkeeping associated with the time tagging of the state input and function output data is different. Similar arguments apply to the impulsive maneuvers shown as segment types 9 and 11.

For any segment, the initial position vector is $\mathbf{r}_0$. The initial velocity vector is $\mathbf{v}_0^-$. The initial velocity impulse, if it exists, is $\Delta\mathbf{v}_0$. The velocity vector after this impulse is

$$\mathbf{v}_0^+ = \mathbf{v}_0^- + \Delta\mathbf{v}_0 \tag{4.2}$$

Due to the velocity impulse, it is necessary to distinguish the state before and after the impulse. The time tags t_0^- and t_0^+ refer to the state before and after the impulse, respectively. The same attributes apply at the t_f node. The initial value of the mass is m_0^{--}. The meaning of the double superscript '$--$' becomes clear after the following description. Before the first velocity impulse, $\Delta\mathbf{v}_0$, a non-maneuver mass discontinuity Δm_0^- is allowed. This is used to account for either a positive or negative mass discontinuity associated with acquiring another spacecraft or discarding a component of the current spacecraft. The velocity impulse itself introduces a mass discontinuity Δm_0 due to the propellant mass consumed. After the impulse, an additional mass discontinuity Δm_0^+ is allowed. For example, Δm_0^+ could be used to represent the stage mass used to perform the impulsive maneuver, in which case Δm_0^+ is a negative quantity. At the t_0 node mass evolves as follows

$$m_0^{--} \longrightarrow m_0^{-+} \longrightarrow m_0^{+-} \longrightarrow m_0^{++}$$

where

$$m_0^{-+} = m_0^{--} + \Delta m_0^- \tag{4.3}$$

$$m_0^{+-} = m_0^{-+} + \Delta m_0 \tag{4.4}$$

$$m_0^{++} = m_0^{+-} + \Delta m_0^+. \tag{4.5}$$

Recall that Δm_0^- and Δm_0^+ are explicit vehicle mass discontinuities that do not depend on the maneuver itself. The impulsive maneuver mass discontinuity is computed from the rocket equation

$$\Delta m_0 = \left(m_0^{-+}\right)\left(e^{-\Delta v_0/c} - 1\right) \tag{4.6}$$

where c is the exhaust velocity of the engine performing the impulsive maneuver. The state vector at the beginning of the segment is

$$\mathbf{x}_0^- = \begin{pmatrix} \mathbf{r}_0 \\ \mathbf{v}_0^- \\ m_0^{--} \end{pmatrix} \tag{4.7}$$

After accounting for the potential state discontinuities at the t_0 node, the initial condition for the state vector that is numerically propagated from t_0 to t_f (if $t_0 \neq t_f$) is

$$\mathbf{x}_0^+ = \begin{pmatrix} \mathbf{r}_0 \\ \mathbf{v}_0^+ \\ m_0^{++} \end{pmatrix} \tag{4.8}$$

The state vector after the propagation and before possible state discontinuities at t_f is

$$\mathbf{x}_f^- = \begin{pmatrix} \mathbf{r}_f \\ \mathbf{v}_f^- \\ m_f^{--} \end{pmatrix} \tag{4.9}$$

The same description regarding the use of an impulsive maneuver and the mass discontinuities at the t_0 node applies to the velocity and mass evolution at the t_f node. Therefore, the final value of the state vector of a segment is

$$\mathbf{x}_f^+ = \begin{pmatrix} \mathbf{r}_f \\ \mathbf{v}_f^+ \\ m_f^{++} \end{pmatrix} \tag{4.10}$$

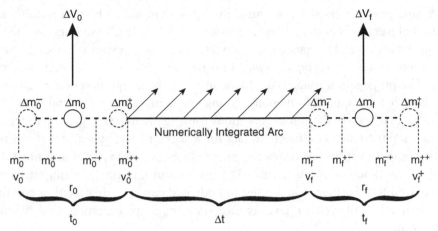

Figure 4.2. The expanded segment showing the evolution of state variables.

In summary, the state elements evolve across the segment as follows

$$
\left\{
\begin{array}{c}
\mathbf{r}_0 \quad \stackrel{integrated}{\Longrightarrow} \quad \mathbf{r}_f \\[4pt]
\mathbf{v}_0^- \quad \longrightarrow \quad \mathbf{v}_0^+ \quad \stackrel{integrated}{\Longrightarrow} \quad \mathbf{v}_f^- \quad \longrightarrow \quad \mathbf{v}_f^+ \\[4pt]
m_0^{--} \to m_0^{-+} \to m_0^{+-} \to m_0^{++} \quad \stackrel{integrated}{\Longrightarrow} \quad m_f^{--} \to m_f^{-+} \to m_f^{+-} \to m_f^{++}
\end{array}
\right\}.
$$

State functions can only be computed from these state vectors and are tagged with the epochs t_0, t_f for position, $t_0^-, t_0^+, t_f^-, t_f^+$ for velocity, and $t_0^{--}, t_0^{-+}, t_0^{+-}, t_0^{++}, t_f^{--}, t_f^{-+}, t_f^{+-}, t_f^{++}$ for mass. Figure 4.2 illustrates the segment structure in detail and shows the evolution of velocity and mass across the segment.

Assuming for the moment that the trajectory arc between t_0 and t_f is purely ballistic, the key independent quantities of the segment are

$$
\left(t_0, \mathbf{r}_0, \mathbf{v}_0^-, \Delta \mathbf{v}_0, m_0^{--}, \Delta m_0^-, \Delta m_0^+, t_f, \Delta \mathbf{v}_f, \Delta m_f^-, \Delta m_f^+ \right)
$$

and the dependent quantities are

$$
\left(\mathbf{v}_0^+, \Delta m_0, m_0^{++}, \mathbf{r}_f, \mathbf{v}_f^-, \mathbf{v}_f^+, m_f^{--}, \Delta m_f, m_f^{++} \right).
$$

Sets of segments are used to construct complete trajectories for one or more spacecraft. Segments can be connected to other segments by forcing the initial state of a segment (or a subset of the initial state) to inherit the state (or a subset of the state) associated with either node of a previous segment. Segments can also be completely disconnected to model independent trajectories. These segments can later be constrained to be connected via continuity constraints to other segments. A mission in this context is defined to contain the set of all segments that are present.

A *base frame* is chosen for a mission and is required to be non-rotating and centered at a celestial body. All state transformations use the base frame as the hub through which all state, maneuver, and state function transformations are made. Each segment has a *state input frame*, a *propagation frame*, and a *function output frame*. The propagation frame is the frame used for the integration of the equations of motion and is required to be non-rotating and centered at a celestial body. The state data that specify the initial conditions for a segment, $(\mathbf{x}_0^-)$, are referenced to the state input frame. The function data for a segment includes all functions that can be computed from the distinct segment states $(\mathbf{x}_0^-, \mathbf{x}_0^+, \mathbf{x}_f^-, \mathbf{x}_f^+)$ and are referenced to the function output frame. The state input and function output frames are required to be centered at a celestial body and can be either fixed or rotating. Each segment is allowed to have its own state input, propagation, and function output frame.

The individual segment impulsive maneuvers $\Delta\mathbf{v}_0$ and $\Delta\mathbf{v}_f$ are each referenced to a unique *impulsive maneuver frame*. The impulsive maneuver frame can be either fixed (non-rotating) or osculating along the trajectory path. In either case the frame is defined relative to a celestial body centered non-rotating or rotating frame. An example of a common osculating impulsive maneuver frame is one with its first axis along the instantaneous velocity vector; the third axis is along the instantaneous angular momentum vector and the second axis completes the right handed system.

A segment vector $\mathbf{s}^i$ is defined for segment i to be a mixed variable type vector containing all of the independent and dependent variables and functions associated with the segment. Additionally it contains all the frame definitions, how and what data is inherited from other segments, what functions need to be evaluated from the state vector at the distinct time tags, and which variables are search variables, etc. This list of information is large, and in this discussion it is limited to only what is relevant. The reduced list includes the state vector at t_0^- and t_f^+ and all of the velocity and mass discontinuities that may be present,

$$
\mathbf{s}^i = \begin{pmatrix} t_0, \mathbf{r}_0^\top, \mathbf{v}_0^{-\top}, m_0^{--}, \Delta m_0^-, \Delta \mathbf{v}_0^\top, \Delta m_0^+, \\ t_f, \Delta m_f^-, \Delta \mathbf{v}_0^\top, \Delta m_f^+, \mathbf{r}_f, \mathbf{v}_f^{++}, m_f^{++}, \ldots \ldots \end{pmatrix}^\top . \tag{4.11}
$$

The mission can be computed uniquely in open loop by processing all of the information contained in the set of segment vectors $\mathbf{s}^i (i = 1, \ldots, n)$ for a mission containing n segments.

Every segment also has a search vector $\mathbf{s}_x^i$ and a function vector $\mathbf{s}_f^i$. $\mathbf{s}_x^i$ is a subset of $\mathbf{s}^i$ and contains only the independent variables that have been tagged to be search variables. $\mathbf{s}_f^i$ is a subset of $\mathbf{s}^i$ and contains only the functions that have been tagged to be constrained or optimized. $\mathbf{s}_f^i$ can contain state functions computed from the segment states $\mathbf{x}_0^-, \mathbf{x}_0^+, \mathbf{x}_f^-, \mathbf{x}_f^+$. For example, $\mathbf{s}_x^i$ could contain the components of $\Delta\mathbf{v}_0$, and $\mathbf{s}_f^i$ could contain a subset of the orbital elements computed from $\mathbf{x}_f^+$.

4.3 Equations of Motion

The state vector for any segment i is governed by the first-order vector equation of motion,

$$\frac{d}{dt}\begin{pmatrix} \mathbf{r} \\ \mathbf{v} \\ m \end{pmatrix}^{i} = \begin{pmatrix} \mathbf{v} \\ \mathbf{g}(\mathbf{r}, \mathbf{v}, m, t) + \dfrac{T(t)}{m}\mathbf{u}(t) \\ -\dfrac{T(t)}{c(t)} \end{pmatrix}^{i} \tag{4.12}$$

where $\mathbf{g}$ is the ballistic acceleration per unit mass, T is the finite engine thrust force, $\mathbf{u}$ is the unit thrust direction, and c is the engine exhaust velocity. T, c, and $\mathbf{u}$ are possible control functions. These equations are evaluated from t_0^+ to t_f^-. The only allowable discontinuities are associated with the control terms.

An example expression for $\mathbf{g}$ is the acceleration acting on a spacecraft due to a main gravitational body c_b and possibly including the gravitational acceleration due to n_b additional celestial bodies whose time dependent positions $\mathbf{r}_j(t)$ with respect to c_b are obtained from a precomputed ephemeris

$$\mathbf{g}(\mathbf{r}, t) = -\frac{Gm_{c_b}}{r^3}\mathbf{r} - G\sum_{j=1}^{n_b} m_j \left(\frac{\mathbf{r} - \mathbf{r}_j(t)}{|\mathbf{r} - \mathbf{r}_j(t)|^3} + \frac{\mathbf{r}_j(t)}{r_j^3(t)} \right) \tag{4.13}$$

where G is the universal constant of gravitation, m_{c_b} is the mass of the main celestial body, and m_j is the mass of celestial body j. Depending on the problem, additional terms accounting for other common accelerations such as atmospheric drag, solar radiation pressure, and non-spherical celestial bodies need to be added to the vector function $\mathbf{g}$. These additional parameters associated with $\mathbf{g}$ also form part of the segment vector $\mathbf{s}^i$. The reduced form for $\mathbf{g}$ used in Equation (4.13) serves only as an example and is used in the example missions to be discussed.

4.4 Finite Burn Control Models

A finite burn arc is based on one of two possible models. The first model is the parameter model that uses a finite set of parameters to describe the time evolution of the control variables. The second model is the optimal control model using the analytical necessary conditions of the problem and thus requires the vector of Lagrange multipliers adjoined to the state vector. The system described can accommodate several types of engine models [2], but only a simple constant exhaust velocity engine model is used to illustrate both methods. In this engine model it is assumed that the exhaust velocity c is known and constant. The control variables are then the thrust force T and the unit thrust direction $\mathbf{u}$ with constraints

$$T_{\min} \leq T(t) \leq T_{\max} \tag{4.14}$$

$$|\mathbf{u}(t)| = 1. \tag{4.15}$$

4.4.1 Thrust Vector Parameter Model

For every finite burn segment it is necessary to determine the value of the thrust $T(t)$ and the unit direction $\mathbf{u}(t)$. It will be assumed that once the value for $T(t)$ is determined, it remains constant for that segment. The reference frame for $\mathbf{u}(t)$ is the *steering frame* which can be specified; alternatively, its basis unit vectors $\hat{\mathbf{u}}_0, \hat{\mathbf{v}}_0, \hat{\mathbf{w}}_0$ can be determined as part of the solution process. If the basis unit vectors are specified, common sets of basis vectors are:

- $\hat{\mathbf{u}}_0, \hat{\mathbf{v}}_0, \hat{\mathbf{w}}_0$ aligned with the basis unit vectors of the propagation frame;
- $\hat{\mathbf{u}}_0, \hat{\mathbf{v}}_0, \hat{\mathbf{w}}_0$ aligned with one of two possible sets of basis unit vectors:
  (a) velocity referenced framed

$$\hat{\mathbf{u}}_0(t) = \frac{\mathbf{v}}{v} \qquad (4.16)$$

  (b) or position referenced frame

$$\hat{\mathbf{u}}_0(t) = \frac{\mathbf{r}}{r} \qquad (4.17)$$

with the remaining basis vectors for both frames given by

$$\hat{\mathbf{v}}_0(t) = \hat{\mathbf{w}}_0 \times \hat{\mathbf{u}}_0 \qquad (4.18)$$

$$\hat{\mathbf{w}}_0(t) = \frac{\mathbf{r} \times \mathbf{v}}{|\mathbf{r} \times \mathbf{v}|} \qquad (4.19)$$

where $\mathbf{r}$ and $\mathbf{v}$ are the instantaneous position and velocity along the trajectory.

If the basis unit vectors are to be determined as part of the solution process they are constrained to be an orthogonal basis

$$|\hat{\mathbf{u}}_0| = |\hat{\mathbf{w}}_0| = 1 \qquad (4.20)$$

$$\hat{\mathbf{w}}_0^\top \hat{\mathbf{u}}_0 = 0 \qquad (4.21)$$

$$\hat{\mathbf{v}}_0 = \hat{\mathbf{w}}_0 \times \hat{\mathbf{u}}_0 \qquad (4.22)$$

Within the steering frame the unit thrust direction is parameterized by the spherical angles α and β,

$$\mathbf{u}_{\hat{u}\hat{v}\hat{w}} = \begin{pmatrix} \cos\beta(t)\cos\alpha(t) \\ \cos\beta(t)\sin\alpha(t) \\ \sin\beta(t) \end{pmatrix} \qquad (4.23)$$

The spherical angles are assumed to be functions of time of the form

$$\alpha(t) = \alpha_0 + \dot{\alpha}_0(t - t_0) + \frac{1}{2}\ddot{\alpha}_0(t - t_0)^2 + a_\alpha \sin(\omega_\alpha(t - t_0) + \phi_\alpha) \qquad (4.24)$$

$$\beta(t) = \beta_0 + \dot{\beta}_0(t - t_0) + \frac{1}{2}\ddot{\beta}_0(t - t_0)^2 + a_\beta \sin(\omega_\beta(t - t_0) + \phi_\beta) \qquad (4.25)$$

where $t_0 \leq t \leq t_f$. The thrust direction unit vector in the propagation frame is

$$\mathbf{u}(t) = [\cos \beta(t) \cos \alpha(t)] \, \hat{\mathbf{u}}_0 + [\cos \beta(t) \sin \alpha(t)] \, \hat{\mathbf{v}}_0 + [\sin \beta(t)] \, \hat{\mathbf{w}}_0. \qquad (4.26)$$

Each of the constants in Eqns. 4.24 and 4.25 can be part of $\mathbf{s}_x^i$. As defined, the functions for the spherical angles α and β admit constant, linear, quadratic, and sinusoidal terms. For most practical applications the constant and linear terms are sufficient. However, for some low-thrust applications, the sinusoidal terms may be important. An example is the oscillation of the thrust direction vector relative to the velocity vector for a low-thrust, long duration, spiral escape or capture trajectory about a celestial body [6].

4.4.2 Thrust Vector Optimal Control Model

An alternate steering profile for the thrust vector is based on Optimal Control Theory [7–11]. In the system under discussion it is only necessary to understand and solve the equations that govern the behavior of the thrust vector. The formulation of the multi-point boundary value problem that results from the complete optimal control solution is not required in the approach described here. This removes the necessity of deriving and including the associated transversality conditions in the solution process.

In the Mayer form of the optimal control problem, the control Hamiltonian is defined as

$$H \equiv \boldsymbol{\lambda}^\top \mathbf{f} \qquad (4.27)$$

where $\boldsymbol{\lambda}$ is the costate vector of Lagrange multipliers adjoined to the state vector and $\mathbf{f}$ is the vector field that governs the evolution of the state vector $\mathbf{x}$

$$\dot{\mathbf{x}} = \mathbf{f}(\mathbf{x}, t, \mathbf{u}_c) \qquad (4.28)$$

where $\mathbf{u}_c$ is a vector of control variables. A first-order necessary condition is the vector differential equation for $\boldsymbol{\lambda}$

$$\dot{\boldsymbol{\lambda}} = -\left(\frac{\partial H}{\partial \mathbf{x}}\right)^\top \qquad (4.29)$$

Additionally, the control vector $\mathbf{u}_c$ is chosen to extremize H at all points on a solution consistent with the control constraints if they are present. Recall the state vector

$$\mathbf{x} = \begin{pmatrix} \mathbf{r}^\top & \mathbf{v}^\top & m \end{pmatrix}^\top. \qquad (4.30)$$

Adjoined to this is the costate vector

$$\boldsymbol{\lambda} = \begin{pmatrix} \boldsymbol{\lambda}_r^\top & \boldsymbol{\lambda}_v^\top & \lambda_m \end{pmatrix}^\top. \qquad (4.31)$$

Using the example force model described in Eqn. 4.12, for a spacecraft with a constant exhaust velocity engine, the vector field is

$$\mathbf{f}(\mathbf{x},t,\mathbf{u}_c) = \begin{pmatrix} \mathbf{v} \\ \mathbf{g}(\mathbf{r},t) + \dfrac{T(t)}{m}\mathbf{u}(t) \\ -\dfrac{T(t)}{c} \end{pmatrix} \tag{4.32}$$

The Hamiltonian is

$$H = \boldsymbol{\lambda}_r^\top \mathbf{v} + \boldsymbol{\lambda}_v^\top \left(\mathbf{g}(\mathbf{r},t) + \frac{T}{m}\mathbf{u} \right) + \lambda_m \left(-\frac{T}{c} \right) \tag{4.33}$$

and the costate vector equations are

$$\dot{\boldsymbol{\lambda}}_r = -\left(\frac{\partial H}{\partial \mathbf{r}} \right)^\top = -\left[\frac{\partial}{\partial \mathbf{r}} \left(\boldsymbol{\lambda}_v^\top \mathbf{g} \right) \right]^\top = \left(\frac{\partial \mathbf{g}}{\partial \mathbf{r}} \right)^\top \boldsymbol{\lambda}_\mathbf{v} \tag{4.34}$$

$$\dot{\boldsymbol{\lambda}}_v = -\left(\frac{\partial H}{\partial \mathbf{v}} \right)^\top = -\boldsymbol{\lambda}_r \tag{4.35}$$

$$\dot{\lambda}_m = -\frac{\partial H}{\partial m} = \frac{T(t)}{m^2}\boldsymbol{\lambda}_v^\top \mathbf{u}. \tag{4.36}$$

Using the form of the ballistic acceleration given in Eqn. 4.13 the gradient term in the equation for the position costate vector of Eqn. 4.34 is

$$\left(\frac{\partial \mathbf{g}}{\partial \mathbf{r}} \right) = -\frac{Gm_{c_b}}{r^5}\left(3\mathbf{r}\mathbf{r}^\top - r^2\mathbf{I} \right) - G\sum_{j=1}^{n_b} m_j \left(\frac{3\left(\mathbf{r} - \mathbf{r}_j(t)\right)\left(\mathbf{r} - \mathbf{r}_j(t)\right)^\top}{\left|\mathbf{r} - \mathbf{r}_j(t)\right|^5} - \frac{\mathbf{I}}{\left|\mathbf{r} - \mathbf{r}_j(t)\right|^3} \right) \tag{4.37}$$

where $\mathbf{I}$ is a 3×3 identity matrix. Extremization of H with respect to the thrust unit vector results in the thrust unit vector steering law [12, 13]

$$\mathbf{u}(t) = \pm\frac{\boldsymbol{\lambda}_v(t)}{\lambda_v(t)} \tag{4.38}$$

and is referred to as the Primer Vector (cf. Chapter 2). The choice of sign depends on whether H is required to be maximized or minimized. In this discussion it is assumed that H is maximized. The assumed engine model admits the well known switching function S which is the coefficient of the thrust term in H after substituting for $\mathbf{u}(t)$,

$$H = \boldsymbol{\lambda}_r^\top \mathbf{v} + \boldsymbol{\lambda}_v^\top \mathbf{g}(\mathbf{r},t) + \left(\frac{\lambda_v}{m} - \frac{\lambda_m}{c} \right) T \tag{4.39}$$

$$S \equiv \left(\frac{\lambda_v}{m} - \frac{\lambda_m}{c} \right). \tag{4.40}$$

Maximization of H with respect to T yields the well known thrust magnitude control law

$$T = \begin{cases} T_{\min} & \text{if} \quad S < 0 \\ T_{\max} & \text{if} \quad S > 0 \end{cases}. \tag{4.41}$$

A switching between the thrust limits occurs instantly when $S = 0$ and $\dot{S} \neq 0$. A singular arc occurs when $S = 0$ and $\dot{S} = 0$ for a finite duration. In this latter case the value of the thrust magnitude is undetermined and a higher order analysis is required to determine the thrust magnitude [9].

In the optimal control model, one of two methods can be used to determine T for a given segment. The first method does not use the switching function and lets the optimization algorithm treat the thrust magnitude as a search variable. If a series of segments are connected sequentially then a trajectory with different values of thrust (some may be coast arcs) is produced. Alternatively, segments can be forced to be either finite burn segments or coast arcs. Since the durations of the segments are search variables, a sequence of sequentially connected segments can be generated representing a string of thrust and coast arcs. Some of these segments may converge to arcs with zero time duration for cases where that particular segment should not exist with the thrust value that was assumed for it.

The second method uses the switching function in one of two ways. A series of segments can be constructed with an assumed thrust and coast switching structure. Again since the durations of these segments are variables the start and end times of the segments can be constrained such that $S = 0$ and $\dot{S} > 0$ if the switch is from T_{min} to T_{max} and $S = 0$ and $\dot{S} < 0$ if the switch is from T_{max} to T_{min}. The simplest way to determine the thrust magnitude, albeit more numerically sensitive, is to monitor the switching function during the numerical integration of the segment and apply the thrust control law in Equation (4.41). This introduces a discontinuity in the equations of motion which may be a problem for most numerical integrators. The DLSODE numerical integration routine from the ODEPACK suite of numerical integration routines [14] has proven to detect the time of this discontinuity accurately and for this reason it has been selected as the main general purpose numerical integration routine in the system.

4.4.2.1 Adjoint Control Transformation

In the optimal control model the costate vector $\lambda(t_0)$ forms part of s_x^i. An estimate for $\lambda(t_0)$ is facilitated by using control related quantities instead of the actual costates. This is accomplished with a simple technique known as an adjoint-control transformation [15]. Using the control optimality condition for the thrust pointing unit vector, the value of the velocity costate vector is

$$\lambda_{v_0} = \lambda_{v_0} \mathbf{u}_0 \tag{4.42}$$

where $\mathbf{u}_0$ is given as a function of the two spherical thrust direction angles α_0 and β_0. The time rate of change of λ_{v_0} is

$$\dot{\lambda}_{v_0} = \dot{\lambda}_{v_0} \mathbf{u}_0 + \lambda_{v_0} \dot{\mathbf{u}}_0. \tag{4.43}$$

The velocity costate vector at t_0 is required to satisfy Eqn. 4.35

$$\dot{\lambda}_{v_0} = -\lambda_{r_0} \tag{4.44}$$

so that the the position costate vector at t_0 becomes

$$\boldsymbol{\lambda}_{r_0} = -\dot{\lambda}_{v_0}\mathbf{u}_0 - \lambda_{v_0}\dot{\mathbf{u}}_0. \tag{4.45}$$

This is the basic transformation and is summarized as follows

$$\begin{pmatrix} \lambda_v \\ \dot{\lambda}_v \\ \alpha \\ \dot{\alpha} \\ \beta \\ \dot{\beta} \\ \lambda_m \end{pmatrix} \implies \begin{pmatrix} \lambda_v \\ \dot{\lambda}_v \\ \mathbf{u} \\ \dot{\mathbf{u}} \\ \lambda_m \end{pmatrix} \implies \begin{pmatrix} \lambda_v \\ \lambda_r \\ \lambda_m \end{pmatrix}. \tag{4.46}$$

The necessity of explicitly specifying all of the components of the costate vector has been removed in favor of specifying the four control related variables $(\alpha, \beta, \dot{\alpha}, \dot{\beta})$ and three costate related variables $(\lambda_v, \dot{\lambda}_v, \lambda_m)$. These remaining variables are still generally non-intuitive. However, further relationships can be derived that remove the necessity of these being specified. These relationships are problem dependent and require knowledge of the transversality conditions that result from the complete formulation of the solution of the optimal control problem [2]. The use of the transformation replaces the explicit costate vector in $\mathbf{s}_x^i$ with the adjoint control variables. Using costate variables or adjoint-control variables without explicitly enforcing the transversality conditions in a parameter optimization solution procedure is referred to here as a hybrid optimal control method and is the basis for all the solutions that use the optimal control model for the thrust vector.

4.5 Solution Methods

All the information necessary to perform an open loop simulation is contained in $\mathbf{s}^i$. If a targeting or optimization problem needs to be solved then it is necessary to cast the given data into a standard format. Let $\mathbf{x}_p$ be the $n \times 1$ problem search vector. It contains all of the elements in $\mathbf{s}_x^i$ $(i = 1, \ldots, n_s)$ for a n_s segment mission. Let $\mathbf{c}$ be the $m \times 1$ problem function vector that will contain the objective function, all of the equality constraints and all of the inequality constraints. It is based on the elements in $\mathbf{s}_f^i$ $(i = 1, \ldots, n_s)$. Let m_{eq} be the number of equality constraints and m_{ineq} be the number of individual inequality constraints so that $m = 1 + m_{eq} + m_{ineq}$. An element j of $\mathbf{s}_f^i$ is required to either

- satisfy an equality constraint

$$\left(\mathbf{s}_f^i\right)_j = f_{eq}$$

- or satisfy a lower and/or an upper bound inequality constraint

$$\left(\mathbf{s}_f^i\right)_j \geq f_l \quad \text{and/or} \quad \left(\mathbf{s}_f^i\right)_j \leq f_u \quad f_u \geq f_l$$

- or be added to the objective function. An element that forms part of the objective function can also have specified lower and/or upper bounds.

The first element of $\mathbf{c}$ is c_1, the objective function, f_{obj}, which is the sum of all of the elements in $\mathbf{s}_f^i$ $(i = 1, \ldots, n_s)$ that have been tagged to be part of the objective function. If no element has been tagged to be part of the objective function, then c_1 is zero. The next part of $\mathbf{c}$ is $\mathbf{c}_{eq}(m_{eq} \times 1)$ which contains all of the equality constraint functions. The last part of $\mathbf{c}$ is $\mathbf{c}_{ineq}(m_{ineq} \times 1)$ which contains all of the inequality constraint functions. All inequality constraints are re-written with a specified upper bound of zero

$$-\left(\mathbf{s}_f^i\right)_j + f_L \leq 0 \tag{4.47}$$

$$\left(\mathbf{s}_f^i\right)_j - f_U \leq 0. \tag{4.48}$$

In summary

$$\mathbf{c} = \begin{pmatrix} f_{obj} \\ \left(\mathbf{c}_{eq}\right)_{m_{eq} \times 1} \\ \left(\mathbf{c}_{ineq}\right)_{m_{ineq} \times 1} \end{pmatrix}_{m \times 1}. \tag{4.49}$$

The elements of $\mathbf{c}$ are assumed to be consistent and independent. For any gradient-based solution method it is necessary to obtain an estimate of the Jacobian matrix $\partial \mathbf{c} / \partial \mathbf{x}_p$. Numerical finite difference methods [16, 17] are the most common and accepted methods used to estimate $\partial \mathbf{c} / \partial \mathbf{x}_p$. For example, the (i, j) element of $\partial \mathbf{c} / \partial \mathbf{x}_p$ can be approximated using a forward difference approximation

$$\frac{\partial c_i}{\partial x_{p_j}} \approx \frac{c_i(x_{p_j} + \Delta x_{p_j}) - c_i(x_{p_j})}{\Delta x_{p_j}} \tag{4.50}$$

or a more accurate central difference approximation

$$\frac{\partial c_i}{\partial x_{p_j}} \approx \frac{c_i(x_{p_j} + \Delta x_{p_j}) - c_i(x_{p_j} - \Delta x_{p_j})}{2\Delta x_{p_j}} \tag{4.51}$$

where Δx_{p_j} is the positive perturbation stepsize for the x_{p_j} element of $\mathbf{x}_p$. A generally accepted rule of thumb for the value of Δx_{p_j} is $10^{-8}|x_{p_j}|$ for a forward difference approximation and $\Delta x_{p_j} = 10^{-4}|x_{p_j}|$ for a central difference approximation [16, 17]. Unfortunately, for many problems it is often necessary to experiment with different values because they are problem and process dependent. The choice for Δx_{p_j} is sometimes based on experience and can be considered more of an art than a science. There are more accurate variational methods based on the use of state transition matrices [18–20] that have proven to be effective for very specific problems, but these are not practical for a generalized system because of the overhead required to derive all of the required relationships between the large number of search variables and constraint functions. The accurate estimation of $\partial \mathbf{c} / \partial \mathbf{x}_p$ is perhaps the single most important calculation for any gradient-based solution method.

4.5.1 Root Finding Problem: $f_{obj} = 0.0, m_{ineq} = 0, m_{eq} = n \geq 1$

This is a nonlinear root finding problem for a system of nonlinear equations

$$\mathbf{c}_{eq}(\mathbf{x}_p) = \mathbf{0} \tag{4.52}$$

which is solved with standard gradient-based nonlinear root finding algorithms [16]. There is no guarantee that a solution exists, and multiple solutions are possible. Many algorithms are available for the solution of this problem. The NS11 and NS12 algorithms from the Harwell Subroutine Library [21] have proven to be very robust and efficient routines for the class of problems considered here. The NS11 algorithm is a Newton-Raphson and steepest descent method which uses Broyden's method for extrapolating the Jacobian array of the system. The NS12 algorithm is based on a trust region method coupled with the Powell dog-leg algorithm.

4.5.2 Mini-Max Problem: $f_{obj} = 0.0, m_{ineq} = 0, m_{eq} \geq 1, n \geq 1$

This is a standard mini-max problem where it is desired to find a local minimum of the function

$$\max \left| \left[\mathbf{c}_{eq}(\mathbf{x}_p) \right]_j \right|, \, j = 1 \ldots m_{eq}. \tag{4.53}$$

If $m_{eq} = n$ then the solution is equivalent to the zero of a set of nonlinear equations as described above. If $m_{eq} > n$ the system is overdetermined and the local minimum of the function can be expected to be nonzero. This is equivalent to nonlinear mini-max data fitting. If $m_{eq} < n$ the system is under determined, the minimum of the function is zero and the solution is the minimum norm solution from the initial value of $\mathbf{x}_p$. The VG11/12 algorithms from the Harwell Subroutine Library [21] have proven to be effective in solving problems under this mode.

4.5.3 Nonlinear Programming Problem: $m_{ineq} \geq 0, 0 \leq m_{eq} < n, n \geq 1$

This is the general problem of nonlinear functional optimization with nonlinear equality and inequality constraints and whose solution methods have been well documented in the literature [17, 22]. The problem is to minimize or maximize the objective function (which can be zero)

$$f_{obj}(\mathbf{x}_p) \tag{4.54}$$

subject to both the equality and inequality constraints

$$\mathbf{c}_{eq}\left(\mathbf{x}_p\right) = \mathbf{0} \qquad m_{eq} \times 1 \tag{4.55}$$

$$\mathbf{c}_{ineq}\left(\mathbf{x}_p\right) \leq \mathbf{0} \qquad m_{ineq} \times 1. \tag{4.56}$$

The sequential quadratic programming (SQP) routines VF13 [21] and SNOPT [23] have proven to be effective in solving the types of problems that can be cast in this system.

4.6 Trajectory Design and Optimization Examples

To illustrate some of the ideas presented in this chapter, two trajectory design examples are presented. These examples capture some of the key elements of the system and are not meant to be comprehensive. It is not possible to formulate one or two examples that make use of all of the ideas presented here. The first example involves the design of a simple free-return lunar flyby trajectory from the Earth. The second example illustrates the procedure used to design a dual spacecraft mission where the first spacecraft is inserted into a lunar orbit and the second spacecraft is required to perform a lunar flyby so that it impacts the lunar north pole approximately half a month later. Both examples will use the VF13 SQP routine in obtaining the feasible and optimal solutions. The procedure is described as an experimental process, and it is the actual procedure used to construct the final solution for both of the examples.

4.6.1 Free-Return Lunar Flyby Mission

Free-return lunar flyby trajectories are well known and have been used since the Apollo era. A new automated and robust procedure not requiring a user-provided initial estimate was recently developed to construct families of these types of trajectories [24]. In the example presented here no special information is assumed and the system will be used without the aid of an initial estimate generating algorithm. The spacecraft is injected impulsively into a trans-lunar trajectory from a low Earth orbit. The spacecraft performs a flyby around the Moon with a specified periapsis radius and returns to the Earth where it is captured impulsively into a low Earth orbit. The objective function is the time of flight.

There are many ways to construct this solution. The method chosen for this example is to divide the trajectory into four segments. The first pair of segments ($S1$ and $S2$) model the departure and capture phases at the Earth, respectively. The Earth departure segment ($S1$) is propagated forward in time and the capture segment ($S2$) is propagated backward in time. The second pair of segments ($S3$ and $S4$) model the hyperbolic flyby at the Moon. A spacecraft state vector is defined at the Moon with a true anomaly of zero and a specified periapsis radius; the remaining orbital elements (e, i, Ω, ω) are search variables. The hyperbolic trajectory ($S4$) is propagated backward in time and represents the lunar approach. The other segment ($S3$) is propagated forward in time and represents the lunar departure. A sketch of this segment model is shown in Figure 4.3. The relevant data associated with each segment is tabulated in Table 4.1. In this table all times are referenced to the reference epoch. The reference epoch is April 3 2009 00:00:00.000 UTC (Coordinated Universal Time).

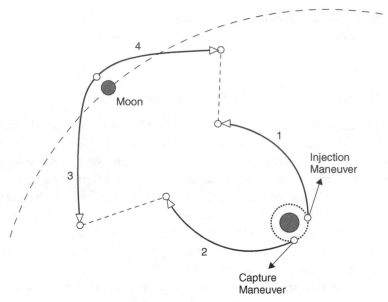

Figure 4.3. Segment Model for the Free-Return Lunar Flyby Mission.

Table 4.1. *Segment Data for the Free-Return Lunar Flyby Mission.*
Starred quantities denote search variables. Definitions: ns = not specified,
na = not applicable. The left double arrow indicates data is inhereted from
the specified node and segment given.

Segment→	S1	S2	S3	S4
t_0 (day)	0.0	8.0*	4.0*	$\Leftarrow t_0$ S3
Δt (day)	1.0*	−1.0*	1.0*	−1.0*
a (km)	6578.0	6578.0	ns	ns
r_p (km)	ns	ns	10000.0	$\Leftarrow t_0^-$ S3
e	0.0	0.0	1.0*	$\Leftarrow t_0^-$ S3
i (deg)	0.0*	45.0*	0.0*	$\Leftarrow t_0^-$ S3
Ω (deg)	0.0*	0.0*	180.0*	$\Leftarrow t_0^-$ S3
ω (deg)	0.0	0.0	0.0*	$\Leftarrow t_0^-$ S3
v (deg)	0.0*	0.0*	0.0	$\Leftarrow t_0^-$ S3
Δv_{0_x} (km/s)	3.2*	−3.2*	na	na

The base frame is $ECJ2000$ (Earth centered $J2000$). The state input and propagation frames for $S1$ and $S2$ are both $ECJ2000$. The state input and propagation frames for $S3$ and $S4$ are both $MCJ2000$ (Moon centered $J2000$). The function output frame for all segments is $ECJ2000$. The estimates provided for some of the segment data are consistent with the time duration of a typical free-return trajectory. The time estimates are based on the duration of a Hohmann-type transfer from a low Earth orbit to the lunar radius which is about four to five days. The departure and capture maneuvers are consistent with the magnitude of an escape maneuver from the Earth from the specified altitude. These are only estimates and they need not be exact. The initial epoch $(t_0)^{S1}$ is fixed. The final time of the trajectory is the initial time of $S2$, $(t_0)^{S2}$, and is the objective function to be minimized. The injection and capture maneuvers at Earth are required to be along the velocity and anti-velocity directions. The directions of these maneuvers in the base frame are controlled by the orientation of the departure and capture orbits. A subset of the angular orbital elements (i, Ω, ω, ν) of these orbits are search variables. Mass variables play no role in this example. The time equality and inequality constraints are

$$\left(t_f\right)^{S1} = \left(t_f\right)^{S4}$$
$$\left(t_f\right)^{S2} = \left(t_f\right)^{S3}$$
$$(\Delta t)^{S1} \geq 1.0$$
$$(\Delta t)^{S2} \leq 1.0$$
$$(\Delta t)^{S3} \geq 1.0$$
$$(\Delta t)^{S4} \leq 1.0$$

and the state equality constraints which are referenced to $ECJ2000$ are

$$\left(\mathbf{r}_f\right)^{S1} = \left(\mathbf{r}_f\right)^{S4}$$
$$\left(\mathbf{v}_f\right)^{S1} = \left(\mathbf{v}_f\right)^{S4}$$
$$\left(\mathbf{r}_f\right)^{S2} = \left(\mathbf{r}_f\right)^{S3}$$
$$\left(\mathbf{v}_f\right)^{S2} = \left(\mathbf{v}_f\right)^{S3} .$$

All segment durations are constrained to be at least one day so that the matching of the state constraints is forced to occur far from the Earth and the Moon. The objective function is

$$f_{obj} = \min \, (t_0)^{S2} .$$

The problem contains 18 search variables, 14 equality constraints, and 4 inequality constraints.

All required gradients are approximated by a forward difference approximation. All time values have a perturbation step size of 1.0×10^{-4} days and a maximum per iteration bound of 1.0 day. All orbital element angular values have a perturbation step size of 1.0×10^{-3} deg and a maximum per iteration bound of 5.0 deg. The Δv magnitude values have a perturbation step size of 1.0×10^{-6} km/s and a maximum per iteration bound of 0.1 km/s. The eccentricity value of $S3$ has a perturbation step size of 1.0×10^{-4} and a maximum per iteration bound of 1.0×10^{-2}.

The initial estimate and final converged solution for the trajectory are shown in Figure 4.4. From this estimate, the solution procedure used the VF13 algorithm with a convergence tolerance of 1.0×10^{-4}. The solution converged in 210 iterations with a minimum flight time of 7.543312 days. The data for the converged optimal solution is tabulated in Table 4.2.

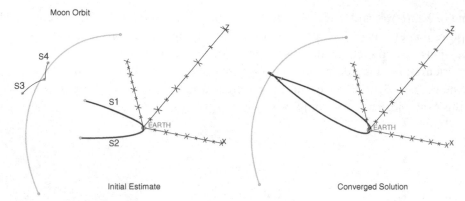

Figure 4.4. Initial estimate and final solution for the Free-Return Lunar Flyby Mission.

Table 4.2. *Converged Segment Data for the Free-Return Lunar Flyby Mission.*

Segment→	S1	S2	S3	S4
t_0 (day)	0.0	7.543312*	3.690074*	$\Leftarrow t_0$ S3
Δt (day)	2.68947*	−2.848924*	1.004313*	−1.000607*
a (km)	6578.0	6578.0	ns	ns
r_p (km)	ns	ns	10000.0	$\Leftarrow t_0^-$ S3
e	0.000	0.000	2.620*	$\Leftarrow t_0^-$ S3
i (deg)	26.676*	27.664*	153.287*	$\Leftarrow t_0^-$ S3
Ω (deg)	351.027*	350.499*	171.634*	$\Leftarrow t_0^-$ S3
ω (deg)	0.000	0.000	12.117*	$\Leftarrow t_0^-$ S3
ν (deg)	354.460*	343.112*	0.000	$\Leftarrow t_0^-$ S3
Δv_{0_x} (km/s)	3.139*	− 3.137*	na	na

4.6.2 Lunar Orbiter and Lunar Impacter Mission

This example illustrates the design and optimization of a lunar bound mission where two spacecraft are launched from a low Earth orbit. This mission is similar in concept to the actual LRO (Lunar Reconnaissance Orbiter) and LCROSS (Lunar Crater Observation and Sensing Satellite) mission launched on June 18, 2009 [25, 5]. The first spacecraft (the Orbiter) is inserted into a lunar polar orbit and the second spacecraft (the Impacter) performs a flyby of the Moon to increase its inclination relative to the Earth-Moon plane so that it impacts the north pole of the Moon exactly vertically about half a month later. It may be possible for the Impacter to collide with the north pole of the Moon without first performing a flyby of the Moon, but the impact angle may then be shallow. The vertical impact requirement makes the mission challenging from a design and performance perspective. It is anticipated that a flyby combined with a maneuver after the flyby is needed to satisfy this requirement. The example is intended to illustrate the use of the system in the formulation, design, and construction of this type of mission. It is not intended to describe the actual LRO/LCROSS mission, even though the Copernicus program was used as the primary mission design and optimization tool for the actual mission.

4.6.2.1 Problem Statement, Data, and Model Setup

The problem description is as follows:

- The Orbiter and the Impacter, each with initial masses of 1000 kg, are attached to a booster rocket in a low circular Earth parking orbit with a radius of 6578.137 km.
- The booster performs a single impulsive maneuver along the velocity vector that injects both spacecraft towards the Moon. This maneuver is the translunar injection (TLI) maneuver. From a fixed inclination parking orbit, the ascending node Ω, argument of periapsis ω, and epoch are search variables.
- The Orbiter performs one impulsive capture maneuver to a lunar orbit. The Orbiter orbit is circular and polar ($i = 90°$) with respect to the lunar equator with a radius of 2000 km.
- The Impacter performs a maneuver sometime after injection but before it encounters the Moon that places it into a hyperbolic lunar flyby trajectory. The flyby is designed to change its inclination relative to the Moon's orbit plane about the Earth to facilitate a collision with the north Pole of the Moon about half a moon revolution later (~14 days later). The impact at the lunar north pole is vertical; this means that the impact angle is 90°. The Impacter is allowed to perform at most one maneuver between the flyby and impact.
- The Impacter separation maneuver must occur no earlier than 6 hours after the injection maneuver performed by the booster. The Impacter separation maneuver is also constrained to occur no later than 12 hours before perilune. The Impacter post flyby maneuver, if needed, is required to occur no earlier than 12 hours after the perilune time and no later than 12 hours before impact.
- The Impacter maneuvers are finite burn maneuvers with a low-thrust engine. The engine is a constant specific impulse engine with specific impulse $Isp = 1000$ seconds and a maximum thrust of 0.5 Newtons.
- The objective function is the sum of the final masses of the Orbiter and the Impacter and is maximized,

$$f_{obj} = \max(m_{f_{Orbiter}} + m_{f_{Impacter}}). \qquad (4.57)$$

The initial estimate of the mission is constructed using a series of segments. Some of the segments are sequentially connected to other segments and some are completely disconnected. Some segments are propagated forward in time and some are propagated backward in time.

Obtaining the optimal impulsive/finite-thrust solution is a four stage process. Stage 1 produces an impulsive solution that satisfies most or all of the constraints. Stage 2 optimizes the impulsive solution by minimizing the sum of the magnitudes of the impulsive maneuvers performed by both the Orbiter and the Impacter. This is equivalent to maximizing the objective function given in Equation (4.57) because both spacecraft use an engine with the same specific impulse. With a converged optimal impulsive solution, Stage 3 converts the Impacter impulsive maneuvers to finite burn maneuvers using the parameter model. This solution is reconverged and optimized with the explicit cost function given in Equation (4.57). Finally, in Stage 4, the thrust vector is reconstructed using the optimal control model and the mission is reconverged and optimized. The final solution is a trajectory that has impulsive maneuvers (for the booster and the Orbiter) and optimal control based finite burn maneuvers for the Impacter. Due to space limitations, Stage 1, which is the most important one, will be described in detail whereas the solution procedure for the remaining stages will only be summarized.

4.6.2.2 Stage 1: Feasible Impulsive Solution

The initial estimate is constructed using 12 segments ($S1, \ldots, S12$). A conceptual sketch of this estimate is shown in Figure 4.5. Some segments have zero time duration and are in some cases just single impulsive maneuvers. The arrows along the propagated trajectory arcs indicate the direction of the time propagation. In this preliminary solution, the Impacter separation maneuver is constrained to occur at least one day after TLI but at least two days before perilune and the Impacter impulsive maneuver after they flyby is constrained to occur at least two days after perilune and at least one day before impact. Later, the actual start and end times of the Impacter finite burn maneuvers will be constrained as dictated in Step 5 in the final solution. The tight constraints on the impulsive maneuver times is intentional and

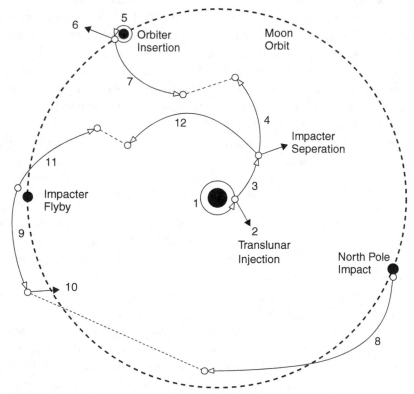

Figure 4.5. Segment Model for the Lunar Orbiter and Impacter Mission.

the reason is to allow a for a padding that may be necessary when these are converted to finite burns.

The reference epoch for this mission is 2011 January 1 00:00:00.000 Barycentric Dynamical Time (TDB). All mission times are given with respect to the reference epoch in days. The base frame is $ECJ2000$. All distance quantities are in kilometers (km), all velocity quantities are in km/s, all mass quantities are in kilograms (kg), and all angular quantities are in degrees. The segment description list follows. Only the pertinent data is listed. All starred quantities are search variables. Only the state input data is listed here. The equality and inequality constraints follow after the segment description list.

- $S1$: Low Earth Parking Orbit for the Booster + Orbiter + Impacter stack. Orbit state is in $ECJ2000$.

$$\begin{pmatrix} t_0^* \\ t_f \end{pmatrix}^{S1} = \begin{pmatrix} 0.0^* \\ 0.0 \end{pmatrix} \qquad \begin{pmatrix} a \\ e \\ i \\ \Omega^* \\ \omega \\ \nu \end{pmatrix}^{S1}_{t_0} = \begin{pmatrix} 6578.137 \\ 0.0 \\ 28.5 \\ 0.0^* \\ 0.0 \\ 0.0 \end{pmatrix}.$$

- $S2$: Translunar Injection Maneuver (TLI). This maneuver is defined in a velocity referenced frame relative to $ECJ2000$

$$\begin{pmatrix} t_0 \\ t_f \end{pmatrix}^{S2} = \begin{pmatrix} (t_f)^{S1} \\ 0.0 \end{pmatrix} \qquad \begin{pmatrix} \mathbf{r}_0 \\ \mathbf{v}_0^- \end{pmatrix}^{S2} = \begin{pmatrix} (\mathbf{r}_f)^{S1} \\ (\mathbf{v}_f^+)^{S1} \end{pmatrix} \qquad \begin{pmatrix} \Delta v_{x_0}^* \\ \Delta v_{y_0} \\ \Delta v_{z_0} \end{pmatrix}^{S2} = \begin{pmatrix} 3.2^* \\ 0.0 \\ 0.0 \end{pmatrix}.$$

- $S3$: Translunar Initial Coast. This segment represents the initial coast for both the Orbiter and the Impacter; at the end of this coast the Impacter performs the separation maneuver which is done in $S12$.

$$\begin{pmatrix} t_0 \\ \Delta t^* \end{pmatrix}^{S3} = \begin{pmatrix} (t_f)^{S2} \\ 1.0^* \end{pmatrix} \qquad \begin{pmatrix} \mathbf{r}_0 \\ \mathbf{v}_0^- \end{pmatrix}^{S3} = \begin{pmatrix} (\mathbf{r}_f)^{S2} \\ (\mathbf{v}_f^+)^{S2} \end{pmatrix}.$$

- $S4$: Translunar Cruise. The Orbiter flies along this segment towards the Moon.

$$\begin{pmatrix} t_0 \\ t_f^* \end{pmatrix}^{S4} = \begin{pmatrix} (t_f)^{S3} \\ 2.5^* \end{pmatrix} \qquad \begin{pmatrix} \mathbf{r}_0 \\ \mathbf{v}_0^- \\ m_0^- \end{pmatrix}^{S4} = \begin{pmatrix} (\mathbf{r}_f)^{S3} \\ (\mathbf{v}_f^+)^{S3} \\ 1000 \end{pmatrix}.$$

- $S5$: Low Lunar Orbit (LLO). The initial condition for the orbit is referenced to $MCJ2000$. The mass of the Orbiter at the Moon is unknown, and hence it is a search variable.

$$\begin{pmatrix} t_0^* \\ \Delta t \end{pmatrix}^{S5} = \begin{pmatrix} 5.0^* \\ 0.0 \end{pmatrix}$$

$$\begin{pmatrix} a \\ e \\ i^* \\ \Omega^* \\ \omega \\ \nu \end{pmatrix}_{t_0}^{S5} = \begin{pmatrix} 2000.0 \\ 0.0 \\ 90.0^* \\ 0.0^* \\ 0.0 \\ 0.0 \end{pmatrix}$$

$$\left(m_0^{--*} \right)^{S5} = \left(1000.0^* \right).$$

- $S6$: Lunar Orbit Capture Maneuver. This segment models the capture maneuver at the Moon for the Orbiter. After the maneuver, the segment is propagated backward in time. The impulsive maneuver is a velocity referenced frame relative to $MCJ2000$.

$$\begin{pmatrix} t_0 \\ \Delta t^* \end{pmatrix}^{S6} = \begin{pmatrix} (t_f)^5 \\ -2.5^* \end{pmatrix} \qquad \begin{pmatrix} \mathbf{r}_0 \\ \mathbf{v}_0^- \\ m_0^{--} \end{pmatrix}^{S6} = \begin{pmatrix} \left(\mathbf{r}_f \right)^{S5} \\ \left(\mathbf{v}_f^+ \right)^{S5} \\ \left(m_f^{++} \right)^{S5} \end{pmatrix}$$

$$\begin{pmatrix} \Delta v_{x_0}^* \\ \Delta v_{y_0} \\ \Delta v_{z_0} \end{pmatrix}^{S6} = \begin{pmatrix} -2.0^* & \text{km/s} \\ 0.0 & \text{km/s} \\ 0.0 & \text{km/s} \end{pmatrix}.$$

- $S7$: Lunar Orbit Capture Approach. This is a simple node segment that inherits the final state from $S6$. It will be constrained to be state continuous with the end of $S4$.

$$\begin{pmatrix} t_0 \\ \Delta t \end{pmatrix}^{S7} = \begin{pmatrix} (t_f)^6 \\ 0.0 \end{pmatrix} \qquad \begin{pmatrix} \mathbf{r}_0 \\ \mathbf{v}_0^- \\ m_0^{--} \end{pmatrix}^{S7} = \begin{pmatrix} \left(\mathbf{r}_f \right)^{S6} \\ \left(\mathbf{v}_f^+ \right)^{S6} \\ \left(m_f^{++} \right)^{S6} \end{pmatrix}.$$

- $S8$: Lunar North Pole Impact. The initial condition is defined relative to a Moon Centered True Equator of Date reference frame. This segment is propagated backward in time. The Impacter final mass is unknown at impact and is therefore a search variable.

$$\begin{pmatrix} t_0^* \\ \Delta t^* \end{pmatrix}^{S8} = \begin{pmatrix} 18.0^* \\ -5.0^* \end{pmatrix} \qquad \begin{pmatrix} r_{x0} \\ r_{y0} \\ r_{z0} \\ v_{x0} \\ v_{y0} \\ v_{z0}^* \end{pmatrix}_{t_0}^{S8} = \begin{pmatrix} 0.0 \\ 0.0 \\ 1738.0 \\ 0.0 \\ 0.0 \\ -3.0^* \end{pmatrix}$$

$$\left(m_0^{--*} \right)^{S8} = \left(1000.0^* \right).$$

- *S9*: Perilune and Forward Flyby. This segment models the hyperbola at the Moon. The initial state is defined in *MCJ*2000.

$$\begin{pmatrix} t_0^* \\ \Delta t^* \end{pmatrix}^{S9} = \begin{pmatrix} 5.0^* \\ 2.0^* \end{pmatrix} \qquad \begin{pmatrix} a^* \\ e^* \\ i^* \\ \Omega^* \\ \omega^* \\ \nu \end{pmatrix}_{t_0}^{S9} = \begin{pmatrix} -10000.0^* \\ 1.5^* \\ 90.0^* \\ 0.0^* \\ 0.0^* \\ 0.0 \end{pmatrix}.$$

- *S10*: Pre-impact Coast and Impulsive Maneuver. The impulsive maneuver is in *ECJ*2000.

$$\begin{pmatrix} t_0 \\ \Delta t^* \end{pmatrix}^{S10} = \begin{pmatrix} (t_f)^{S9} \\ -6.0^* \end{pmatrix} \quad \begin{pmatrix} \mathbf{r}_0 \\ \mathbf{v}_0^- \\ m_0^{--} \end{pmatrix}^{S10} = \begin{pmatrix} (\mathbf{r}_f)^{S9} \\ (\mathbf{v}_f^+)^{S9} \\ (m_f^{++})^{S9} \end{pmatrix} \quad \begin{pmatrix} \Delta v_{x_f}^* \\ \Delta v_{y_f}^* \\ \Delta v_{z_f}^* \end{pmatrix}^{S10} = \begin{pmatrix} 0.0^* \\ 0.0^* \\ 0.0^* \end{pmatrix}.$$

- *S11*: Perilune and Backward Flyby.

$$\begin{pmatrix} t_0 \\ \Delta t^* \end{pmatrix}^{S10} = \begin{pmatrix} (t_0)^{S9} \\ -2.0^* \end{pmatrix} \quad \begin{pmatrix} \mathbf{r}_0 \\ \mathbf{v}_0^- \\ m_0^{--} \end{pmatrix}^{S10} = \begin{pmatrix} (\mathbf{r}_0)^{S9} \\ (\mathbf{v}_0^-)^{S9} \\ (m_f^{++})^{S12} \end{pmatrix}.$$

- *S12*: Impacter Separation and Coast. The impulsive maneuver is defined in *ECJ*2000.

$$\begin{pmatrix} t_0 \\ \Delta t^* \end{pmatrix}^{S12} = \begin{pmatrix} (t_f)^{S3} \\ 2.0^* \end{pmatrix} \quad \begin{pmatrix} \mathbf{r}_0 \\ \mathbf{v}_0^- \\ m_0^{--} \end{pmatrix}^{S12} = \begin{pmatrix} (\mathbf{r}_f)^{S3} \\ (\mathbf{v}_f^+)^{S3} \\ 1000 \text{ kg} \end{pmatrix} \quad \begin{pmatrix} \Delta v_{x_0}^* \\ \Delta v_{y_0}^* \\ \Delta v_{z_0}^* \end{pmatrix}^{S12} = \begin{pmatrix} 0.0^* \\ 0.0^* \\ 0.0^* \end{pmatrix}.$$

The open loop mission based on these segment definitions is shown in Figure 4.6. By any standard the initial estimate provided can be considered a bad initial guess.

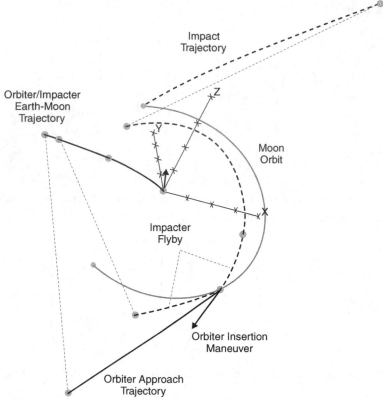

Figure 4.6. Initial estimate for the impulsive Lunar Orbiter and Impacter Mission. The straight dotted lines connect the node points that need to be connected in time, position, and velocity in the feasible solution.

The inequality constraint functions for the segment durations are (all values are in days)

$$(\Delta t)^{S3} \geq 1.0$$

$$(\Delta t)^{S4} \geq 0.0$$

$$(\Delta t)^{S6} \leq -1.0$$

$$(\Delta t)^{S8} \leq -1.0$$

$$(\Delta t)^{S9} \geq 2.0$$

$$(\Delta t)^{S10} \geq 0.0$$

$$(\Delta t)^{S11} \leq -2.0$$

$$(\Delta t)^{S12} \geq 0.0$$

In Stages 3 and 4 of the solution procedure, these constraints are adjusted accordingly to meet the maneuver time constraints associated with the Impacter as described in item 5 of the problem description. The equality constraint functions are

$$
\begin{pmatrix} t_f \\ \mathbf{r}_f \\ \mathbf{v}_f^+ \\ m_f^{++} \end{pmatrix}^{S7} = \begin{pmatrix} t_f \\ \mathbf{r}_f \\ \mathbf{v}_f^+ \\ m_f^{++} \end{pmatrix}^{S4}
$$

$$
\left(m_f^{++} \right)^{S9} = \left(m_f^{++} \right)^{S8}
$$

$$
\begin{pmatrix} t_f \\ \mathbf{r}_f \\ \mathbf{v}_f^+ \\ m_f^{++} \end{pmatrix}^{S10} = \begin{pmatrix} t_f \\ \mathbf{r}_f \\ \mathbf{v}_f^+ \\ m_f^{++} \end{pmatrix}^{S8}
$$

$$
\begin{pmatrix} t_f \\ \mathbf{r}_f \\ \mathbf{v}_f^+ \\ m_f^{++} \end{pmatrix}^{S12} = \begin{pmatrix} t_f \\ \mathbf{r}_f \\ \mathbf{v}_f^+ \\ m_f^{++} \end{pmatrix}^{S11}
$$

The problem as stated has 31 search variables, 25 equality constraints, and 8 inequality constraints. A feasible solution is first obtained using the VF13 algorithm with the cost function set to zero. The converged solution that results required 102 iterations and is shown in Figure 4.7. The search variable values for the both the initial estimate and feasible solutions are listed in Table 4.3. The data listed has been rounded to six significant figures. Note that the results for some of the angular quantities are not modulated since these data are extracted as-is from the final solution iterate generated by the system.

4.6.2.3 Stage 2: Optimal Impulsive Solution

In this stage, the feasible impulsive solution is optimized. The objective function to minimize is the sum of the impulsive maneuvers made by both the Orbiter and the Impacter. This is equivalent to maximizing the sum of the final masses of both spacecraft. The Orbiter makes a single insertion maneuver at the Moon. The Impacter makes a separation maneuver before the lunar flyby and an impulsive maneuver prior to impact. Based on the segment model described above, the objective function is explicitly

$$
f_{obj} = \min \left[\left(\Delta v_0 \right)^{S6} + \left(\Delta v_f \right)^{S10} + \left(\Delta v_0 \right)^{S12} \right].
$$

The value of the objective function of the impulsive feasible solution is 2.850 km/s. The optimized result required 138 iterations and yielded an objective function value

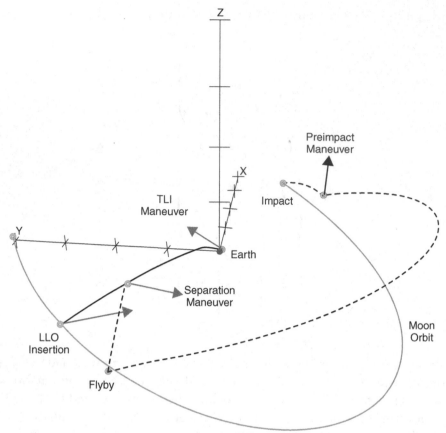

Figure 4.7. Feasible solution for the impulsive Lunar Orbiter and Impacter Mission. The total velocity maneuver impulse required is 2.850 km/s.

of 1.187 km/s. The optimized impulsive solution is shown in Figure 4.8. Using the rocket equation and a specific impulse value of 1000 sec. for all maneuvers results in a combined final mass for the Orbiter and Impacter of 1882.954 kg. This value is an upper limit for any finite burn solution and the quality of any finite burn solution is measured by how close it can get to this value. For this example, obtaining the optimal impulsive solution is the most important phase in arriving at the final finite burn solution. The results for the optimal impulsive solution are tabulated in Table 4.3. The last few rows of Table 4.3 lists the magnitudes of the impulses performed by the Orbiter and the Impacter along with their associated final masses. It is the sum of their final masses that will be maximized in the finite burn solutions to follow.

4.6.2.4 Stage 3: Finite Burn Solution using the Parameter Model
In this stage the optimal impulsive solution is used as the basis for a solution that has the Impacter performing low-thrust maneuvers. The applicable segment duration constraints are set to produce a solution that satisfies the maneuver start and end time constraints as listed in Step 5 of the problem description. Also, in the impulsive solution described in the previous sections, a mistake was made in specifying

Table 4.3. *Impulsive Lunar Orbiter and Impacter Mission Data.*

	Variable	Estimate Values	Feasible Values	Optimal Values	Units
1	$(t_0)^{S1}$	0.000	$-0.833156e+01$	$-0.974303e+01$	day
2	$(\Omega)^{S1}$	0.000	$+0.3437234e+03$	$+0.345934e+03$	deg
3	$(\Delta v_{x_0})^{S2}$	3.200	$+0.321438e+01$	$+0.312655e+01$	km/s
4	$(\Delta t)^{S3}$	1.000	$+0.100000e+01$	$+0.100000e+01$	day
5	$(t_f)^{S4}$	2.500	$-0.733156e+01$	$-0.637816e+01$	day
6	$(t_0)^{S5}$	5.000	$-0.633156e+01$	$-0.537816e+01$	day
7	$(i)^{S5}$	90.000	$+0.200787e+02$	$+0.387285e+02$	deg
8	$(\Omega)^{S5}$	0.000	$+0.420167e+03$	$+0.403068e+03$	deg
9	$(m_0^{--})^{S5}$	1000.000	$+0.886661e+03$	$+0.921474e+03$	kg
10	$(\Delta t)^{S6}$	-2.500	$-0.100000e+01$	$-0.100000e+01$	day
11	$(\Delta v_{x_0})^{S6}$	-2.000	$-0.117966e+01$	$-0.801997e+00$	km/s
12	$(t_0)^{S8}$	18.000	$+0.831994e+01$	$+0.805582e+01$	day
13	$(\Delta t)^{S8}$	-5.000	$-0.120323e+01$	$-0.100000e+01$	day
14	$(v_{z_0})^{S8}$	-3.000	$-0.231773e+01$	$-0.246334e+01$	km/s
15	$(m_0^{--})^{S8}$	1000.000	$+0.8434035e+03$	$+0.961480e+03$	kg
16	$(t_0)^{S9}$	5.000	$-0.531623e+01$	$-0.532782e+01$	day
17	$(\Delta t)^{S9}$	2.000	$+0.124189e+02$	$+0.200000e+01$	day
18	$(a)^{S9}$	-10000.000	$-0.101072e+05$	$-0.699220e+04$	km
19	$(e)^{S9}$	1.500	$+0.155698e+01$	$+0.174934e+01$	–
20	$(i)^{S9}$	90.000	$+0.894873e+02$	$+0.922372e+02$	deg
21	$(\Omega)^{S9}$	0.000	$+0.477525e+03$	$+0.437921e+03$	deg
22	$(\omega)^{S9}$	0.000	$+0.324308e+03$	$+0.326995e+03$	deg
23	$(\Delta t)^{S10}$	6.000	$+0.139905e-01$	$+0.103836e+02$	day
24	$(\Delta v_{x_f})^{S10}$	0.000	$-0.209686e-01$	$-0.404867e-01$	km/s
25	$(\Delta v_{y_f})^{S10}$	0.000	$-0.690054e-01$	$+0.316228e+00$	km/s
26	$(\Delta v_{z_f})^{S10}$	0.000	$+0.827989e+00$	$+0.152771e+00$	km/s
27	$(\Delta t)^{S11}$	-2.000	$-0.200000e+01$	$-0.200000e+01$	day
28	$(\Delta t)^{S12}$	2.000	$+0.153349e-01$	$+0.141521e+01$	day
29	$(\Delta v_{x_0})^{S12}$	0.000	$+0.314938e+00$	$-0.113901e-01$	km/s
30	$(\Delta v_{y_0})^{S12}$	0.000	$-0.706849e+00$	$+0.512349e-02$	km/s
31	$(\Delta v_{z_0})^{S12}$	0.000	$-0.324319e+00$	$-0.291293e-01$	km/s
n/a	$\Delta v_{\text{Orbiter}}$	2.000	$+0.117967e+01$	$+0.8019976e+00$	km/s
n/a	$\Delta v_{\text{Impacter}}$	0.000	$+0.167017e+01$	$+0.385216e+00$	km/s
n/a	Δv_{total}	2.000	$+0.284984e+01$	$+0.118721e+01$	km/s
n/a	$m_{f_{\text{Orbiter}}} + m_{f_{\text{Impacter}}}$	2000.000	$+0.173006e+04$	$+0.188295e+04$	kg

the Orbiter's lunar capture orbit. The requirements dictated a lunar polar orbit. In the impulsive solution, the orientation of the lunar capture orbit was referenced to $MCJ2000$ and the inclination was allowed to be a search variable. This is corrected in this stage by fixing the inclination to be 90° in a Moon centered true equator of data reference frame.

The Impacter impulses are converted to finite burns using the rocket equation to estimate the start and end times of the maneuver. In this conversion gravity losses

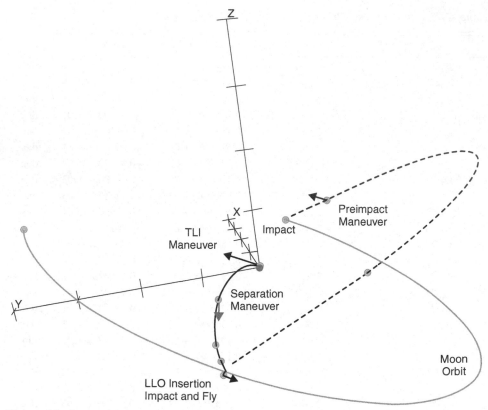

Figure 4.8. Optimal impulsive solution for the Lunar Orbiter and Impacter Mission. The objective function value is 1.187 km/s which corresponds to a combined final of 1885.784 kg for the Orbiter and the Impacter.

are assumed to be zero. The thrust direction is estimated to be along the impulsive maneuver direction. The thrust direction is initially assumed to be held inertially fixed. Following convergence, the thrust direction is allowed to change by allowing it to have a constant linear rate in both the right ascension and declination variables. The thrust steering frame is chosen to be $ECJ2000$. Recall that the times and components of the impulsive maneuvers for the Impacter are search variables. In the finite burn model, the start and end times of the maneuvers, the two spherical angles, and their rates are now search variables. The thrust level for the finite burn maneuvers is $T_{max} = 0.5$ Newtons. The impulsive to finite burn conversion is simple and is described as follows. Assume the impulsive maneuver is $\Delta \mathbf{v}$ and occurs at t_i. The estimated start time t_s and end time t_e are

$$t_s = t_i - \frac{\Delta t}{2} \qquad t_e = t_i + \frac{\Delta t}{2} \tag{4.58}$$

where

$$\Delta t = \frac{m_e - m_s}{\dot{m}} \tag{4.59}$$

where m_s is the mass before the impulse which is known and m_e is the mass after the impulse

$$m_e = m_s e^{-(\Delta v / I_{sp}/g_0)} \tag{4.60}$$

where g_0 is the value of the surface gravity at the Earth's surface, and $\dot{m}$ is the mass flow rate

$$\dot{m} = -\frac{T_{\max}}{I_{sp}g_0}. \tag{4.61}$$

The thrust vector direction is estimated to be along the impulsive maneuver unit vector

$$\hat{\mathbf{u}} = \Delta \mathbf{v}/\Delta v. \tag{4.62}$$

The spherical angles α and β are computed from $\hat{\mathbf{u}}$. For each of the Impacter maneuvers, the linear terms $\dot{\alpha}$ and $\dot{\beta}$ are included as search variables for each of the maneuvers and their initial values are estimated to be 0.0 deg/day.

Using a similar segment model as before, the two impulsive segments are replaced with two finite burn segments. The optimal solution converged with a final objective function value of 1858.075 kg and is shown in Figure 4.9. The optimal Impacter finite burn maneuver data for this case is listed in Table 4.4. Recall that the optimal impulsive solution had a final objective function value of 1882.954 kg, which, as expected, can not be achieved with a finite burn solution.

4.6.2.5 Stage 4: Finite Burn Solution using the Optimal Control Model

For this solution, the two finite burns performed by the Impacter are converted to optimal control finite burns via an adjoint control transformation. By optimal control

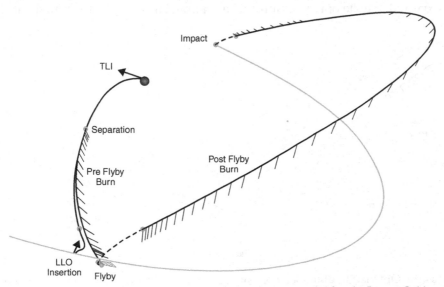

Figure 4.9. Optimal finite burn solution using the parameter model for the Lunar Orbiter and Impacter Mission. The objective function value is 1858.075 kg.

Table 4.4. *Impacter maneuver data for the parameter model finite burn.*

Impacter Maneuver→	Finite Burn 1 ($S11$)	Finite Burn 2 ($S13$)
t_0 (day)	−8.456392	−0.468141
Δt (day)	3.198987	11.177308
t_f (day)	−5.257405	6.495897
α (deg)	−122.805	120.562
β (deg)	−81.388	−68.974
$\dot{\alpha}$ (deg/day)	7.608592	−4.683136
$\dot{\beta}$ (deg/day)	16.490283	9.376933
m_0^{--} (kg)	1000.000	985.908
m_f^{++} (kg)	985.908	936.670

in this context, it is meant that the thrust vector will follow the primer vector as dictated by the costate differential equations. As presented previously the transversality conditions are not enforced and this is intentional. Additionally, for this example, the switching function will not be used to determine whether a segment should be a coast arc or a thrust arc. Because of this the mass costate variable λ_m plays no role in the solution process. For each finite burn segment the magnitude of the primer vector is set to unity and this sets the scaling for the costate vector. The adjoint control variables $\dot{\lambda}_v, \alpha, \beta, \dot{\alpha}, \dot{\beta}$ form part of the list of search variables and these replace the parameter model based thrust vector variables used in the previous solution. The initial estimate for $\dot{\lambda}_v$ is taken to be 0.0. The angular quantities and their rates, along with the estimate for the start and end times of the maneuvers are taken from the previous solution that used the parameter model for the finite burns. The implementation of this procedure yields a solution only slightly better in performance with respect to the previous solution so that both solutions are nearly indistinguishable. The final converged solution yields a value for the objective function of 1858.424 kg. It was expected that the optimal control solution would have offered a more significant

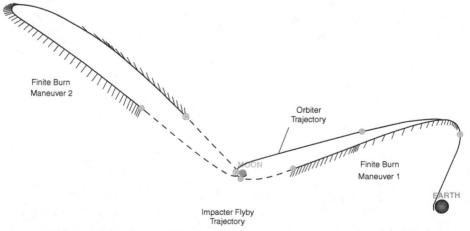

Figure 4.10. Optimal finite burn solution using the optimal control model for the Lunar Orbiter and Impacter Mission. The objective function value is 1858.424 kg. The trajectory is shown in a MCJ2000 frame.

improvement over the more-constrained finite burn model of Stage 3, but it was not the case for this specific example. The converged solution is shown in Figure 4.10. The optimal Impacter finite burn maneuver data for the parameter model is listed in Table 4.5. Figure 4.11 shows a near Moon view of the same trajectory showing the Impacter flyby and vertical north pole impact. Table 4.6 summarizes and compares the objective function for the three main solutions; these are the optimal impulsive solution, and the two finite burn solutions.

Table 4.5. *Impacter maneuver data for the optimal control finite burn model.*

Impacter Maneuver→	Finite Burn 1 ($S11$)	Finite Burn 2 ($S13$)
t_0 (day)	−8.477456	−4.265911
Δt (day)	3.212911	11.084013
t_f (day)	−5.264544	6.818102
λ_v	$0.839533E - 05$	−0.164419
α (deg)	−109.474	113.160
β (deg)	−106.052	−62.349
$\dot\alpha$ (deg/day)	19.111	1.907
$\dot\beta$ (deg/day)	99.654	6.518
m_0^{--} (kg)	1000.000	985.847
m_f^{++} (kg)	985.847	937.020

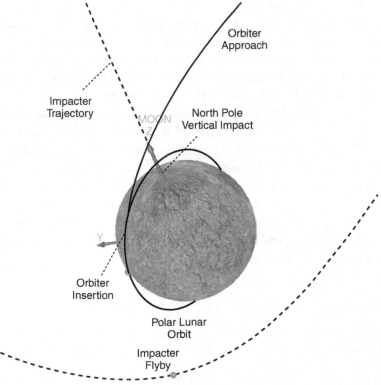

Figure 4.11. Near Moon view of the optimal finite burn solution using the optimal control model for the Lunar Orbiter and Impacter Mission.

Table 4.6. *Performance comparison between the optimized impulsive solution, the parameter model finite burn, and the optimal control finite burn solutions.*

Parameter	Impulsive	Parameter Model	Optimal Control Model
$m_{f_{\text{Orbiter}}}$ (kg)	921.474	921.405	921.405
$m_{f_{\text{Impacter}}}$ (kg)	961.480	936.670	937.020
total (kg)	1882.954	1858.075	1858.424

4.7 Concluding Remarks

The key elements associated with a general trajectory optimization system have been presented. It is a system that can be used to analyze a large range of problems. The system facilitates the solutions to these problems because it uses a segment-based trajectory model as the building block to construct simple to complex trajectories. With the trajectory model in place, the fundamental process that follows is the identification of the set of search variables, the set of functions to be satisfied, and the function to be extremized, in the case of an optimization problem. Depending on how these are identified, the process is followed by casting the problem into one of the three solution methods: nonlinear root finding, a mini-max problem, or a parameter optimization problem. The emphasis is placed on the modeling and casting process and less so on the actual numerical methods used in the solution processes which are assumed to be well understood.

Two example missions were used to illustrate some of the ideas presented. It is clear that the approach taken to solve these examples with the described trajectory model is only one of several possible strategies. It is the responsibility of the analyst to identify the best strategy that produces a valid solution with the least amount of guessing and with a high level of robustness and efficiency.

The system described then can be considered as an experimental toolbox that requires the analyst to be an integral part of the overall solution process. For very specific problems, algorithms for initial estimate generators can be developed so that the analyst is removed from the solution process (see for example [20, 24, 26, 27]). However, for a general system that solves general problems, the analyst will always be in the solution process loop. The goal then, which may not be achievable for a while, is to develop a system intelligent enough so that it does not need an analyst to provide an initial estimate and interact with it in order to obtain a converged solution for any spacecraft trajectory optimization problem.

REFERENCES

[1] Ocampo, C. (2003) An Architecture for a Generalized Trajectory Design and Optimization System, in *Libration Point Orbits and Applications*, World Scientific, Publishing Co., Singapore.

[2] Ocampo, C. (2004) Finite Burn Maneuver Modeling for a Generalized Spacecraft Trajectory Design and Optimization System, *Astrodynamics, Space Missions, and Chaos, Annals of the New York Academy of Sciences*, Vol. 1017, p. 210–233.

[3] Williams, J. et. al. (2009) Global Performance Characterization of the Three Burn Trans-Earth Injection Maneuver Sequence over the Lunar Nodal Cycle, AAS 09–380 Proceedings of the 2009 AAS/AIAA Astrodynamics Specialist Conference, Pittsburgh, PA.

[4] Garn, M. et. al. (2008) NASA's Planned Return to the Moon: Global Access and Anytime Return Requirement Implications on the Lunar Orbit Insertion Burns, Proceedings of the 2008 AIAA Guidance, Navigation and Control Conference, Honolulu, HI.

[5] Lunar Crater Observation and Sensing Satellite (LCROSS) Official Website: http://lcross.arc.nasa.gov/

[6] Ranieri, C. L., Ocampo, C. (2009) Indirect Optimization of Three-Dimensional, Finite-Burning Interplanetary Transfers Including Spiral Dynamics, Journal of Guidance, Control, and Dynamics, Vol. 32, No. 2.

[7] Leitmann, G. (1966) *An Introduction to Optimal Control*, McGraw-Hill, New York, NY.

[8] Bryson A.E. Jr. and Ho Y.C. (1975) *Applied Optimal Control*, Hemisphere Publishing Corporation, Revised Printing, New York, NY.

[9] Vincent T.L and Grantham W.J. (1997) *Nonlinear and Optimal Control Systems*, John Wiley and Sons, New York, NY.

[10] Hull, D.G.(2003) *Optimal Control Theory for Applications,* Springer Verlag, New York, NY.

[11] Pontryagin, L.S., et al. (1964) *The Mathematical Theory of Optimal Processes*, Pergamon Press, New York, NY.

[12] Lawden, D.E. (1963) *Optimal Trajectories for Space Navigation*, London Butterworths, London.

[13] Marec, J.P. (1979) *Optimal Space Trajectories*, Elsevier Scientific Publishing Company, New York, NY.

[14] ODEPACK: Serial Fortran Solvers for Initial Value Problems https://computation.llnl.gov/casc/odepack/odepack_home.html

[15] Dixon, L.C., Bartholomew-Biggs, M.C. (1981) Adjoint Control Transformations for Solving Practical Optimal Control Problems, *Optimal Control Applications and Methods*, 2, 365–381.

[16] Dennis, J.E., Schnabel, R.B. (1983) *Numerical Methods for Unconstrained Optimization and Nonlinear Equations*, Prentice-Hall, Inc., Englewood Cliffs, NJ.

[17] Gill, P.E. et. al., (1998) *Practical Optimization*, Cambridge University Press, San Diego, CA.

[18] Zimmer, S., Ocampo, C. (2005) Use of Analytical Gradients to Calculate Optimal Gravity-Assist Trajectories, *Journal of Guidance, Control, and Dynamics*, 28, No. 2, 324–332.

[19] Zimmer, S., Ocampo, C. (2005) Analytical Gradients for Gravity Assist Trajectories Using Constant Specific Impulse Engines, Journal of Guidance, Control, and Dynamics, 28, No. 4, 753–760.

[20] Ocampo, C. et al. (2009) Variational Equations for a Generalized Trajectory Model, Proceedings of the AAS/AIAA Spaceflight Mechanics Conference, Savannah, GA.

[21] Harwell Subroutine Library (HSL) and Archive, http://www.cse.scitech.ac.uk/nag/hsl/

[22] Nocedal, J., Wright, S.J. (1999) *Numerical Optimization*, Springer-Verlag, New York, NY.

[23] Stanford Business Software http://www.sbsi-sol-optimize.com/ asp/sol_product_snopt.html

[24] Jesick, M., Ocampo, C. (2009) Automated Lunar Free-Return Trajectory Generation, Proceedings of the 19th AAS/AIAA Space Flight Mechanics Meeting, Savannah, GA.

[25] Lunar Reconnaissance Orbiter (LRO) Official Website: http://lro.gsfc.nasa.gov/

[26] Jesick, M., Ocampo, C. (2009) Lunar Orbit Insertion from a Fixed Free-Return, Proceedings of the 2009 AAS/AIAA Astrodynamics Specialist Conference, Pittsburgh, PA.

[27] Ocampo, C. et al. (2009) Initial Trajectory Model for a Multi-Maneuver Moon to Earth Abort Sequence, Proceedings of the 19th AAS/AIAA Space Flight Mechanics Meeting, Savannah, GA.

Low-Thrust Trajectory Optimization Using Orbital Averaging and Control Parameterization

Craig A. Kluever
Department of Mechanical & Aerospace Engineering,
University of Missouri-Columbia

5.1 Introduction and Background

It is well known that spacecraft propelled by low-thrust electric propulsion (EP) can potentially deliver a greater payload fraction compared to vehicles propelled by conventional chemical propulsion. The increase in payload fraction for EP systems is due to its much higher specific impulse (I_{sp}) or engine exhaust velocity, which is often an order of magnitude greater than the I_{sp} for a chemical system. However, optimizing low-thrust orbit transfers is a challenging problem due to the low control authority of the EP system and the existence of long powered arcs and subsequent multiple orbital revolutions. Therefore, obtaining optimal transfers is sometimes tedious and time consuming. In his seminal paper, Edelbaum presented analytical solutions for optimizing continuous-thrust transfers between inclined circular Earth orbits [1]. These results serve as an excellent preliminary design tool for estimating ΔV and transfer time for low-thrust missions with continuous thrust and quasi-circular transfers. Real solar electric propulsion (SEP) spacecraft, however, experience periods of zero thrust during passage through the Earth's shadow, and this major effect is not accommodated in Edelbaum's analysis. Colasurdo and Casalino [2] have extended Edelbaum's analysis and developed an approximate analytic technique for computing optimal quasi-circular transfers with the inclusion of the Earth's shadow. Only coplanar transfers are considered, and the thrust-steering is constrained so that the orbit remains circular in the presence of the Earth's shadow. Kechichian [3] also developed an analytical method for obtaining coplanar orbit-raising maneuvers in the presence of Earth shadow where eccentricity is constrained to remain zero. Both References 2 and 3 develop suboptimal solutions for the coplanar circle-to-circle transfer problem with Earth-shadow arcs, since steering the thrust vector to maintain zero eccentricity ultimately leads to steering losses compared to the minimum-time transfer.

SEPSPOT [4] is a widely used program for computing optimal Earth-orbit transfers using SEP, and it obtains the optimal trajectory by using a calculus of variations approach and a shooting method or the so-called "indirect method." However, the problem is sensitive to the initial guesses for the costate variables, and oftentimes a converged solution can only be obtained by using known solutions for the initial guess. Kluever and Oleson [5] and Ilgen [6] have developed methods for obtaining optimal Earth-orbit transfers using a direct optimization approach. In Reference 5, weighting functions for various control laws serve as optimization parameters in the nonlinear programming (NLP) problem. Ilgen uses a "hybrid" approach where the costate histories are directly parameterized by piecewise linear functions, and the optimal transfer is obtained by using NLP methods. Jenkin [7] applied Ilgen's trajectory optimization program to perform trade studies for orbit-raising missions that combine chemical- and electric-propulsion stages. Scheel and Conway [8] used a direct-transcription method for obtaining optimal, many-revolution, planar low-thrust Earth-orbit transfers for the case of continuous thrust. In this approach, both the state and control time histories are discretized, and the optimal transfer is determined using NLP methods.

In this chapter, we present a direct method for obtaining optimal Earth-orbit transfers. The approach presented here is related to the technique in Reference 5, where the control time history is parameterized and the subsequent constrained parameter-optimization problem is solved using NLP methods. The thrust-steering parameterization is based on the necessary conditions from optimal control theory, and the resulting NLP problem has relatively few free design variables (on the order of 10) for low-thrust transfers that require hundreds of days. Orbital averaging methods are used to quickly and efficiently compute multiple powered trajectories during the optimization process. The optimization technique is also able to easily accommodate other control variables, such as specific impulse modulation during the orbit transfer. Numerical results are presented for a range of minimum-time and minimum-propellant problems, with three-dimensional, Earth-orbit transfers in the presence of Earth-shadow arcs.

5.2 Low-Thrust Trajectory Optimization

5.2.1 Problem Statement

A general problem statement for low-thrust trajectory optimization can be stated as follows: determine the optimal transfer time t_f and optimal thrust-direction program $\mathbf{u}(t), 0 \leq t \leq t_f$, that minimize the performance index

$$J = J\left(\mathbf{x}(t_f), t_f\right) \tag{5.1}$$

subject to the equations of motion

$$\dot{\mathbf{x}} = \mathbf{f}(t, \mathbf{x}, \mathbf{u}) \tag{5.2}$$

and the terminal state constraints

$$\boldsymbol{\psi}\left[\mathbf{x}(t_f), t_f\right] = \mathbf{0}. \tag{5.3}$$

The 7×1 state vector for the general problem is comprised of the classical orbital elements and spacecraft mass, $\mathbf{x} = [a \quad e \quad i \quad \Omega \quad \omega \quad \theta \quad m]^T$. The equations of motion (5.2) are governed by the Gauss form of Lagrange's planetary equations [9], plus the appropriate differential equation for mass-flow rate

$$\frac{da}{dt} = \frac{2a^2 v}{\mu} a_t \tag{5.4}$$

$$\frac{de}{dt} = \frac{1}{v} \left[2 (e + \cos \theta) a_t + \frac{r}{a} a_n \sin \theta \right] \tag{5.5}$$

$$\frac{di}{dt} = \frac{r}{h} a_h \cos (\omega + \theta) \tag{5.6}$$

$$\frac{d\Omega}{dt} = \frac{r}{h \sin i} a_h \sin (\omega + \theta) \tag{5.7}$$

$$\frac{d\omega}{dt} = \frac{1}{ev} \left[2a_t \sin \theta - \left(2e + \frac{r}{a} \cos \theta \right) a_n \right] - \frac{r}{h \sin i} a_h \sin (\omega + \theta) \cos i \tag{5.8}$$

$$\frac{d\theta}{dt} = \frac{h}{r^2} - \frac{1}{ev} \left[2a_t \sin \theta - \left(2e + \frac{r}{a} \cos \theta \right) a_n \right] \tag{5.9}$$

$$\frac{dm}{dt} = \frac{-2\eta P}{\left(g I_{\mathrm{sp}} \right)^2} \tag{5.10}$$

where $[a_n \quad a_t \quad a_h]^T$ represent the perturbing acceleration components. Traditionally, the perturbing forces (or accelerations) for the variational equations are expressed in an orthogonal radial-horizontal (RSW) frame, where in-plane unit vectors R and S are along the local radial and horizontal directions, respectively, and the out-of-plane unit vector W is along the angular momentum direction. Here we use an orthogonal normal-tangential (NTH) coordinate frame where the perturbing acceleration component a_n lies in the osculating orbital plane and is normal to the velocity vector (where the radial outward direction is positive), a_t is the component along the instantaneous velocity vector, and a_h is along the osculating angular momentum direction. Therefore, rotating the RSW frame about the negative angular momentum direction (orbit normal) by the flight-path angle will produce the NTH frame. Perturbing accelerations are due to low-thrust propulsive forces and Earth-oblateness (J_2) effects. Terminal state constraints (5.3) represent a desired target orbit after the transfer is completed.

Our approach is to use a direct optimization method to solve the trajectory optimization problem. One potential direct-solution technique is to parameterize the state $\mathbf{x}(t)$ and/or control profiles $\mathbf{u}(t)$ and use nonlinear programming to solve the subsequent parameter optimization problem. Scheel and Conway [8] have successfully used collocation methods to parameterize the states and controls in order to solve low-thrust orbit transfer problems. However, the nature of low-thrust flight mechanics complicates this approach since the thrust-to-weight ratio is typically on the order of 10^{-3} to 10^{-5}, and therefore the desired orbit transfer usually requires a trip time of tens or hundreds of days and subsequently hundreds or thousands of orbital revolutions. Hence a collocation method can lead to large NLP problems, with hundreds or thousands of free variables and constraints. Since one of our principal goals is to develop a low-thrust trajectory optimization method that can quickly solve a wide range of problems, we choose to only parameterize the control profile $\mathbf{u}(t)$ with as few optimization variables as possible, and numerically integrate the state equations (5.4–5.10) as efficiently as possible.

The computation load of any method that relies on orbit propagation based on accurate numerical integration of the full variational equations (5.4–5.9) would be considerable. For example, a time step on the order of minutes would be required to accurately capture the periodic fluctuations of the orbital elements during a transfer that might require a trip time of six months or more. The computational load of propagating the orbit is significantly reduced by using the method of orbital averaging [4]. Because the five orbital elements a, e, i, Ω, and ω vary slowly with time due to the small perturbing accelerations, each element's *average* time rate-of-change can be computed, and subsequently the orbital transfer can be propagated ahead in time by using large integration time steps on the order of several days. A consequence of using the orbital averaging method is that knowledge of the "fast" orbital element (for example, true anomaly θ) is lost, and therefore the exact location of the spacecraft in the osculating orbit cannot be determined. Orbital averaging determines each element's *mean* time rate-of-change, which is done by calculating the incremental change in an orbital element over a single revolution and dividing by the orbital period. For example, the averaged time rate for the mean state vector due to thrust acceleration is

$$\dot{\bar{\mathbf{x}}} = \frac{1}{T_p} \int_{E_{ex}}^{E_{en}} \frac{d\mathbf{x}}{dt} \frac{dt}{dE} dE \qquad (5.11)$$

where $\bar{\mathbf{x}} = \begin{bmatrix} \bar{a} & \bar{e} & \bar{i} & \bar{\Omega} & \bar{\omega} & \bar{m} \end{bmatrix}^T$ is the mean state vector (note that true anomaly has been removed), and T_p is the orbital period. The overbar indicates the mean or averaged value. The first five elements of the integrand term $d\mathbf{x}/dt$ are computed by evaluating the variational equations (5.4–5.8) with the orbital elements fixed at their mean values over a single orbit; the mass-flow rate equation (5.10) is used for the last element of $d\mathbf{x}/dt$. The time-rate of eccentric anomaly is used to change the independent variable from time to position variable E

$$\frac{dE}{dt} = \frac{na}{r} \qquad (5.12)$$

where $n = \sqrt{\mu/a^3}$ and the limits of integration for the orbital-averaging integral are the Earth-shadow exit angle E_{ex} and Earth-shadow entrance angle E_{en}, respectively. We assume that thruster power (and hence thrust) is zero when the spacecraft is in the Earth's shadow. If no shadowing conditions exist for a particular osculating orbit, then the integration limits are $E_{ex} = -\pi$ and $E_{en} = \pi$.

Earth-oblateness effects are modeled by including the secular rates of change for longitude of the ascending node and argument of perigee, which are determined by averaging the J_2 zonal harmonic perturbation over one orbital period [10]

$$\dot{\bar{\Omega}}_{J2} = \frac{-3nR_E^2 J_2}{2a^2(1-e^2)^2}\cos i \tag{5.13}$$

$$\dot{\bar{\omega}}_{J2} = \frac{3nR_E^2 J_2}{4a^2(1-e^2)^2}(4 - 5\sin^2 i). \tag{5.14}$$

These secular rates from oblateness are added to the mean rates of change for Ω and ω caused by thrust, that is, the fourth and fifth elements of $\dot{\bar{\mathbf{x}}}$ in Equation (5.11). There-fore, it should be noted that the perturbing acceleration components $[a_n \quad a_t \quad a_h]^T$ in Equations (5.4–5.8) and hence the orbital averaging integral (5.11) are only due to the propulsive thrust.

5.2.2 Thrust-Steering Control Laws

Evaluating the orbital-averaging integral (5.11) requires knowledge of the thrust acceleration perturbations in the NTH orbital coordinate frame as a function of eccentric anomaly E. Furthermore, our direct optimization approach requires that we (somehow) parameterize the thrust-steering direction $\mathbf{u}(t)$ over the entire orbit transfer. In addition, it is advantageous to parameterize $\mathbf{u}(t)$ with as few optimization parameters as possible in order to pose a small-scale NLP problem that can be quickly and efficiently solved. One technique is to use the thrust-steering control laws that can be derived by applying the necessary conditions from optimal control theory. In the derivation that follows, keep in mind that we do not intend to develop all first-order necessary conditions and solve a two-point boundary-value problem; instead, we only wish to use the structure of the optimal control in order to efficiently parameterize the thrust direction.

We begin by defining the Hamiltonian function for the optimal control problem

$$H(\mathbf{x}, \mathbf{u}, \boldsymbol{\lambda}) = \lambda_a \frac{da}{dt} + \lambda_e \frac{de}{dt} + \lambda_i \frac{di}{dt} \tag{5.15}$$

where Equations (5.4–5.6) represent the three differential equations for orbital elements a, e, and i. Three costate variables $\boldsymbol{\lambda} = [\lambda_a \quad \lambda_e \quad \lambda_i]^T$ are introduced. We choose to only include three time-rate equations in H because our orbit-transfer problems typically only involve target conditions for semi-major axis, eccentricity, and inclination (keep in mind that trajectory propagation is per-formed using Equation (5.11), which involves orbital averaging and all five classical orbital elements). The thrust acceleration components in the normal-tangent orbital frame are

$$[a_n \quad a_t \quad a_h]^T = a_T [\sin\alpha\cos\beta \quad \cos\alpha\cos\beta \quad \sin\beta]^T = a_T\mathbf{u} \tag{5.16}$$

where the in-plane (pitch) thrust-direction steering angle α is measured from the velocity vector to the projection of the thrust vector onto the orbital plane, and the out-of-plane (yaw) steering angle β is measured from the orbital plane to the thrust vector. Thrust acceleration a_T is a function of input power P, thruster efficiency η, specific impulse I_{sp}, Earth's gravitational acceleration at sea level g, and spacecraft mass m

$$a_T = \frac{2\eta P}{mgI_{sp}}. \tag{5.17}$$

The optimality condition can be used to derive the structure of the optimal control laws. Optimal pitch steering is determined from the derivative of H with respect to α

$$\frac{\partial H}{\partial \alpha} = -\lambda_a \frac{2a^2 v}{\mu} a_T \sin\alpha \cos\beta + \lambda_e \frac{a_T}{v} \left[-2(e + \cos\theta)\sin\alpha + \frac{r}{a}\sin\theta\cos\alpha \right] \cos\beta = 0. \tag{5.18}$$

After some algebraic manipulations, Equation (5.18) can be written as

$$\tan\alpha^* = \frac{\lambda_e \dfrac{r}{a} \sin\theta}{2 \left[\lambda_a \dfrac{a^2 v^2}{\mu} + \lambda_e (e + \cos\theta) \right]}. \tag{5.19}$$

At this point, the derivative of Equation (5.18) must be computed in order to determine the proper signs for control law (5.19) so that $\partial^2 H / \partial\alpha^2 > 0$ and the Hamiltonian H is minimized. Analysis of the second partial derivative leads to

$$\sin\alpha^* = \frac{-\lambda_e \dfrac{r}{a} \sin\theta}{\sqrt{4 \left[\lambda_a \dfrac{a^2 v^2}{\mu} + \lambda_e (e + \cos\theta) \right]^2 + \lambda_e^2 \dfrac{r^2}{a^2} \sin^2\theta}} \tag{5.20}$$

$$\cos\alpha^* = \frac{-2 \left[\lambda_a \dfrac{a^2 v^2}{\mu} + \lambda_e (e + \cos\theta) \right]}{\sqrt{4 \left[\lambda_a \dfrac{a^2 v^2}{\mu} + \lambda_e (e + \cos\theta) \right]^2 + \lambda_e^2 \dfrac{r^2}{a^2} \sin^2\theta}}. \tag{5.21}$$

Equations (5.20) and (5.21) constitute the optimal steering law used to parameterize the pitch angle profile for the orbital averaging integral (5.11). Recall that orbital elements a, e, and i are held constant over the averaging integral, and that eccentric anomaly E is the integration variable. True anomaly θ can easily be determined from E

$$\cos\theta = \frac{\cos E - e}{1 - e\cos E}$$
$$\sin\theta = \frac{\sin E\sqrt{1 - e^2}}{1 - e\cos E}. \tag{5.22}$$

Radius and velocity magnitudes can be computed from the trajectory equation and
energy integral, respectively

$$r = \frac{a(1 - e^2)}{1 + e \cos \theta}$$

$$\xi = \frac{-\mu}{2a} = \frac{v^2}{2} - \frac{\mu}{r}.$$

(5.23)

Equation (5.19) can be simplified for the case of "quasi-circular" orbits, where eccentricity remains small throughout the transfer. Therefore, if we set $e = 0$, $a = r$ and
$v = \sqrt{\mu/a}$ (circular orbital speed), Equation (5.19) becomes

$$\tan \alpha^* = \frac{\lambda_e \sin \theta}{2 \left(\lambda_a a + \lambda_e \cos \theta\right)}.$$

(5.24)

Edelbaum [1] derived this control law for quasi-circular low-thrust orbit transfers
with continuous thrust (that is, no periods of zero thrust due to the Earth's shadow);
however, Edelbaum set $\lambda_a a = 1$ in his analysis.

Optimal yaw steering is determined from the derivative of H with respect to β

$$\frac{\partial H}{\partial \beta} = -\lambda_a \frac{2a^2 v}{\mu} a_T \cos \alpha \sin \beta - \lambda_e \frac{a_T}{v} \left[2 \left(e + \cos \theta\right) \cos \alpha + \frac{r}{a} \sin \theta \sin \alpha\right] \sin \beta$$

$$+ \lambda_i \frac{a_T r}{h} \cos \left(\omega + \theta\right) \cos \beta = 0.$$

(5.25)

After some algebraic manipulations, Equation (5.25) can be written as

$$\tan \beta^* = \frac{\lambda_i \frac{rv}{h} \cos \left(\omega + \theta\right)}{\lambda_a \frac{2a^2 v^2}{\mu} \cos \alpha^* + \lambda_e \left[2 \left(e + \cos \theta\right) \cos \alpha^* + \frac{r}{a} \sin \theta \sin \alpha^*\right]}.$$

(5.26)

After checking the second partial derivative in order to ensure that $\partial^2 H / \partial \beta^2 > 0$,
we find

$$\sin \beta^* = \frac{-\lambda_i \frac{rv}{h} \cos \left(\omega + \theta\right)}{d}$$

(5.27)

$$\cos \beta^* = \frac{-\lambda_a \frac{2a^2 v^2}{\mu} \cos \alpha^* - \lambda_e \left[2 \left(e + \cos \theta\right) \cos \alpha^* + \frac{r}{a} \sin \theta \sin \alpha^*\right]}{d}$$

(5.28)

where the denominator term d is

$$d = \sqrt{\lambda_i^2 \frac{r^2 v^2}{h^2} \cos^2 \left(\omega + \theta\right) + \lambda_a^2 \frac{4a^4 v^4}{\mu^2} \cos^2 \alpha^* + \lambda_e^2 \left[2 \left(e + \cos \theta\right) \cos \alpha^* + \frac{r}{a} \sin \theta \sin \alpha^*\right]^2}.$$

(5.29)

Equations (5.27–5.29) constitute the optimal steering law used to parameterize the
yaw angle profile for the orbital averaging integral (5.11). The in-plane normal and

tangent components ($\sin \alpha^*$ and $\cos \alpha^*$) are determined by Equations (5.20) and (5.21), respectively.

Edelbaum derived a similar yaw-steering law for quasi-circular orbit transfers with inclination change. Because tangent steering is optimal for quasi-circular transfers (that is, align the in-plane thrust component with the velocity vector), we can set $\lambda_e = 0$ so that $\alpha^* = 0$ for every position in the orbit [see Equations (5.20) and (5.21)]. Therefore, Equation (5.26) can be reduced to

$$\tan \beta^* = \frac{\lambda_i \cos(\omega + \theta)}{2\lambda_a a}. \tag{5.30}$$

If we set the magnitude of $\lambda_a a$ to unity, Equation (5.30) becomes Edelbaum's optimal yaw-steering control law for quasi-circular transfers [1]. Equations (5.20), (5.21) and (5.27–5.29) are the "complete" optimal pitch and yaw control laws, whereas Equations (5.24) and (5.30) are the optimal pitch and yaw controls for "quasi-circular" orbit transfers where eccentricity remains nearly zero.

An alternative method can be applied to derive the optimal control laws. To begin, write the three differential equations for the classical orbital elements in a matrix-vector format

$$\dot{\mathbf{z}} = a_T M \mathbf{u} = a_T \begin{bmatrix} 0 & \dfrac{2a^2 v}{\mu} & 0 \\ \dfrac{r \sin \theta}{av} & \dfrac{2(e + \cos \theta)}{v} & 0 \\ 0 & 0 & \dfrac{r \cos(\omega + \theta)}{h} \end{bmatrix} \mathbf{u} \tag{5.31}$$

where $\mathbf{z} = [a \; e \; i]^T$ and $\mathbf{u} = [\sin \alpha \cos \beta \quad \cos \alpha \cos \beta \quad \sin \beta]^T$ is the unit vector in the thrust direction. Now, the Hamiltonian can be written as

$$H(\mathbf{z}, \mathbf{u}, \lambda) = a_T \lambda^T M \mathbf{u} \tag{5.32}$$

which is equivalent to Equation (5.15), the previous definition of the Hamiltonian. Clearly, the control law

$$\mathbf{u} = \frac{-M^T \lambda}{\|M^T \lambda\|} \tag{5.33}$$

minimizes the Hamiltonian H. Equation (5.33) is equivalent to the pitch and yaw control laws summarized by Equations (5.20), (5.21), and (5.27–5.29).

5.2.3 Analysis of the Control Laws

Before we apply the thrust-steering laws to the orbital averaging integral (5.11) in order to solve the optimal orbit-transfer problem, it is instructive to analyze the various optimal control laws. Both the "complete" and "quasi-circular" control laws demonstrate a feedback structure, where the pitch and yaw steering angles depend on the osculating orbital elements and the spacecraft's current position in the orbit (true anomaly θ). In addition, the three costate variables $\lambda = [\lambda_a \quad \lambda_e \quad \lambda_i]^T$ act as "influence coefficients" or "weighting functions" for each control law. For example, setting a particular costate variable to a "large" magnitude relative to the other costates (for example, $|\lambda_i| >> \lambda_a$ and $|\lambda_i| >> \lambda_e$) results in a steering program that maximizes the magnitude of the *average* rate-of-change for the corresponding orbital element (for example, $\overline{di}/dt$). Furthermore, setting a costate value to zero (or a small number) causes a zero (or small) *average* rate-of-change for the corresponding orbital element. Pitch steering laws (5.20) and (5.21) clearly show that if the eccentricity costate (λ_e) is zero and the semi-major axis costate (λ_a) is negative, then pitch angle α is zero over the orbital revolution and the in-plane thrust remains aligned with the velocity vector (that is, "tangent steering"). Tangent steering provides the maximum instantaneous increase in energy (semi-major axis), and for the case of continuous thrust (no Earth-shadow effect) and a nearly circular orbit, tangent steering results in a zero net change in eccentricity over one orbital revolution.

Figure 5.1 shows the optimal pitch steering program for a low-Earth orbit with $a = 1.1568$ Earth radii (7,378 km), $e = 0.05$, and $\lambda_a = -0.8645$ (that is, $\lambda_a a = -1$) for $\lambda_e = [0, 0.5, 1, 1.06, 2, 10^6]$, while Figures 5.2a–5.2d show the thrust direction over a single orbit for eccentricity costates of 0, 1, 2, and 10^6 (the orbit is direct, and perigee is located on the right-hand sides of Figures 5.2a–5.2d). Note that all pitch steering programs exhibit tangent steering ($\alpha = 0$) at apogee ($\theta = 180$ deg), which results in $de/dt < 0$, and that pitch steering is "anti-tangent" ($\alpha = 180$ deg) at perigee when $\lambda_e \geq 1.06$. Hence, as the eccentricity costate is increased, the steering effort is

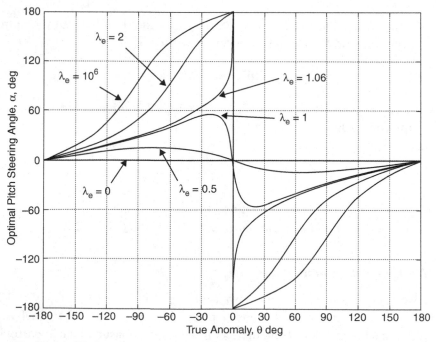

Figure 5.1. Pitch-steering program over one orbital revolution for range of λ_e.

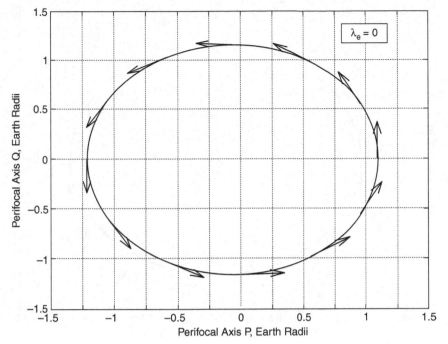

Figure 5.2a. Pitch-steering direction for $\lambda_e = 0$ (tangent steering).

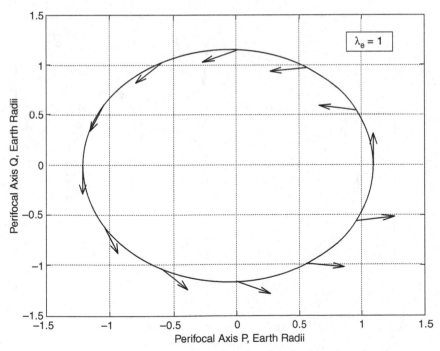

Figure 5.2b. Pitch-steering direction for $\lambda_e = 1$.

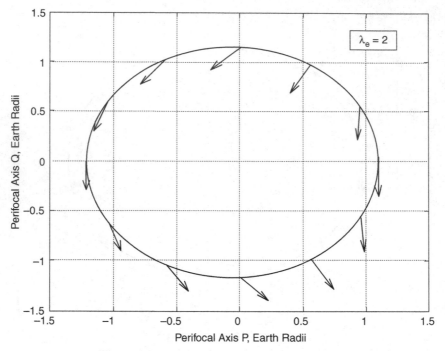

Figure 5.2c. Pitch-steering direction for $\lambda_e = 2$.

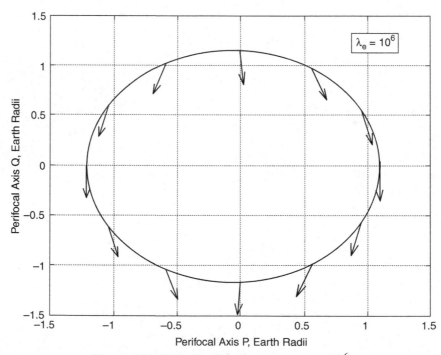

Figure 5.2d. Pitch-steering direction for $\lambda_e = 10^6$.

concentrated on reducing eccentricity. We can determine the net changes in elements a and e over a single orbit by substituting the optimal pitch-steering program into Equations (5.4) and (5.5) and evaluating the orbital-averaging integral (5.11) with integration limits of $[-\pi, \pi]$ (that is, assume continuous thrust). The net changes in semi-major axis and eccentricity are always positive and negative, respectively, for $\lambda_a = -0.8645$ and $\lambda_e \geq 0$. It is interesting to note that the semi-major axis rate da/dt remains positive everywhere in the orbit until the eccentricity costate is increased to approximately $\lambda_e = 1.0526$. When $\lambda_e = 1.0526$, the thrust direction instantly switches from pointing radially outward ($\alpha = 90$ deg) to radially inward ($\alpha = -90$ deg) as the spacecraft passes through perigee; hence $da/dt = 0$ at perigee passage since the tangential acceleration component is zero. When the eccentricity costate λ_e is greater than 1.214, the eccentricity rate de/dt is negative at every location in the orbit. It is important to note that the limiting values presented here only apply to a specific orbit ($a = 1.1568$ Re and $e = 0.05$) and a specific value $\lambda_a = -0.8645$; however, the same general trends can be observed for other combinations of these parameters.

A range of yaw steering programs can be determined using Equations (5.27–5.29) for a range of values for the inclination costate λ_i. Figure 5.3 shows the optimal yaw steering program for a circular low-Earth orbit with $a = 1.1568$ Earth radii (7,378 km), $e = 0$, $\omega = 0$, $\lambda_a = -0.8645$ (that is, $\lambda_a a = -1$), and $\lambda_e = 0$ for $\lambda_i = [0, 1, 2, 4, 10, 10^6]$. Note that the optimal yaw angle always passes through zero at ± 90 deg from the nodal crossings since $di/dt = 0$ at these orbital locations (the "anti-nodes") regardless of the magnitude of the out-of-plane thrust acceleration component [see Equation (5.6)]. Figure 5.3 also clearly shows that the yaw steering

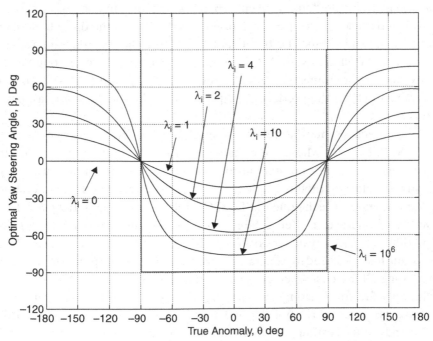

Figure 5.3. Yaw-steering program over one orbital revolution for range of $\lambda_i (\omega = 0)$.

angle is symmetric about the orbital plane. A symmetric yaw program occurs only if the eccentricity and eccentricity costate are both zero. If the orbit is eccentric and $\lambda_e = 0$ (that is, little emphasis on reducing eccentricity), then the yaw-steering amplitude is greatest at apogee, since di/dt is maximized at this orbital location. Figure 5.3 also clearly shows that yaw-steering amplitude (and hence the magnitude of di/dt) increases with increasing inclination costate, and that when λ_i is an extremely large positive number then the thrust vector is always normal to the orbit plane ($\beta = \pm 90$ deg). The net change in inclination from the orbital averaging integral is always negative for all positive values of the inclination costate; furthermore, when $\lambda_i > 0$, the instantaneous inclination rate di/dt is negative at every point in the orbit, except at the anti-nodes.

5.2.4 Solution Method

Recall that this chapter focuses on solving low-thrust orbit transfer problems using control parameterization coupled with a direction optimization technique. Sections 5.2.1 and 5.2.2 have outlined the general problem statement, and the control laws as derived from optimal control theory. The optimal control laws for in-plane thrust steering, Equations (5.20) and (5.21), and out-of-plane thrust steering, Equations (5.27–5.29), all depend on the current orbital elements, position in the orbit, and the costate vector $\lambda = [\lambda_a \quad \lambda_e \quad \lambda_i]^T$. Section 5.2.3 described how the pitch and yaw steering programs vary by adjusting the costates. Therefore, the control parameters for our trajectory optimization problem are the time histories of the costate variables. We choose to parameterize $a\lambda_a(a)$, $\lambda_e(a)$, and $\lambda_i(a)$ with linear interpolation among n discrete "nodes" evenly spaced along semi-major axis, $a \in [a_0, a_f]$. The "design parameters" or free optimization variables are the discrete nodal values for $a\lambda_a$, λ_e, and λ_i, and the free final time, t_f. We choose to parameterize the product $a\lambda_a$ (instead of costate λ_a) since this product appears in the optimal pitch and yaw steering programs [see Equations (5.20), (5.21), and (5.27–5.29)]. Recall that Edelbaum [1] set the product $a\lambda_a$ equal to unity for the case of quasi-circular transfers.

Our approach is somewhat related to the classical indirect approach used in the program SEPSPOT [4] where the costate differential equations are derived from the Hamiltonian, the mean costate histories are determined by numerical integration, and a shooting method is employed to meet the desired boundary conditions. However, the indirect technique of SEPSPOT relies on a good initial guess for the costate vector and hence often exhibits poor convergence properties. Our direct method offers distinct advantages over the SEPSPOT (indirect) approach: (1) only three costate histories need to be parameterized with linear interpolation, as opposed to integrating all five costate equations; (2) as shown in Section 5.2.3, the user has some intuitive feel for selecting initial nodal guesses for the costates since the magnitude and sign has a direct relationship with the time-rate of the corresponding orbital element; and (3) the costate profiles are selected by the user and do not depend on numerical integration of the adjoint system, which aids convergence properties. It is likely that if the NLP problem is expanded to include parameterization of all five costates with a fine grid of nodal values, then the resulting solution may approach the corresponding indirect solution produced by SEPSPOT. However, the computational time will be greatly increased and the convergence properties will likely be degraded for the expanded problem.

We replace the original optimal control problem with a nonlinear programming problem, which is solved using a gradient-based optimization technique. Sequential quadratic programming (SQP) is used to solve the constrained parameter optimization problem. The SQP algorithm used here is *fmincon* from the Matlab® optimization toolbox, which computes the gradients using a finite-difference method. Terminal states constraints are enforced through SQP equality constraints.

5.3 Numerical Results

Several optimal low-thrust orbit transfers are presented in this section. Before proceeding with the examples, a few details regarding the numerical simulation are in order. The averaged state equations (5.11) are numerically integrated using a second-order Euler method (or Huen's method) in order to determine the averaged or mean state trajectories. The fixed integration time step is determined by dividing the time axis into 40 equal steps (therefore, the time-step is 5 days if the transfer time is 200 days). Evaluation of the orbital-averaging integral [the right-hand side of Equation (5.11)] is required at each time step, and trapezoidal-rule integration is used to integrate dx/dE with 20 equally spaced steps between the integration limits E_{ex} and E_{en}. Finally, Earth-shadow exit/entrance angles (E_{ex} and E_{en}) are determined by finding the intersection between the osculating elliptical orbit and a cylindrical shadow model; this procedure requires the solution of a fourth-order polynomial in $\cos E$ and sorting out spurious roots (see Neta and Vallado [11] for details).

5.3.1 Minimum-Time LEO-GEO Transfer

The first case involves finding the low-thrust transfer from circular low-Earth orbit (LEO) to circular geostationary Earth orbit (GEO) with minimum transfer time, and the orbital conditions and vehicle parameters are taken from Kluever and Oleson [5]. Hence, the performance index in Equation (5.1) is $J = t_f$. The initial orbital elements for LEO are $a_0 = 6,927$ km (1.086 Re), $e_0 = 0$, $i_0 = 28.5$ deg, $\Omega_0 = \omega_0 = 0$ deg, and the target elements for GEO are $a_f = 42,164$ km, $e_f = 0$, and $i_f = 0$ deg (therefore, the terminal state constraint vector $\psi[\mathbf{x}(t_f), t_f]$ in Equation (5.3) is 3×1, which is enforced by three equality constraints for the SQP algorithm). Initial mass in LEO is 1,200 kg, and the constant SEP parameters are $I_{\text{sp}} = 3,300$ s, $P = 10$ kW, and $\eta = 65\%$. These SEP parameters are representative of a xenon ion thruster system. Departure date is required for Earth-shadow calculations, and it is arbitrarily set at January 1, 2000.

Edelbaum's analytical method for optimizing continuous-thrust, quasi-circular transfers is used to provide good initial guesses for the inclination costates (yaw steering program) and final end-time t_f. A good initial guess for $\lambda_i(a)$ is determined from Kechichian's extension [12] to Edelbaum's original formulation [1]. Kechichian [12] derived analytic formulas for the initial and final yaw amplitudes based on the required plane change, the initial and final circular orbital speeds, and transfer time. The corresponding initial and final inclination costates are easily determined from Equation (5.30)

$$
\lambda_i(a_0) = 2 \tan \beta_0
$$
$$
\lambda_i(a_f) = 2 \tan \beta_f
$$
(5.34)

where β_0 and β_f are the initial and final yaw-steering amplitudes, respectively, as determined by Kechichian's method. Initial guesses for the inclination costates nodes are linearly spaced between $\lambda_i(a_0)$ and $\lambda_i(a_f)$. Transfer time for Edelbaum's continuous-thrust, quasi-circular transfer can be analytically computed from the required propellant mass (computed from ΔV and the rocket equation), and the constant mass-flow rate. Because Edelbaum's solution assumes no shadow-eclipse periods, we set the initial guess for transfer time (with Earth shadow) to 1.2 times the continuous-thrust transfer time. Finally, the initial guess for the nodes for the product $a\lambda_a$ is set to -1 (following Edelbaum's technique), and the initial guess for the eccentricity costate nodes is simply a linear distribution between two arbitrary small values (that is, the initial trajectory essentially uses tangent steering for the pitch-steering program).

For the first LEO-GEO solution, we use three nodes for $a\lambda_a$, λ_e, and λ_i, and therefore the NLP problem has a total of 10 optimization parameters (final time t_f is the tenth parameter). Table 5.1 shows a comparison between the minimum-time LEO-GEO transfers computed by our control-parameterization method and by using SEPSPOT. Unfortunately, the ΔV and final mass values are not available for the SEPSPOT solution. The minimum transfer time differs by only 0.2 days, or 0.1%.

Figures 5.4 and 5.5 show the time histories of a, e, and i for the minimum-time LEO-GEO transfer. Clearly, all three elements are simultaneously adjusted during the orbit transfer until they meet their desired targets. Figures 5.6 and 5.7 present the optimal costates and the corresponding (maximum) pitch- and yaw-steering amplitudes, respectively. Note that the optimal costates λ_e and λ_i (as determined by the NLP solver) are nearly linear with semi-major axis, whereas the product $a\lambda_a$ remains close to -1. The initial rise in eccentricity in Figure 5.5 is due to the Earth-shadow effect; because apogee lies in the shadow as the spacecraft spirals away from circular LEO, the low-amplitude pitch steering program (see Figure 5.7) produces a positive

Table 5.1. *Minimum-time LEO-GEO transfer solutions*

Optimization method	Minimum t_f (days)	ΔV (km/s)	Mass ratio m_f/m_0
Control parameterization	199.0	5.675	0.8392
SEPSPOT	198.8	N/A	N/A

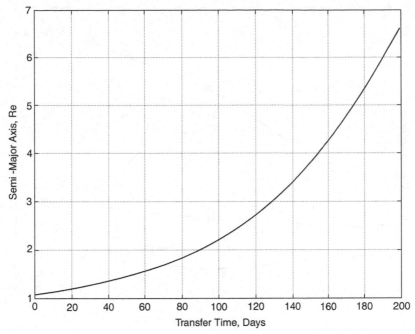

Figure 5.4. Semi-major axis versus time for minimum-time LEO-GEO transfer.

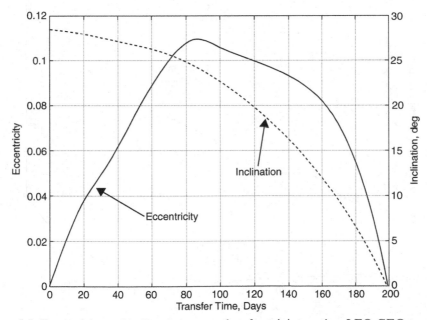

Figure 5.5. Eccentricity and inclination versus time for minimum-time LEO-GEO transfer.

eccentricity rate. Figure 5.8 shows that the angular arc of the Earth shadow is about 134 deg in LEO and eventually goes to zero at about 91 days into the transfer, and therefore the remainder of the orbit transfer involves continuous thrust. This exit from shadowing periods at 91 days is seen in Figure 5.5, since the eccentricity steadily

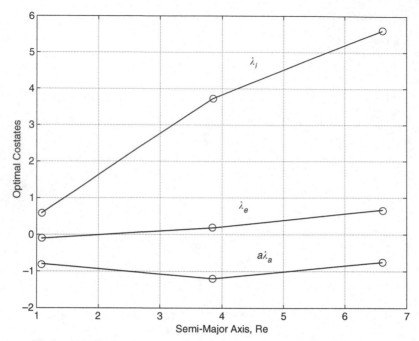

Figure 5.6. Optimal costates for minimum-time LEO-GEO transfer.

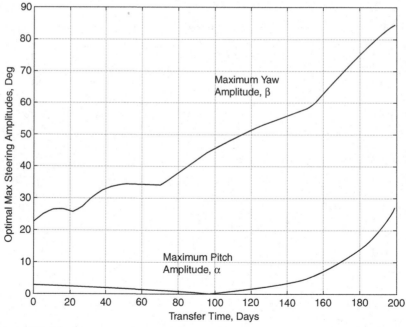

Figure 5.7. Maximum steering amplitudes for minimum-time LEO-GEO transfer.

decreases once apogee is no longer eclipsed. It is interesting to note that the optimal pitch steering program essentially becomes tangent steering ($\alpha = 0$) as the spacecraft no longer experiences eclipse periods (see Figure 5.7). Therefore, semi-major axis (or energy) rate is maximized while the orbit is "naturally" circularized. Pitch amplitude

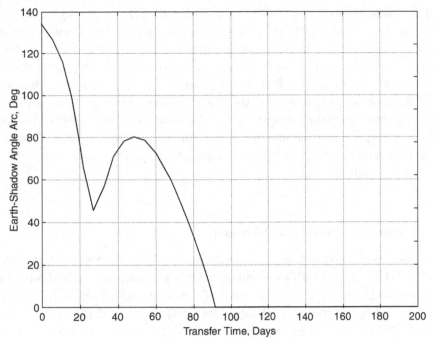

Figure 5.8. Earth-shadow angle arc for minimum-time LEO-GEO transfer.

is increased at the end of the transfer in order to complete the circularization (however, pitch amplitude remains small), while yaw amplitude steadily increases during
the transfer since it is more efficient to perform plane changes at lower velocity.
The initial and final yaw amplitudes are 23 deg and 84 deg (see Figure 5.7); the yaw
amplitudes predicted by the analytic Edelbaum/Kechichian method are $\beta_0 = 22$ deg
and $\beta_f = 60$ deg, respectively.

Next, we analyzed the effect of the number of control parameters (costate nodes)
on both solution accuracy and the computational load. Table 5.2 presents the optimal
transfer time and run-time for several minimum-time LEO-GEO solutions. Run-
time is the "real time," or "wall time" as computed by the Matlab tic/toc commands.
All numerical computations were performed on a Pentium M laptop with a 2 GHz
processor. Initially the two-node solution is obtained with an initial guess derived
from the analytic Edelbaum/Kechichian solution, and then each subsequent problem
uses the previous optimal solution as its initial guess (it should be noted that the three

Table 5.2. *Minimum-time LEO-GEO transfers with an increasing number of costate nodes*

Number of nodes	Number of SQP variables	Minimum t_f (days)	Run-time (sec)	Number of SQP iterations	Time/iteration (sec/iteration)
2	7	199.0401	33.68	23	1.46
3	10	198.9901	59.83	30	1.99
5	16	198.9898	3.01	1	3.01
9	28	198.9898	5.47	1	5.47

costate profiles always have the same number of nodes). The extra nodes are placed linearly at the midpoints of the prior solution, so that the initial guess produces a trajectory identical to the one from the previous converged solution. Table 5.2 shows that very little performance is gained by increasing the number of nodes, at least for this LEO-GEO problem. Furthermore, when the number of nodes (for each costate variable) reaches five, the problem converges after a single iteration without any improvement in transfer time. Therefore, for this standard LEO-GEO problem, a simple linear parameterization of the two costates (that is, seven total optimization parameters) produces a sufficiently accurate optimal transfer with very minimal computational load.

5.3.2 Minimum-Time GTO-GEO Transfer

The second case involves finding the low-thrust transfer from geostationary-transfer orbit (GTO) to circular GEO with minimum transfer time. The initial orbital elements for GTO are $a_0 = 24{,}364$ km (3.820 Re), $e_0 = 0.7306$, $i_0 = 28.5$ deg, $\Omega_0 = \omega_0 = 0$ deg, and the target elements for GEO are $a_f = 42{,}164$ km, $e_f = 0$, and $i_f = 0$ deg. Initial mass in LEO is 1,200 kg, and the SEP parameters are $I_{sp} = 1{,}800$ s, $P = 5$ kW, and $\eta = 55\%$. This particular set of SEP parameters is representative of a Hall-effect thruster. Departure date is arbitrarily set at March 22, 2000, so that the initial perigee direction is approximately aligned with the Earth-Sun vector.

The minimum-time transfer is readily obtained by using simple linear profiles as initial guesses for the costate profiles (note that the analytic Edelbaum solution cannot be used to provide initial guesses for flight time or yaw-steering amplitudes). A series of solutions were obtained for a range of number of costates nodes as summarized in Table 5.3. The two-node case converged in 23 iterations; trials with additional nodes converged in 40–60 iterations. Clearly there is no advantage to be gained by increasing the number of costate nodes, since the two-node solution is within 0.2 days (or 0.2%) of the five-node solution.

Figures 5.9 and 5.10 show the time histories of a, e, and i for the minimum-time GTO-GEO transfer with five nodes for each costate variable. Note that eccentricity (Figure 5.10) steadily decreases despite having the apogee of the GTO initially in the Earth's shadow. Figure 5.11 presents the time histories of the maximum steering amplitudes. Note that maximum pitch-steering amplitude steadily increases and reaches 90 deg (inward/outward radial steering at perigee) when transfer time is

Table 5.3. *Minimum-time GTO-GEO solutions*

Number of nodes	Number of SQP variables	Minimum t_f (days)	Run-time/iteration (sec/iteration)
2	7	118.494	2.13
3	10	118.415	2.79
4	13	118.375	3.60
5	16	118.357	4.66

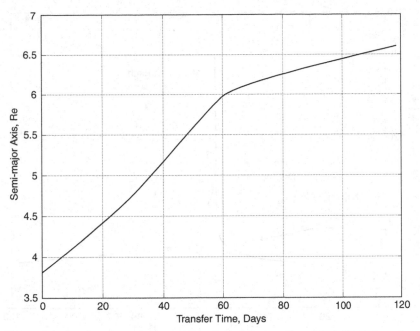

Figure 5.9. Semi-major axis versus time for minimum-time GTO-GEO transfer.

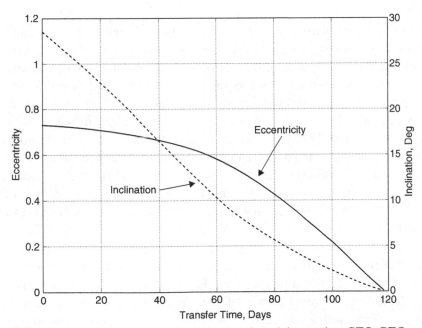

Figure 5.10. Eccentricity and inclination versus time for minimum-time GTO-GEO transfer.

approximately 55 days. At this same instant, the maximum yaw amplitude is also 90 deg, and also occurs at perigee. Therefore, at $t = 55$ days, the radial thrust command at perigee is not a wasted effort, since all of the thrust is normal to the orbit plane (that is, yaw amplitude is 90 deg). Immediately after 55 days, the maximum pitch-steering amplitude is 180 deg (antitangent steering), which occurs at perigee

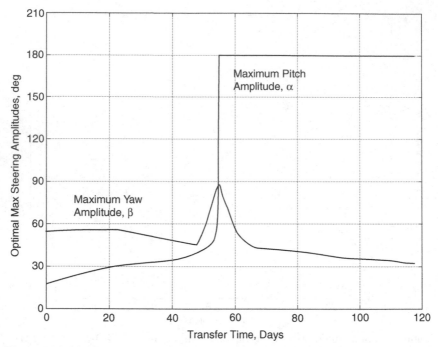

Figure 5.11. Maximum steering amplitudes for minimum-time GTO-GEO transfer.

(similar to Figure 5.1 with $\lambda_e > 1.06$). After 55 days, the maximum yaw amplitude decreases. It is interesting to note that the maximum yaw amplitude is generally larger during the initial phase of the transfer, where the maximum yaw amplitude occurs near apogee. Figure 5.12 shows the time history of the Earth-shadow angle arc. Note that the initial shadow arc angle is small (about 18 deg), since the calendar date for the start of the transfer is March 22 and $\Omega_0 = 0$ deg. Earth-shadow periods do not exist after $t = 57$ days, which corresponds to the transition to antitangent steering (maximum $\alpha = 180$ deg, see Figure 5.11) at each perigee pass.

The minimum-time GTO-GEO transfer was obtained for the case where initial argument of perigee is $\omega_0 = 180$ deg (starting calendar date and Ω_0 remain unchanged). Therefore, apogee is now initially in sunlight and perigee is in the Earth's shadow. The minimum-time transfer (using five costate nodes) is 112.11 days, which is more than 6 days faster than the previous solution where perigee is initially in sunlight. For this case, with apogee initially in sunlight, the initial Earth-shadow arc angle is 107 deg (centered at perigee), and Earth-shadow periods do not exist after 70 days. Despite the larger shadow angle, the actual time spent in eclipse is smaller for the second case with a perigee shadow, since perigee speed is significantly larger than apogee speed.

5.3.3 Minimum-Propellant LEO-GEO Transfer

The third case involves finding the low-thrust LEO-GEO transfer that minimizes total propellant usage instead of trip time. We use the same initial LEO, target

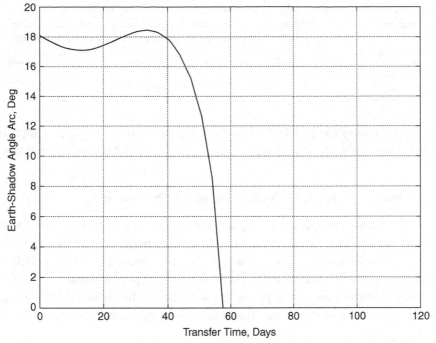

Figure 5.12. Earth-shadow angle arc for minimum-time GTO-GEO transfer.

GEO, and EP spacecraft parameters as case 1 (minimum-time LEO-GEO transfer) described in Section 5.3.1. The only difference is that the performance index J to be minimized, Equation (5.1), is now the negative of the final mass

$$J = -m(t_f). \tag{5.35}$$

As before, transfer time t_f is a free design variable. For this problem one additional trajectory propagation is required during the SQP search in order to compute the performance index (final mass).

The minimum-propellant problem is readily solved using the direct optimization method with three nodes for each costate profile (10 total optimization variables), and Table 5.4 compares the minimum-propellant and minimum-time solutions. Note that the minimum-propellant solution increases the final delivered mass to GEO by about 0.6 kg with a corresponding increase in transfer time of 1.2 days. The two trajectory solutions are nearly identical; the slight difference in transfer time is due to slight differences in the Earth-shadow profile during the LEO-GEO transfer. The minimum-propellant solution seeks an Earth-shadow profile that minimizes the so-called "geometry losses" associated with thrusting during nonoptimal orbit locations of the transfer [13]. Computation time (run time per iteration) is approximately doubled for the minimum-propellant problem compared to the minimum-time problem. This case demonstrates the direct optimization method's ability to easily accommodate a different performance index; in practice, a mission designer would likely

Table 5.4. *Minimum-propellant and minimum-time LEO-GEO solutions*

Performance index	Propellant mass m_{prop} (kg)	Transfer time t_f (days)	Run-time/iteration (sec/iteration)
Minimum propellant	192.40	200.16	3.96
Minimum time	193.01	198.99	1.99

solve the minimum-time problem when electric power and I_{sp} are constant (that is, constant mass-flow rate) since this problem requires half the computational time.

5.3.4 Minimum-Propellant LEO-GEO Transfer with Variable I_{sp}

The fourth and final numerical example involves finding the minimum-propellant, low-thrust LEO-GEO transfer with variable specific impulse (I_{sp}). Input power is assumed to be constant at 10 kW, and we use the same initial LEO, target GEO, and EP spacecraft parameters as described in Section 5.3.1. Specific impulse is allowed to vary during the orbit transfer, and its profile is defined by linear interpolation among five nodes equally spaced between the initial and target semi-major axis. Thruster efficiency is modeled as a simple function of I_{sp}

$$\eta = \frac{b_1 V_{ex}^2}{b_2^2 + V_{ex}^2} \tag{5.36}$$

where $V_{ex} = gI_{sp}$ is the engine exhaust speed, and $b_1 = 0.73$ and $b_2 = 10,400$ m/s are engine parameters [14] used to model a typical ion thruster efficiency variation with I_{sp}. Specific impulse is constrained to be between 2,000 and 5,000 s by using the "box" limits on the free SQP variables, and therefore maximum efficiency is 70% (at $I_{sp} = 5,000$ s), while minimum efficiency is 57% (at $I_{sp} = 2,000$ s). Because propellant usage is always minimized by using the highest possible I_{sp}, the transfer time must be fixed for each minimum-propellant problem. Hence, the subsequent NLP problem has 14 free optimization variables: three nodes each for the three costates, and five nodes for I_{sp}.

Several minimum-propellant (maximum final mass) problems are solved using the direct optimization method for a range of fixed transfer times. Figure 5.13 shows the maximum final spacecraft mass for the various transfer times, where each solution is indicated by a symbol. Figure 5.14 shows the optimal I_{sp} profile for selected transfer times (all I_{sp} profiles are plotted with semi-major axis as the independent variable). Specific impulse approaches the lower boundary (2,000 s) in order to produce a higher thrust magnitude as transfer time is reduced; the converse is true as transfer time is increased. Figure 5.13 shows that final mass increases with transfer time and reaches a maximum value at a transfer time of about 300 days when the optimal I_{sp} profile is constant at 5,000 s. The smallest final mass value shown in Fig. 5.13 demonstrates the case where the resulting optimal I_{sp} profile is constant at its lower bound of 2,000 s, and the corresponding transfer time of 132.4 days is the fastest

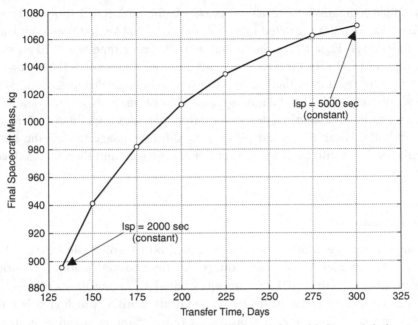

Figure 5.13. Minimum-propellant LEO-GEO transfers with I_{sp} modulation.

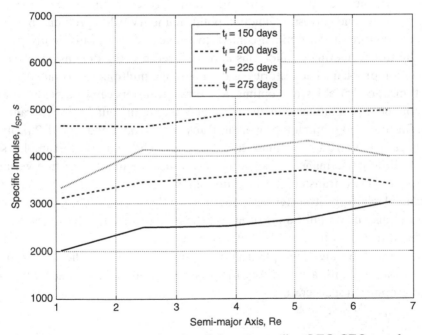

Figure 5.14. Optimal I_{sp} profiles for minimum-propellant LEO-GEO transfers.

possible LEO-GEO transfer for this particular SEP system. Specific impulse cannot be modulated for improved efficiency when transfer time is fixed at 132.4 days; if I_{sp} is increased above the lower limit of 2,000 s then the thrust decreases and the spacecraft cannot complete the LEO-GEO transfer in the given trip time.

As a final comparison, note that the previous minimum-time LEO-GEO transfer with fixed $I_{sp} = 3,300$ s resulted in a final mass of 1,007 kg and minimum transfer time of 199 days. Figure 5.13 shows that when transfer time is 199 days, varying I_{sp} during the LEO-GEO transfer will result in a final mass of 1,012 kg, or a 5-kg (0.5%) improvement over the fixed-I_{sp} transfer. Although this mass improvement is essentially negligible, modulating specific impulse may provide advantages for scenarios that involve solar cell degradation from particles trapped in the Van Allen radiation belts. Using the lowest possible I_{sp} during passage though the heart of the radiation belts will maximize the rate of energy gain and therefore reduce the power loss.

5.4 Conclusions

This chapter presents a direct optimization method for solving low-thrust, Earth-orbit transfer problems. The technique relies on two key features: (1) orbital averaging and (2) control parameterization. Orbital averaging allows relatively efficient numerical propagation of the low-thrust orbit transfer, which greatly reduces the run time of the optimization process. Control parameterization utilizes the optimality condition (derived from the necessary conditions from optimal control theory) to define the thrust-direction program. However, the method makes no attempt to solve the corresponding two-point boundary value problem, but rather uses the structure of the optimal control to conveniently and efficiently parameterize the thrust-steering profile. This approach reduces the dimensionality of the optimization problem. The histories of the Lagrange multipliers (costates) serve as the optimization variables, and optimal low-thrust transfers can be quickly obtained using nonlinear programming methods, namely sequential quadratic programming. Minimum-time and minimum-propellant transfers are obtained for NLP problems that typically have 10 free optimization variables. The numerical examples show that very little performance is gained by increasing the number of free variables, even for low-thrust transfers that require hundreds of days. Earth-shadow effects are included in all orbit-transfer cases presented in this chapter. The specific impulse profile can also be optimized for orbit transfers with fixed transfer times. In addition, this direct optimization method demonstrates rapid convergence in all test cases attempted. Because this optimization technique is fast and efficient, it provides mission designers with a useful tool for performing preliminary trade studies for electric-propulsion spacecraft.

Nomenclature

a $\quad\quad\quad$ = semi-major axis, km

a_T $\quad\quad\quad$ = thrust acceleration, m/s^2

a_n, a_t, a_h = perturbing accelerations in NTH frame, m/s^2

e	=	eccentricity
E	=	eccentric anomaly, rad
$\mathbf{f}$	=	vector of orbital equations of motion
g	=	Earth's gravitational acceleration at sea level, m/s^2
h	=	angular momentum, km^2/s
H	=	Hamiltonian function
I_{sp}	=	specific impulse, s
i	=	inclination, rad
J	=	performance index
J_2	=	Earth oblateness constant
m	=	spacecraft mass, kg
n	=	mean motion, rad/s
P	=	input power, kW
r	=	orbital radius, km
R_E	=	Earth radius, km
t	=	time, s
$\mathbf{u}$	=	unit vector in thrust direction
v	=	orbital velocity magnitude, km/s
$\mathbf{x}$	=	state vector of classical orbital elements and mass $[a,e,i,\Omega,\omega,\theta,m]^T$
$\mathbf{z}$	=	state vector of orbital elements, $[a,e,i]^T$
α	=	in-plane pitch steering angle, rad
β	=	out-of-plane yaw steering angle, rad
ΔV	=	velocity increment, km/s
ξ	=	orbital energy, km^2/s^2
η	=	thruster efficiency
λ	=	costate vector associated with orbital elements, $[\lambda_a \quad \lambda_e \quad \lambda_i]^T$
λ_a	=	costate associated with semi-major axis
λ_e	=	costate associated with eccentricity
λ_i	=	costate associated with inclination
μ	=	Earth's gravitational constant, km^3/s^2
θ	=	true anomaly, rad
ψ	=	vector of terminal state constraints
Ω	=	longitude of the ascending node, rad
ω	=	argument of periapsis, rad

Subscripts

0	=	initial value
en	=	Earth-shadow entrance
ex	=	Earth-shadow exit
f	=	final value
max	=	maximum value
min	=	minimum value

REFERENCES

[1] Edelbaum, T. N. (1961) Propulsion Requirements for Controllable Satellites, *ARS Journal*, **31**, 1079–1089.

[2] Colasurdo, G., and Casalino, L. (2004) *Optimal Low-Thrust Maneuvers in Presence of Earth Shadow*, AIAA/AAS Astrodynamics Specialist Conference, AIAA Paper 2004-5087, Providence, RI.

[3] Kechichian, J. A. (1998) Low-Thrust Eccentricity-Constrained Orbit Raising, *Journal of Spacecraft and Rockets,* **35**, No. 3, 327–335.

[4] Sackett, L. L., Malchow, H. L., and Edelbaum, T. N. (1975) *Solar Electric Geocentric Transfer with Attitude Constraints: Analysis*, NASA CR-134927.

[5] Kluever, C. A., and Oleson, S. R. (1998) Direct Approach for Computing Near-Optimal Low-Thrust Earth-Orbit Transfers, *Journal of Spacecraft and Rockets,* **35**, No. 4, 509–515.

[6] Ilgen, M. R. (1994) Hybrid Method for Computing Optimal Low Thrust OTV Trajectories, *Advances in the Astronautical Sciences*, **87**, No. 2, 941–958.

[7] Jenkin, A. B. (2004) *Representative Mission Trade Studies for Low-Thrust Transfers to Geosynchronous Orbit*, AIAA/AAS Astrodynamics Specialist Conference, AIAA Paper 2004-5086, Providence, RI.

[8] Scheel, W. A., and Conway, B. A. (1994) Optimization of Very-Low-Thrust, Many-Revolution Spacecraft Trajectories, *Journal of Guidance, Control, and Dynamics,* **17**, No. 6, 1185–1192.

[9] Battin, R. H. (1987) *An Introduction to the Mathematics and Methods of Astrodynamics*, AIAA Education Series, AIAA, Washington, DC, 488–489.

[10] Vallado, D. A. (1997) *Fundamentals of Astrodynamics and Applications*, McGraw-Hill, New York, 579–583.

[11] Neta, B., and Vallado, D. (1998) On Satellite Umbra/Penumbra Entry and Exit Positions, *Journal of the Astronautical Sciences*, **46**, No. 1, 91–104.

[12] Kechichian, J. A. (1997) Reformulation of Edelbaum's Low-Thrust Transfer Problem Using Optimal Control Theory, *Journal of Guidance, Control, and Dynamics,* **20**, No. 5, 988–994.

[13] Pollard, J. E., and Janson, S. W. (1996) *Spacecraft Electric Propulsion Applications*, Aerospace Corporation, Report No. ATR-96 (8201)-1.

[14] Kluever, C. A. (2005) Geostationary Orbit Transfers Using Solar Electric Propulsion with Specific Impulse Modulation, *Journal of Spacecraft and Rockets*, **41**, No. 3, 461–466.

6   Analytic Representations of Optimal Low-Thrust Transfer in Circular Orbit

Jean A. Kéchichian

The Aerospace Corporation, El Segundo, CA

6.1 Introduction

The analysis of optimal low-thrust orbit transfer using averaging methods has benefited from the contributions in [1]–[5]. In particular the minimum-time transfer between inclined circular orbits was solved in [1] by rotating the orbit plane around the relative line of nodes of initial and final orbits using a piecewise-constant out-of-plane thrust angle switching signs at the relative antinodes while simultaneously changing the orbit size from its initial to its final required value. The original analysis in [1] was further reformulated in [6] in order to arrive at expressions that depict the evolutions of the transfer parameters in a uniform manner valid for all transfers. Further contributions using both analytic and numerical approaches were also carried out in [7] and [8] to solve increasingly more difficult problems, such as by considering realistic constraints and orbit precession due to the important second zonal harmonic J_2, while also considering various averaging schemes for rapid computation of the solutions.

The first part of this chapter presents the reformulated Edelbaum problem of the minimum-time low-thrust transfer between inclined circular orbits by further extending it in order to constrain the intermediate orbits during the transfer to remain below a given altitude. The minimum-time problem involving an inequality constraint on the orbital velocity is shown to be equivalent to one involving an equality constraint in terms of the thrust yaw angle representing the control variable that is optimized resulting in the minimum-time solution. The transfers are shown to comprise an initial unconstrained arc followed by a constrained arc where the altitude remains constant while pure inclination change is effected, and terminating by a final unconstrained arc with simultaneous altitude and inclination changes as during the first arc.

Edelbaum's optimal low-thrust orbit transfer problem between inclined circular orbits [1] is reformulated within the framework of optimal control theory [6]. This is done in order to cast the transfer problem as a minimum-time problem between given noncoplanar circular orbits and obtain a single analytic expression for the orbital inclination involving a single inverse-tangent function uniformly valid for all transfers. Edelbaum's original treatment considered the time-constrained inclination

139

maximization with velocity as the independent variable allowing the use of the theory of maxima.

Because the independent variable used by Edelbaum is double-valued for some transfers, two expressions for the inclination change involving inverse-sine functions were needed to describe all possible transfers. Besides achieving this simplification of uniform validity through the use of the optimal control theory formalism, additional expressions for the initial value of the control parameter needed for a given transfer, as well as its functional dependency on time, are also derived [6] using inverse-tangent functions without any quadrant ambiguity.

This linearized theory [1] and [6] averages out the spacecraft position and reduces the Lagrange Planetary Equations to a set of two coupled differential equations for the relative inclination i and orbital velocity V inasmuch as the orbit is assumed to remain circular during the transfer. These two differential equations describe the evolution of the orbit size and orientation without regard to the actual position of the spacecraft. The out-of-plane or thrust vector yaw angle β is the control variable that is optimized during the transfer to minimize transit time.

For large relative inclination changes, the intermediate or current circular orbits during the transfer can grow to very large values due to the unconstrained nature of the transfer because the extra effort spent in growing the orbit size to those large values and then shrinking it back to its final value is more than compensated by spending much less effort in rotating the orbit plane at those higher altitudes versus the less efficient lower altitudes. It is within this framework that the analysis of Reference 6 is extended by also constraining the maximum altitude reached during the transfer to a user-defined value.

Because the altitude or semi-major axis is related to orbital velocity, the problem is now cast as a minimum-time problem with an inequality constraint on functions of the state variables, which is reduced in this case to $S = V_{limit} - V \leq 0$. The control β appearing in the first derivative of S, namely $\dot{S}$, the problem is now of the type involving an equality constraint on functions of the control and state variables [9].

A variable Lagrange multiplier $\mu(t)$ is used to adjoin the constraint $\dot{S}$ to augment the original Hamiltonian of the unconstrained problem, leading to an optimal solution involving three phases during the constrained transfer, with the second phase holding the altitude constant at the limiting altitude while purely rotating the orbit plane, whereas the initial and final legs expand and shrink the orbit respectively from its initial size to the limiting size and from that limiting size to its final size while simultaneously contributing to the overall orbit plane rotation in an optimal manner as well. References [10] and [11] are representative of the many excellent texts published recently on the theory of optimization and its applications, while an example on the use of the averaging technique for rapid integration in optimal orbit transfer is depicted in [3].

For the sake of completeness and clarity of the presentation, the analysis of the unconstrained transfer that appeared in [6] in great detail is shown briefly in the next section because the subsequent section dealing with the constrained transfer is based on this first analysis. The analysis of the constrained transfer is based on [12] and [13] where the necessary conditions for an extremal solution with a state variable inequality constraint are fully derived. It is shown in [12] that one or more functions of the state and time must satisfy equality constraints at the beginning of an inequalty constraint boundary arc. These constraints are responsible for the presence of discontinuities in the Lagrange multipliers at the constraint boundary entry point.

The completely closed-form analysis of the constrained transfer is presented with all the necessary details for rapid implementation on a computer. These analytic theories are of great benefit for preliminary spacecraft systems analyses and design optimization applications, as well as for future autonomous onboard guidance implementations.

This chapter later revisits the original analytic formulation of [1] and [6], and in anticipation of folding the J_2 perturbation, by considering a split-sequence transfer where either the inclination variable or the right ascension of the ascending node is controlled first while simultaneously adjusting the orbit size in an optimal way. The original (V, i^*) analytic theory, where V is the varying orbital velocity in circular orbit and i^* the varying relative inclination measured at the relative line of nodes, is now replaced by the sequence (V, i), (V, Ω) or (V, Ω), (V, i) where i is now the varying equatorial inclination, such that analytic descriptions of all the pertinent parameters are still achieved even though the combined two-step transfer is now near-optimal because i and Ω are not allowed to change simultaneously.

The remaining part of this chapter dwells on the numerical solution of fixed-time continuous-thrust transfers whereby the final equatorial inclination must be maximized for general circular orbits. A specific averaging scheme that rotates the orbit plane around the equatorial line of nodes is considered and the node is allowed to precess due to J_2. Thus the thrust acceleration is used only to change the orbital inclination while optimally varying the orbit size in order to carry out the transfer in an efficient manner. These various simplifications and assumptions lead to robust computer codes that converge rather easily to the optimal solutions of interest, and in the purely analytic theory of the first part lead to near-optimal solutions that are generated essentially in no time.

6.2 The Optimal Unconstrained Transfer

6.2.1 The Linearized Reduced Equations of Motion

The thrust acceleration vector $\mathbf{f}$ is assumed not to have any radial component $f_n = 0$, and with the thrust yaw angle β defining the angle made by the thrust acceleration vector with the current orbit plane, the tangential and out-of-plane components of $\mathbf{f}$ are simply given by $f_t = fc_\beta$ and $f_h = fs_\beta$, with $f = |\mathbf{f}|$ the acceleration magnitude, such that the simplified equations of motion can be written as

$$\dot{a} = \frac{2af_t}{V} \tag{6.1}$$

$$\dot{i} = \frac{c_\theta f_h}{V} \tag{6.2}$$

$$\dot{\theta} = n \tag{6.3}$$

These equations are obtained directly from the full set of the Gaussian form of the Lagrange planetary equations for near-circular orbits [6] after setting $e = 0$ and $\alpha = \omega + M = \omega + \theta^* = \theta = nt$ and holding β piecewise constant switching sign at the orbital antinodes such that the $f_h s_\alpha$ terms will be eliminated from the original equations because their net contribution will be zero. Here $n = (\mu/a^3)^{1/2}$ is the orbit mean motion, a, i, and θ are the orbit semimajor axis, inclination and orbital position, with α, ω, M, and θ^* standing for the mean angular position, argument of perigee, mean anomaly, and true anomaly respectively. V and μ are the orbital velocity also equal to $na = (\mu/a)^{1/2}$ and the gravity constant respectively. From the energy equation $\frac{V^2}{2} - \frac{\mu}{r} = -\frac{\mu}{2a}$ with $r = a$, r being the radial distance, Equation (6.1) can be written as

$$\dot{V} = -f_t = -fc_\beta. \tag{6.4}$$

6.2.2 The Averaging Out of the Orbital Position

The angular position θ can effectively be averaged out in Equation (6.2) by integrating this equation with respect to θ while holding β, f, and V constant such that

$$\dot{\bar{i}} = \frac{1}{T} \int_0^T \dot{i}\, dt = \frac{n}{2\pi} \int_0^{2\pi} \dot{i}\, \frac{r^2}{h}\, d\theta^*$$

and because $r = a$, and the angular momentum h is equal to $na^2(1 - e^2)^{1/2} = na^2$, the above averaged inclination rate is reduced to

$$\begin{aligned}
\dot{\bar{i}} = \overline{\left(\frac{di}{dt}\right)} &= \frac{1}{2\pi} \int_0^{2\pi} \left(\frac{di}{dt}\right) d\theta \\
&= \frac{2fs_\beta}{2\pi V} \int_{-\pi/2}^{\pi/2} c_\theta\, d\theta \\
&= \frac{2fs_\beta}{\pi V}.
\end{aligned} \tag{6.5}$$

β can now be considered as a continuous function of time due to this averaging operation. T is of course the current orbit period at the time of the averaging, and i can now be considered as the relative inclination between the current orbit and the initial orbit itself, and Equation (6.5) is written for convenience as

$$\dot{i} = \frac{2fs_\beta}{\pi V}. \tag{6.6}$$

6.2.3 The Hamiltonian and Euler-Lagrange Equations

The two state variables being i and V, with time t the independent variable and the thrust yaw angle β the control variable, the optimal transfer between two given circular orbits (i_o, V_o) and (i_f, V_f) with the thrust acceleration continuously on is formulated as a minimum time problem with the variational Hamiltonian

$$H = 1 + \lambda_i \left(\frac{2fs_\beta}{\pi V} \right) - \lambda_V fc_\beta. \tag{6.7}$$

The performance index being given by $J = \int_{t_0}^{t_f} L dt$ with $L = 1$. The Euler-Lagrange equations are given by

$$\dot{\lambda}_V = -\frac{\partial H}{\partial V} = \frac{2}{\pi} \frac{fs_\beta}{V^2} \lambda_i \tag{6.8}$$

$$\dot{\lambda}_i = -\frac{\partial H}{\partial i} = 0. \tag{6.9}$$

λ_i is therefore constant, and the optimal control law is obtained from the optimality condition

$$\frac{\partial H}{\partial \beta} = \lambda_i \frac{2}{\pi} \frac{f}{V} c_\beta + f\lambda_V s_\beta = 0 \tag{6.10}$$

such that

$$\tan \beta = -\frac{2}{\pi} \frac{\lambda_i}{V\lambda_V}. \tag{6.11}$$

6.2.4 The Analytic Form of the State and Control Variables

Because the Hamiltonian is not an explicit function of time, it is constant and equal to zero due to the transversatility condition $H_f = 0$ at the end time t_f. Therefore, λ_i and λ_V can be obtained from Equation (6.10) and the following expression of the Hamiltonian

$$H = 0 = 1 + \frac{2}{\pi} \frac{fs_\beta}{V} \lambda_i - fc_\beta \lambda_V \tag{6.12}$$

such that

$$\lambda_i = -\frac{\pi s_\beta V}{2f} = const \tag{6.13}$$

$$\lambda_V = \frac{c_\beta}{f}. \tag{6.14}$$

f being assumed constant, Equation (6.13) yields $Vs_\beta = V_0 s_{\beta_0}$, which allows for the integration of $\dot{V}$ in equation (6.4) as in [6], that is

$$V = \left(V_0^2 + f^2 t^2 - 2ftV_0 c_{\beta_0} \right)^{1/2}. \tag{6.15}$$

The control variable is also obtained explicitly as a function of time by observing that

$$V = \frac{V_0 s_{\beta_0}}{s_\beta} = V_0 s_{\beta_0} \frac{(1 + \tan^2 \beta)}{\tan \beta}$$

$$\frac{d}{dt}(\tan \beta) = \frac{2}{\pi} \lambda_i \frac{\dot{V}\lambda_V + V\dot{\lambda}_V}{V^2 \lambda_V^2} = \frac{d \tan \beta}{d\beta}\dot{\beta} = \frac{\dot{\beta}}{c_\beta^2}$$

which yields with the use of Equations (6.13), (6.14), as well as (6.4) and (6.8), $\dot{\beta} = \frac{f s_\beta}{V} = \frac{f s_\beta^2}{V_0 s_{\beta_0}}$ and after integration to

$$\tan \beta = \frac{V_0 s_{\beta_0}}{V_0 c_{\beta_0} - ft}. \tag{6.16}$$

Using this explicit control law, and Equation (6.15) for V, the inclination Equation (6.6) can also be integrated between the limits 0 and i yielding

$$\Delta i = \frac{2}{\pi}\left[\tan^{-1}\left(\frac{ft - V_0 c_{\beta_0}}{V_0 s_{\beta_0}}\right) + \frac{\pi}{2} - \beta_0\right]. \tag{6.17}$$

Finally, the multiplier λ_V is also given explicitly in terms of t after integrating $\dot{\lambda}_V$ in Equation (6.8) and using Equation (6.13) for λ_i. From $\dot{\lambda}_V = -\frac{V_0^2 s_{\beta_0}^2}{V^3}$ and with V given by Equation (6.15), the integration yields

$$\lambda_V = \frac{V_0 c_{\beta_0} - ft}{fV} \tag{6.18}$$

with $(\lambda_V)_0 = \frac{c_{\beta_0}}{f}$ and because $\lambda_V = \frac{c_\beta}{f}$ in Equation (6.14)

$$c_\beta = \frac{V_0 c_{\beta_0} - ft}{V} \tag{6.19}$$

The total ΔV needed for the transfer is given by [6]

$$\Delta V = V_0 c_{\beta_0} - \frac{V_0 s_{\beta_0}}{\tan\left[\frac{\pi}{2}\Delta i + \beta_0\right]} \tag{6.20}$$

$$\Delta V = V_0 c_{\beta_0} \pm \left(V^2 - V_0^2 s_{\beta_0}^2\right)^{1/2}$$

and the initial value β_0 needed in all the relevant equations that describe the transfer parameters as a function of time, obtained from Equation (6.17) or $\Delta i = \frac{2}{\pi}(\beta - \beta_0)$ such that

$$\tan \beta_0 = \frac{\sin\left(\frac{\pi}{2}\Delta i\right)}{\frac{V_0}{V} - \cos\left(\frac{\pi}{2}\Delta i\right)}. \tag{6.21}$$

The derivations leading to the expressions in (6.20) and (6.21) are not repeated here but they are shown in [6] with all the needed details. Given Δi_t, the total inclination change desired, as well as V_0 and V_f, the value of β_0 is obtained from Equation (6.21) with $V = V_f, \Delta i = \Delta i_t$, then the ΔV_t is evaluated from Equation (6.20) yielding $t_f = \frac{\Delta V_t}{f}$ the transfer time.

The running values of V, Δi, β, and λ_V are readily available from Equations (6.15)–(6.18) with Equation (6.16) written as

$$\beta = \tan^{-1}\left(\frac{V_0 s_{\beta_0}}{V_0 c_{\beta_0} - ft}\right). \tag{6.22}$$

6.3 The Optimal Transfer with Altitude Constraints

Because the semi-major axis can grow to very large values during a transfer that requires a large inclination change, it is important to constrain its value with a limiting a_{lim} such that $a \leq a_{lim}$ at all times during the transfer. This translates into $V \geq V_{lim}$ for the related variable V. This problem is therefore an optimal control problem, that is, minimum-time, with an inequality constraint on functions of the state variables [9] which is simply written here as

$$S(V, i, t) = V_{lim} - V \leq 0. \tag{6.23}$$

This constraint is of the type $S(x, t) \leq 0$ with x standing for the state variables. The equations of motion are still given by Equations (6.4) and (6.5), such that

$$\dot{S} = -\dot{V} = fc_\beta. \tag{6.24}$$

6.3.1 The Augmented Hamiltonian

This first time derivative is explicitly dependent on the control β, therefore S is a first-order state variable inequality constraint, and $\dot{S} = S^{(q)}$ now plays the same role as an equality constraint on functions of the control and state variables $C(\mathbf{x}, \mathbf{u}, t) = 0$, where $\mathbf{u}$ represents the control variables and q is the order of the inequality constraint with $S^{(q)}$ standing for the qth time derivative of S. Here $q = 1$, and the augmented Hamiltonian $H = L + \boldsymbol{\lambda}^T \mathbf{f}^* (\mathbf{x}, \mathbf{u}, t) + \mu S^{(q)}$ can be written as

$$H = 1 + \lambda_i \left(\frac{2fs_\beta}{\pi V}\right) - \lambda_V fc_\beta + \mu fc_\beta \tag{6.25}$$

with $\mathbf{f}^*$ standing for the constraint differential Equations (6.4) and (6.5), and $\mu(t)$ a time-dependent Lagrange multiplier used to adjoin the $\dot{S}$ constraint. The quantities μ and $\mathbf{u}$ are generally solved from the optimality condition $H_u = \frac{\partial H}{\partial \mathbf{u}} = \boldsymbol{\lambda}^T \mathbf{f}_u^* + L_u + \mu C_u = 0$ and $C(\mathbf{x}, \mathbf{u}, t) = 0$.

6.3.2 The Euler-Lagrange Equations and the q Tangency Constraints

The Euler-Lagrange equations are such that

$$\dot{\boldsymbol{\lambda}}^T = -H_x = -L_x - \boldsymbol{\lambda}^T \mathbf{f}_x^* - \mu S_x^{(q)}; \quad S^{(q)} = 0 \text{ on } S = 0 \tag{6.26}$$

$$\dot{\boldsymbol{\lambda}}^T = -H_x = -L_x - \boldsymbol{\lambda}^T \mathbf{f}_x^*; \qquad\qquad \mu = 0 \text{ for } S < 0. \tag{6.27}$$

Furthermore, $\mu(t) \geq 0$ on $S = 0$ for a minimizing solution because the minimizing control $\mathbf{u}(t)$ requires $\delta J = \boldsymbol{\lambda}^T(t_0)\delta \mathbf{x}(t_0) + \int_{t_0}^{t_f} \frac{\partial H}{\partial \mathbf{u}} \delta \mathbf{u} dt \geq 0$ for all admissible $\delta \mathbf{u}(t)$ in the fixed time subproblem, with the minimum-time problem itself being a particular problem of the set of fixed-time problems.

Finally, the trajectory entering the constraint boundary must satisfy the well-known q tangency constraints

$$\mathbf{N}(\mathbf{x},t) = \begin{vmatrix} S(\mathbf{x},t) \\ S^{(1)}(\mathbf{x},t) \\ \vdots \\ S^{(q-1)}(\mathbf{x},t) \end{vmatrix} = 0 \tag{6.28}$$

which are also satisfied at the exit from the boundary $S = 0$. The Lagrange multipliers are in general discontinuous at the junction points linking the constrained and unconstrained arcs [9], and choosing the entry point at time t_1 to satisfy the tangency constraints, the multipliers $\boldsymbol{\lambda}$ and Hamiltonian H are in general discontinuous at t_1 and continuous at the exit point at time t_2. It is shown in [12] that the Lagrange multipliers, also called influence functions, are not unique on a state variable inequality constraint boundary, meaning that the jump at t_1 is not unique and that the jump at t_2 is determined by the jump at t_1. However, a particular choice of the jump at t_1 will result in the corresponding multiplier being continuous at the exit corner of t_2.

6.3.3 The Jump Conditions at the Constraint Entry Point

The jump conditions at t_1 are then given by

$$\boldsymbol{\lambda}^T(t_1^-) = \boldsymbol{\lambda}^T(t_1^+) + \boldsymbol{\pi}^{*T} \left. \frac{\partial \mathbf{N}}{\partial \mathbf{x}} \right|_{t_1} \tag{6.29}$$

$$H(t_1^-) = H(t_1^+) - \boldsymbol{\pi}^{*T} \left. \frac{\partial \mathbf{N}}{\partial t} \right|_{t_1} \tag{6.30}$$

where $\boldsymbol{\pi}^*$ is a q vector of constant Lagrange multipliers introduced to enforce the satisfaction of the q constraints in Equation (6.28).

Furthermore, the control $\mathbf{u}$ can be discontinuous at both t_1 and t_2, resulting in corners. Therefore since $\mathbf{N}(V,i,t) = (V_{lim} - V) = S = 0$, Equations (6.29) and (6.30) yield

$$\lambda_i(t_1^-) = \lambda_i(t_1^+) + \pi_1^* \left. \frac{\partial (V_{lim} - V)}{\partial i} \right|_{t_1} \tag{6.31}$$

$$\lambda_V\left(t_1^-\right) = \lambda_V\left(t_1^+\right) + \pi_2^* \left.\frac{\partial\left(V_{lim} - V\right)}{\partial V}\right|_{t_1} \tag{6.32}$$

$$H\left(t_1^-\right) = H\left(t_1^+\right) - \pi_3^* \left.\frac{\partial\left(V_{lim} - V\right)}{\partial t}\right|_{t_1} \tag{6.33}$$

or in view of Equation (6.25)

$$\lambda_i\left(t_1^-\right) = \lambda_i\left(t_1^+\right) \tag{6.34}$$

$$\lambda_V\left(t_1^-\right) = \lambda_V\left(t_1^+\right) - \pi^* \tag{6.35}$$

$$H\left(t_1^-\right) = H\left(t_1^+\right). \tag{6.36}$$

Here π^* stands for π_2^*, introducing a jump or discontinuity in λ_V at t_1. However, both λ_i and H are continuous at t_1 because H is not an explicit function of time and $\mathbf{N}$ is not an explicit function of the relative inclination. The Euler-Lagrange Equations (6.26) and (6.27) yield with $L = 1$

$$\dot{\lambda}_i = -\lambda_i \frac{\partial\left(\frac{di}{dt}\right)}{\partial i} - \lambda_V \frac{\partial\left(\frac{dV}{dt}\right)}{\partial i} - \mu \frac{\partial\left(fc_\beta\right)}{\partial i}$$

$$\dot{\lambda}_i = 0 \tag{6.37}$$

$$\dot{\lambda}_V = -\lambda_i \frac{\partial\left(\frac{di}{dt}\right)}{\partial V} - \lambda_V \frac{\partial\left(\frac{dV}{dt}\right)}{\partial V} - \mu \frac{\partial\left(fc_\beta\right)}{\partial V}$$

$$\dot{\lambda}_V = \frac{2}{\pi}\frac{fs_\beta}{V^2}\lambda_i \tag{6.38}$$

on the constraint boundary $S = 0$; these equations for $\dot{\lambda}_i$ and $\dot{\lambda}_V$ being then identical to the differential Equations (6.9) and (6.8), respectively, which are valid on the unconstrained arcs. This is due to the fact that $\dot{S}$ is not a function of the states i and V. The optimal control β on the constraint boundary is obtained from the optimality condition

$$\frac{\partial H}{\partial \beta} = \lambda_i \frac{2}{\pi}\frac{f}{V}c_\beta + f\lambda_V s_\beta - \mu f s_\beta = 0 \tag{6.39}$$

yielding

$$\tan\beta = \frac{2}{\pi}\frac{\lambda_i}{V\left(\mu - \lambda_V\right)}. \tag{6.40}$$

The transversality condition $H_f = 0$ for the minimum-time transfer from t_0 to t_f with t_0 fixed and t_f free, holds true that is, $H = 0$ for all t because H is not an explicit function of time. Therefore

$$H = 1 + \lambda_i \frac{2fs_\beta}{\pi V} - \lambda_V\left(fc_\beta\right) + \mu\left(t\right)fc_\beta = 0. \tag{6.41}$$

Since λ_i is constant throughout, and because $V = V_{lim}$ and $\beta = \frac{\pi}{2}$ on the constrained arc, Equation (6.13) for λ_i on the unconstrained arcs yields the constant λ_i value as

$$\lambda_i = -\frac{\pi}{2} \frac{V_{lim}}{f}. \tag{6.42}$$

Because $V = V_{lim}$ on the constrained arc, $\dot{V} = -fc_\beta = 0$ leading to $\beta = \frac{\pi}{2}$ as stated above. In an identical manner with $\beta = \frac{\pi}{2}$, the Hamiltonian in Equation (6.41) yields

$$H = 1 + \lambda_i \frac{2}{\pi} \frac{f}{V_{lim}} = 0$$

or as in Equation (6.42)

$$\lambda_i = -\frac{\pi}{2} \frac{V_{lim}}{f}.$$

It is also true from $\dot{i} = \frac{2}{\pi} \frac{f}{V} s_\beta$, that $\dot{i} = \frac{2}{\pi} \frac{f}{V_{lim}}$ on a constrained arc because $\beta = \frac{\pi}{2}$. Therefore, $\dot{i} = $ constant and $\dot{V} = 0$, indicating a pure inclination change linear in time when $S = 0$. Equations (6.14) and (6.42) provide the initial value of the control β, that is, β_0 at time zero, since

$$\lambda_i = -\frac{\pi}{2} \frac{V s_\beta}{f} = -\frac{\pi}{2} \frac{V_{lim}}{f}$$

or

$$V s_\beta = V_{lim} \tag{6.43}$$

and at t_0, $V_0 s_{\beta_0} = V_{lim}$ yielding

$$s_{\beta_0} = \frac{V_{lim}}{V_0}. \tag{6.44}$$

The initial value of λ_V, or $(\lambda_V)_0$, is now easily obtained from the control law on the unconstrained arc starting at t_0, namely Equation (6.11), by substituting λ_i from Equation (6.42)

$$\tan \beta_0 = -\frac{2}{\pi} \left(-\frac{\pi}{2} \frac{V_{lim}}{f} \right) \frac{1}{V_0 (\lambda_V)_0} = \frac{V_{lim}}{f V_0 (\lambda_V)_0}$$

such that

$$(\lambda_V)_0 = \frac{f V_0}{V_{lim}} \tan \beta_0. \tag{6.45}$$

6.3.4 The Non-Existence of a Corner at the Constraint Entry and Exit Points

Because $V(t_1^-) = V(t_1^+) = V_{lim}$ at t_1, and due to the constancy of Vs_β on the unconstrained arcs

$$Vs_\beta = V_0 s_{\beta_0} = V\left(t_1^-\right) s_{\beta_1^-} = V_{lim}$$

yielding $s_{\beta_1^-} = 1$ or

$$\beta_1^- = \frac{\pi}{2}. \tag{6.46}$$

And due to $\beta_1^+ = \frac{\pi}{2}$ as the constrained arc starts, it follows that there does not exist a corner at t_1, such that there is no discontinuity in the control β at t_1. From Equation (6.16) at time t_1, and with $\beta_1^- = \frac{\pi}{2}$

$$\tan \beta_1^- = \frac{V_0 s_{\beta_0}}{V_0 c_{\beta_0} - ft_1} = \infty$$

or

$$V_0 c_{\beta_0} = ft_1$$

and

$$t_1 = \frac{V_0 c_{\beta_0}}{f} \tag{6.47}$$

The time t_1 at the first junction can also be obtained from the velocity expression in Equation (6.15)

$$V_0^2 + f^2 t_1^2 - 2ft_1 V_0 c_{\beta_0} = V_{lim}^2$$

which yields t_1 as

$$t_1 = \frac{V_0 c_{\beta_0} \pm \sqrt{V_0^2 c_{\beta_0}^2 - (V_0^2 - V_{lim}^2)}}{f}$$

However,

$$V_0^2 c_{\beta_0}^2 = V_0^2 \left(1 - s_{\beta_0}^2\right) = V_0^2 \left(1 - \frac{V_{lim}^2}{V_0^2}\right) = V_0^2 - V_{lim}^2$$

where s_{β_0} in Equation (6.44) has been used. Therefore, time $t_1 = \frac{V_0 c_{\beta_0}}{f}$ as in Equation (6.47) is known because V_0 is given and β_0 is known from Equation (6.44) for given V_{lim}. From $\lambda_V = \frac{c_\beta}{f}$ on the first unconstrained arc from t_0 to t_1, and in view of $\beta_1^- = \frac{\pi}{2}$, it follows that $\lambda_V(t_1^-) = 0$. From t_1 to t_2 on the constrained arc, $\beta = \frac{\pi}{2}$, $V = V_{lim}$ and $\lambda_i = -\frac{\pi}{2}\frac{V_{lim}}{f}$, resulting in $\dot{\lambda}_V = \frac{2}{\pi}\frac{f}{V^2}s_\beta\lambda_i = -\frac{1}{V_{lim}} = C$, a constant.

$\lambda_V(t)$ is therefore a linearly decreasing function in the interval (t_1, t_2), and due to the fact that λ_V does not experience any discontinuity at t_2 where $\lambda_V(t) = \frac{c_\beta}{f}$ is again valid on the unconstrained arc until t_f, and in view of $\beta_2^- = \beta_2^+ = \frac{\pi}{2}$, $\lambda_V(t_2^-) = \lambda_V(t_2^+) = \lambda_V(t_2) = 0$.

$\beta_2^+ = \beta(t_2^+) = \frac{\pi}{2}$ because $\lambda_i\left(t_2^-\right) = \lambda_i\left(t_2^+\right) = -\frac{\pi}{2}\frac{V_{lim}}{f} = -\frac{\pi}{2}s_{\beta_2^+}\frac{V_{lim}}{f}$ implying $\beta_2^+ = \frac{\pi}{2}$ and the absence of any corner at t_2 since the control β has no discontinuity there. There exist no jumps in either the λ's or H at t_2, that is

$$\lambda_i\left(t_2^-\right) = \lambda_i\left(t_2^+\right)$$
$$\lambda_V\left(t_2^-\right) = \lambda_V\left(t_2^+\right)$$
$$H\left(t_2^-\right) = H\left(t_2^+\right)$$

and due to the linear behavior of λ_V in the (t_1, t_2) interval

$$\lambda_V(t_2) - \lambda_V\left(t_1^+\right) = C(t_2 - t_1) = -\frac{1}{V_{lim}}(t_2 - t_1) = -\pi^* \qquad (6.48)$$

such that the jump in λ_V at t_1 is given by

$$\pi^* = \left(t_2 - \frac{V_0 c_{\beta_0}}{f}\right)\bigg/V_{lim} \qquad (6.49)$$

because $t_1 = \frac{V_0 c_{\beta_0}}{f}$ is already known. π^* is therefore a function of t_2 still to be determined.

In summary, $\lambda_V(t_1^-) = 0$, $\lambda_V(t_1^+) = \pi^*$, and $\lambda_V(t_2^-) = \lambda_V(t_2^+) = 0$. The control law on the constrained arc in the interval (t_1, t_2) given by Equation (6.40) yields the value of the multiplier $\mu(t) = \lambda_V(t)$ because $\beta = \frac{\pi}{2}$, $\tan\beta = \infty$ such that

$$\tan\beta = \frac{2}{\pi}\frac{\lambda_i}{V(\mu - \lambda_V)}$$

requires

$$\mu(t) = \lambda_V(t) = \lambda_V\left(t_1^+\right) + C(t - t_1)$$

or

$$\mu(t) = \lambda_V(t) = \pi^* - \frac{1}{V_{lim}}(t - t_1) \qquad (6.50)$$

where π^* is given in Equation (6.49) in terms of t_2. The remaining unknowns t_2 and t_f can now be resolved from the end conditions i_f and V_f, which are given quantities.

6.3.5 Evaluation of the Constraint Arc Exit and Final Transfer Times

Applying Equation (6.17) at time t_1 will yield the value of i_1 at the end of the first unconstrained arc such as with $t_1 = \frac{V_0 c_{\beta_0}}{f}$

$$\Delta i = i - i_0 = \frac{2}{\pi} \left[\tan^{-1} \left(\frac{ft - V_0 c_{\beta_0}}{V_0 s_{\beta_0}} \right) + \frac{\pi}{2} - \beta_0 \right]$$

$$i = i_0 + \frac{2}{\pi} \left[\tan^{-1} \left(\frac{ft - V_0 c_{\beta_0}}{V_0 s_{\beta_0}} \right) + \frac{\pi}{2} - \beta_0 \right] \tag{6.51}$$

$$i_1 = i_0 + \frac{2}{\pi} \left(\frac{\pi}{2} - \beta_0 \right) \tag{6.52}$$

with β_0 known from Equation (6.44), that is, $s_{\beta_0} = \frac{V_{lim}}{V_0}$. Equation (6.51) depicts the evolution of i as a function of time in the interval $(0, t_1)$, while Equation (6.52) is identical to the $\Delta i = \frac{2}{\pi} (\beta - \beta_0)$ Equation of Reference 6 with $\beta = \beta_1^- = \frac{\pi}{2}$ here.

On the constrained arc between t_1 and t_2, $\dot{i} = \frac{2f}{\pi V_{lim}} = const$ such that

$$i = i_1 + \frac{2f}{\pi V_{lim}} (t - t_1) \tag{6.53}$$

and at time t_2

$$i_2 = i_1 + \frac{2f}{\pi V_{lim}} (t_2 - t_1)$$

$$i_2 = i_0 + \frac{2}{\pi} \left(\frac{\pi}{2} - \beta_0 \right) + \frac{2f}{\pi V_{lim}} \left(t_2 - \frac{V_0 c_{\beta_0}}{f} \right). \tag{6.54}$$

Finally, in the interval (t_2, t_f), a further application of Equation (6.17) on this unconstrained arc yields

$$\Delta i = i - i_2 = \frac{2}{\pi} \left[\tan^{-1} \left(\frac{f(t - t_2) - V_2 c_{\beta_2^+}}{V_2 s_{\beta_2^+}} \right) + \frac{\pi}{2} - \beta_2^+ \right]$$

with $V_2 = V_{lim}$ and $\beta_2^+ = \beta_2^- = \beta_2 = \frac{\pi}{2}$, such that

$$\Delta i = i - i_2 = \frac{2}{\pi} \left\{ \tan^{-1} \left[\frac{f(t - t_2)}{V_{lim}} \right] \right\}^2. \tag{6.55}$$

Therefore, the evolution of i on this interval is given by

$$i = i_0 + \frac{2}{\pi} \left(\frac{\pi}{2} - \beta_0 \right) + \frac{2f}{\pi V_{lim}} \left(t_2 - \frac{V_0 c_{\beta_0}}{f} \right) + \frac{2}{\pi} \left\{ \tan^{-1} \left[\frac{f(t - t_2)}{V_{lim}} \right] \right\}. \tag{6.56}$$

Similarly, Equation (6.15) yields the variation of V on the unconstrained arcs in the intervals $(0, t_1)$ and (t_2, t_f) respectively as

$$V = \left(V_0^2 + f^2 t^2 - 2ftV_0 c_{\beta_0}\right)^{1/2}$$

$$V = \left[V_2^2 + f^2 (t - t_2)^2 - 2f (t - t_2) V_2 c_{\beta_2^+}\right]^{1/2} \tag{6.57}$$

once again with $V_2 = V_{lim}$ and $\beta_2^+ = \frac{\pi}{2}$. At time t_f, Equation (6.57) yields

$$V_f^2 = V_{lim}^2 + f^2 \left(t_f - t_2\right)^2$$

yielding

$$\left(t_f - t_2\right) = \frac{\sqrt{(V_f - V_{lim})(V_f + V_{lim})}}{f} \geq 0. \tag{6.58}$$

Equation (6.56) evaluated at t_f, along with the use of $(t_f - t_2)$ above, allows for the resolution of t_2

$$t_2 = \frac{\pi V_{lim}}{2f} \left\{ i_f - \frac{2}{\pi} \left(\frac{\pi}{2} - \beta_0\right) - \frac{2}{\pi} \tan^{-1} \left[\frac{\sqrt{(V_f - V_{lim})(V_f + V_{lim})}}{V_{lim}}\right] \right\} + \frac{V_0 c_{\beta_0}}{f}. \tag{6.59}$$

Thus t_2 is given in terms of known quantities, which in turn yields the remaining unknown t_f from Equation (6.58)

$$t_f = t_2 + \frac{\sqrt{(V_f - V_{lim})(V_f + V_{lim})}}{f} \geq 0. \tag{6.60}$$

Equation (6.49) solves for π^*, the jump in λ_V at t_1, such that $\mu(t) = \lambda_V(t) = \pi^* - \frac{1}{V_{lim}}\left(t - \frac{V_0 c_{\beta_0}}{f}\right)$ is now known on the constrained arc. This $\lambda_V(t)$ multiplier is already known on the first and third arcs from Equation (6.14)

$$\lambda_V(t) = \frac{c_\beta}{f}$$

with β given as a function of time t by Equation (6.16) for the two arcs respectively as

$$\tan \beta = \frac{V_0 s_{\beta_0}}{V_0 c_{\beta_0} - ft} = \frac{V_{lim}}{V_0 c_{\beta_0} - ft} \tag{6.61}$$

and

$$\tan \beta = -\frac{V_{lim}}{f(t - t_2)} \tag{6.62}$$

the latter expression derived from

$$\tan \beta = \frac{V_2 s_{\beta_2^+}}{V_2 c_{\beta_2^+} - f(t - t_2)}$$

with $V_2 = V_{lim}$ and $\beta_2^+ = \frac{\pi}{2}$.

A transfer from a low orbit to a higher orbit with a large enough inclination change or from (i_0, V_0) to (i_f, V_f) without any constraints on the intermediate V values, will show a certain V_{min} at $t = t^*$. From Equation (6.15) for V, $\dot{V} = 0$ yields $t^* = \frac{V_0 c_{\beta_0}}{f}$, which in turn yields

$$V_{min} = \left(V_0^2 - 2V_0 f t^* c_{\beta_0} + f^2 t^{*2} \right)^{1/2} = V_0 s_{\beta_0}. \tag{6.63}$$

First the unconstrained ΔV is obtained from Edelbaum's equation [1]

$$\Delta V = \left[V_0^2 + 2V_0 V_f \cos \left(\frac{\pi}{2} \Delta i_f \right) + V_f^2 \right]^{1/2} \tag{6.64}$$

or Equation (6.20) with $\Delta i_f = i_f - i_0$, leading to the transfer time $t_f = \frac{\Delta V}{f}$. We must have $t^* \leq t_f$ for a constrained transfer to be possible because otherwise $V_{min} < V_f$ is reached after t_f such that $V_{lim} < V_f$ or $a_{lim} > a_f$ is never reached. A constrained transfer requires

$$V_{min} \leq V_{lim} \leq V_f$$

or

$$a_{max} \geq a_{lim} \geq a_f$$

because if $V_{lim} > V_f$ or equivalently $a_{lim} < a_f$, then the final value a_f can never be reached and the transfer will not be achieved. On the other hand, if $V_{lim} < V_{min}$ or $a_{lim} > a_{max}$, then the constraining value is never reached and the transfer is achieved unconstrained. Therefore, $V_{min} \leq V_{lim} \leq V_f$ must be satisfied by an appropriate choice of V_{lim} for given V_{min} and V_f.

In the limiting case $t^* = t_f$, $V_{min} = V_f$ or $a_{max} = a_f$ and the only acceptable value of a_{lim} is a_{max} itself, that is, $V_{lim} = V_{min}$ is only reached at $t = t^* = t_f$ and the transfer will remain unconstrained. Finally when $t^* = t_f$, then $V_{min} = V_f$ or $a_{max} = a_f$ and if a_{lim} is chosen such that $a_{lim} \leq a_{max}$, that is, $V_{lim} \geq V_{min}$, then the transfer is effectively constrained.

As an example, let $a_0 = 6,563$ km or $V_0 = \sqrt{\frac{\mu}{a_0}} = 7.793241$ km/s, $i_0 = 0$, $a_f = 6,878$ km, or $V_f = \sqrt{\frac{\mu}{a_f}} = 7.612692$ km/s, $i_f = 69.607$ deg, and let $a_{lim} = 7,000$ km in one case and $a_{lim} = 10,000$ km in another case.

With the gravity constant $\mu = 398,601.3$ km^3/s^2 and $f = 3.5 \times 10^{-8}$ km/s^2 for the continuous constant acceleration value, these two cases are generated together

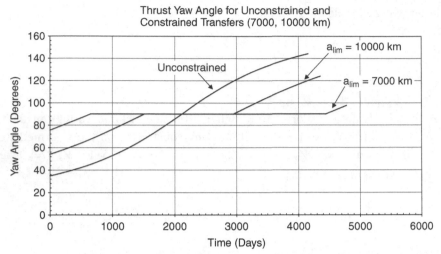

Figure 6.1. Thrust yaw angle for unconstrained and constrained transfers (7,000 km, 10,000 km).

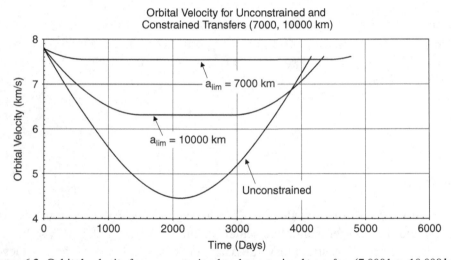

Figure 6.2. Orbital velocity for unconstrained and constrained transfers (7,000 km, 10,000 km).

with the unconstrained transfer where no limit is set for the intermediate semimajor axis values. Figure 6.1 shows the thrust yaw control angle β as a function of time for all three transfers. The constrained arcs where $\beta = 90$ deg get longer as the a_{lim} gets smaller. As expected, the unconstrained transfer results in the shortest transit time at 4,156.418 days, while the constrained transfer with $a_{lim} = 10,000$ km requires 4,355.975 days, and the one with $a_{lim} = 7,000$ km requiring an even longer trip time of 4,777.683 days. Figures 2 and 3 show the evolution of the orbital velocity and semi-major axis in time for the three transfers, with the velocity and semimajor axis reaching their minimum and maximum respectively on the unconstrained transfer at about $t = 2,110$ days with $V = 4.453$ km/s, $a = 20,094$ km.

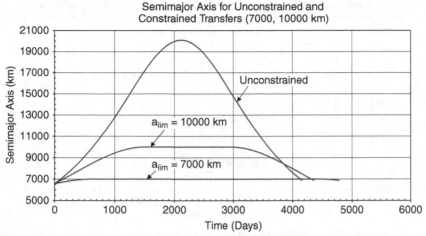

Figure 6.3. Semimajor axis for unconstrained and constrained transfers (7,000 km, 10,000 km).

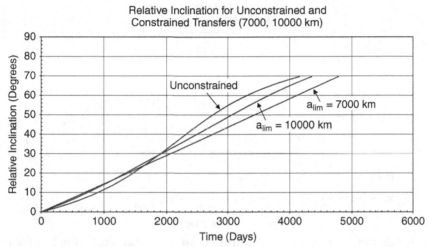

Figure 6.4. Relative inclination for unconstrained and constrained transfers (7,000 km, 10,000 km).

For the constrained transfers, the semi-major axis never exceeds the 7,000 km and 10,000 km limits as seen in Figure 6.3.

Figure 6.4 shows the relative inclination histories for the three transfers with linear buildup during the constrained arcs, while Figures 5 and 6 depict the evolutions of the multipliers λ_V and λ_i, with the jumps in λ_V clearly visible at time t_1, the junction between the end of the first unconstrained leg and the start of the constrained arc for the two constrained transfers. The λ_V history of the unconstrained transfer is, of course, continuous throughout. Figure 6.7 shows the variations of a versus i, once again showing that the a_{lim} constraints are effectively enforced during the transfers. Finally, Figure 6.8 shows the effective constancy of the Hamiltonian at the zero mark with the small departures from zero due to machine round off. Unlike the unconstrained arcs where λ_V is computed from the knowledge of the control angle

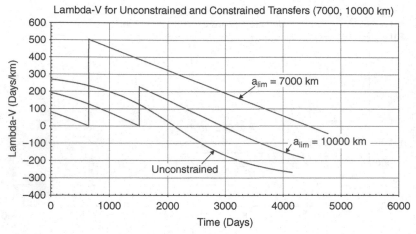

Figure 6.5. Lambda-V for unconstrained and constrained transfers (7,000 km, 10,000 km).

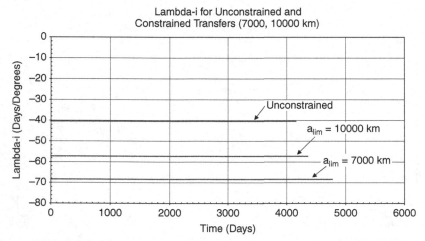

Figure 6.6. Lambda-i for unconstrained and constrained transfers (7,000 km, 10,000 km).

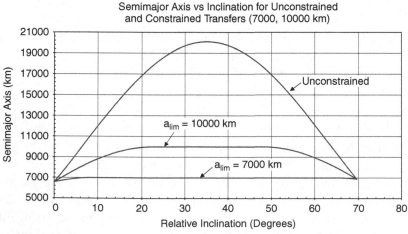

Figure 6.7. Semimajor axis versus inclination for unconstrained and constrained transfers (7,000 km, 10,000 km).

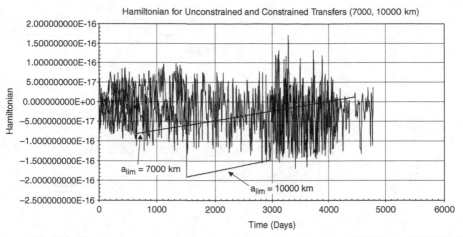

Figure 6.8. Hamiltonian for unconstrained and constrained transfers (7,000 km, 10,000 km).

β, that is, $\lambda_V = \frac{c_\beta}{f}$ with β itself computed by way of an inverse tangent function, the constrained arcs make effective use of the linear drop in λ_V as a function of time to calculate λ_V itself, and because λ_i is constant throughout, the linear segments appear in Figure 6.8 during the constrained arcs.

The total ΔV needed for each transfer is only dependent on the orbit parameters and the a_{lim} constraint, but not on the acceleration level.

ΔV = 12.569009 km/s for the unconstrained case, and 13.172470 km/s and 14.447714 km/s for the constrained a_{lim} = 10,000 km and 7,000 km, respectively. If f is set to 3.5×10^{-6} km/s^2, or 100 times larger than before, then the transfer times become 41.564, 43.559, and 47.776 days, respectively, or 100 times faster transfers with the same corresponding ΔV values shown above. The shapes of the various figures remain unchanged except that all the intermediate states V, i, as well as the control β, are reached 100 times faster than with the lower acceleration.

The use of this analytic theory in quickly generating whole sets or families of transfers is shown in Figure 6.9 where transfers from a_0 = 7,000 km, i_0 = 0 deg to a range of final orbits with a_f spanning from 7,000 km to 15,000 km and a common i_f = 70 deg are generated for various a_{lim} values. Note that the acceleration f only affects the transfer times but not the ΔV requirements that are only dependent on the initial and final orbit parameters. Each curve in Figure 6.9 cannot extend obviously beyond a_f = a_{lim}, and as the value of a_{lim} is relaxed to higher values, the ΔV requirements themselves relax further because the transfers start to approach the unconstrained solutions which, of course, are overall minimizing. Figure 6.9 is essentially generated in a single run in a matter of a few seconds or less, and it can be repeated for other values of say i_f to generate a further family of transfers and ΔV requirements.

6.4 The Split-Sequence Transfers

A pair of practical problems in optimal continuous-thrust transfer in general circular orbit are analyzed next within the context of analytic averaging for rapid computations leading to near-optimal solutions.

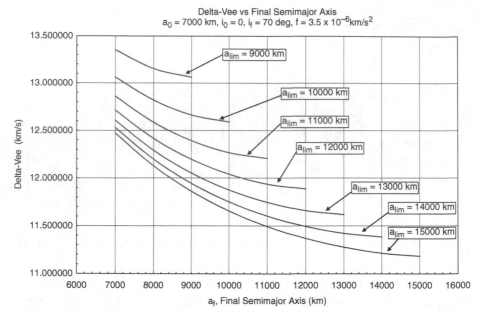

Figure 6.9. ΔV requirements for a family of constrained transfers with various limiting altitudes.

The first problem addresses the minimum-time transfer between inclined circular orbits by proposing an analytic solution based on a split-sequence strategy in which the equatorial inclination and node controls are done separately by optimally selecting the intermediate orbit size at the sequence switch point that results in the minimum-time transfer. The consideration of the equatorial inclination and node state variables besides the orbital velocity variable is needed to further account for the important J_2 perturbation that precesses the orbit plane during the transfer, unlike the thrust-only case in which it is sufficient to consider the relative inclination and velocity variables, thus reducing the dimensionality of the system equations. Further extensions of the split-sequence strategy with analytic J_2 effect are thus possible for equal computational ease. The second problem addresses the maximization of the equatorial inclination in fixed time by adopting a particular thrust-averaging scheme that controls only the inclination and velocity variables, leaving the node at the mercy of the J_2 precession, providing robust fast-converging codes that lead to efficient near-optimal solutions.

Example transfers for both sets of problems are solved showing near-optimal features as far as transfer time is concerned by directly comparing the solutions to "exact" purely numerical counterparts that rely on precision integration of the raw unaveraged system dynamics with continuously varying thrust vector orientation in three-dimensional space.

6.4.1 Analytic (V, i) Transfer

The analysis of the (V, i) transfer is identical to the one shown in Section 6.2. If the equator is ignored, then it is sufficient to consider the relative inclination i

instead of the equatorial inclination. However, when the J_2 perturbation must also be accounted for, then i must represent the equatorial inclination and the averaging of the thrust contribution must switch the sign of β at the antinodes of the equatorial line of nodes instead. The i expression in Equation (6.5) stays unchanged because the integration limits are still $-\frac{\pi}{2}, \frac{\pi}{2}$, that is

$$\dot{i} = \frac{1}{2\pi} \int_0^{2\pi} \left(\frac{di}{dt}\right) d\theta = \frac{2fs_\beta}{2\pi V} \int_{-\pi/2}^{\pi/2} c_\theta \, d\theta = \frac{2fs_\beta}{\pi V} \tag{6.65}$$

with θ the angular position measured from the equator and where $\frac{di}{dt} = \frac{f_h c_\alpha}{V}$ is the unaveraged rate with α the mean angular position equal to θ because $\alpha = \omega + M = \omega + \theta^* = \theta$ in circular orbit where θ^* is the true anomaly. Therefore the (V, i) analysis is still valid for orbit rotation around the equatorial line of nodes instead of the current line of nodes provided that the β sign switch is carried out at the equatorial antinodes. Note that the tilde over $\tilde{i}, \tilde{V}$ is deleted for ease of writing as $\dot{i}, \dot{V}$.

6.4.2 Analytic (V, Ω) transfer

If the transfer is from (V_0, Ω_0) to (V_f, Ω_f) with no change in i, then the averaging out of the angular position from the unaveraged Ω equation is carried out with the $0, \pi$ integration limits, that is, switching the β sign at the ascending and descending nodes, such that using $\dot{\Omega} = \frac{f_h s_\alpha}{V s_i}$

$$\dot{\Omega} = \frac{1}{2\pi} \times 2 \int_0^\pi \frac{f s_\beta s_\theta}{V s_i} d\theta = \frac{2 f s_\beta}{\pi V s_i}. \tag{6.66}$$

Removing the tilde once again, the differential equations of interest read as

$$\dot{V} = -f c_\beta \tag{6.67}$$

$$\dot{\Omega} = \frac{2 f s_\beta}{\pi V s_i}. \tag{6.68}$$

The Hamiltonian is now written as

$$H = 1 + \lambda_V \left(-f c_\beta\right) + \lambda_\Omega \frac{2 f s_\beta}{\pi V s_i} \tag{6.69}$$

with adjoint equations

$$\dot{\lambda}_V = -\partial H / \partial V = \frac{2 f s_\beta}{\pi V^2 s_i} \lambda_\Omega \tag{6.70}$$

$$\dot{\lambda}_\Omega = -\partial H / \partial \Omega = 0 \tag{6.71}$$

such that $\lambda_\Omega = const$ and the optimal control variable obtained from $\partial H / \partial \beta = 0$ leading to

$$\tan \beta = -\frac{2}{\pi V s_i} \frac{\lambda_\Omega}{\lambda_V}. \tag{6.72}$$

The transversality condition $H_f = H = 0$ (H not explicit function of time) together with $\partial H / \partial \beta = 0$ allows to solve for λ_V and λ_Ω this time around as

$$\lambda_V = \frac{c_\beta}{f} \tag{6.73}$$

$$\lambda_\Omega = -\frac{\pi V s_\beta s_i}{2f} \tag{6.74}$$

and because λ_Ω is constant, it follows as in the (V, i) analysis that $V s_\beta = const = V_0 s_{\beta_0}$ because in Equation (6.74) s_i remains constant. As in the (V, i) analysis, and due to the constancy of $V s_\beta$, the analytic form of V is obtained from

$$\dot{V} = -f c_\beta = -f(\pm)\sqrt{1 - s_\beta^2} = \mp \frac{f\sqrt{V^2 - V^2 s_\beta^2}}{V} = \mp \frac{f\sqrt{V^2 - V_0^2 s_{\beta_0}^2}}{V}$$

which yields

$$V = \left(V_0^2 + f^2 t^2 - 2ft V_0 c_{\beta_0}\right)^{1/2}. \tag{6.75}$$

Also in view of i constant, $\frac{d}{dt}(\tan \beta) = \frac{d}{d\beta}(\tan \beta)\dot{\beta} = \frac{\dot{\beta}}{c_\beta^2}$ and using $\tan \beta$ from Equation (6.72), we get $\dot{\beta} = \frac{f s_\beta}{V} = \frac{f s_\beta^2}{V_0 s_{\beta_0}}$ integrating to

$$\tan \beta = \frac{V_0 s_{\beta_0}}{V_0 c_{\beta_0} - ft}. \tag{6.76}$$

Note that in the (V, i) analysis $\tan \beta = -\frac{2}{\pi} \frac{\lambda_i}{V \lambda_V}$, and after forming $\frac{d}{dt} \tan \beta$ and using $\dot{\lambda}_i = 0, \lambda_V$ from Equation (6.8) and $\dot{V} = -f c_\beta$, the same expression for $\dot{\beta}$ is arrived at, as $\dot{\beta} = \frac{f s_\beta}{V}$ such that the running value of β is identical to Equation (6.76) above. From Equation (6.70), it follows that $\dot{\lambda}_V = -\frac{s_\beta^2}{V} = -\frac{V^2 s_\beta^2}{V^3} = -\frac{V_0^2 s_{\beta_0}^2}{V^3}$ and in view of $V = f(t)$ given in Equation (6.75), λ_V is integrated as

$$\lambda_V = \frac{V_0 c_{\beta_0} - ft}{fV} \tag{6.77}$$

such that from $\lambda_V = \frac{c_\beta}{f}$, the control is given as

$$c_\beta = \frac{V_0 c_{\beta_0} - ft}{V}. \tag{6.78}$$

Finally, the current value of Ω or rather the change $\Delta\Omega = \Omega - \Omega_0$ is obtained from $\dot{\Omega}$ in Equation (6.68) by observing that i is constant such that with $s_\beta = V_0 s_{\beta_0}/V$ and V from Equation (6.75), we get

$$\Delta\Omega = \frac{2}{\pi s_i}\left[\tan^{-1}\left(\frac{ft - V_0 c_{\beta_0}}{V_0 s_{\beta_0}}\right) + \frac{\pi}{2} - \beta_0\right]. \tag{6.79}$$

This expression is identical in its form to the Δi expression in Equation (6.17) except that we have the s_i factor in $\Delta \Omega$ such that $s_i \Delta \Omega$ is equivalent to Δi. In other words, the same amount of ΔV will either rotate the orbit by Δi or by $s_i \Delta \Omega$. Thus by analogy with the analysis in [6], the initial value of β_0 is now given by

$$\tan \beta_0 = \frac{\sin \left(\frac{\pi}{2} s_i \Delta \Omega_t \right)}{\frac{V_0}{V_f} - \cos \left(\frac{\pi}{2} s_i \Delta \Omega_t \right)} \tag{6.80}$$

instead of Equation (6.21) with the understanding that s_i is constant because only Ω changes are allowed in this (V, Ω) analysis. Here $\Delta \Omega_t$ is the total $\Delta \Omega$ change required by the transfer. Also by analogy with the (V, i) analysis in [6], we have for the (V, i) case

$$\Delta V = \left[V_0^2 - 2VV_0 \cos \left(\frac{\pi}{2} \Delta i \right) + V^2 \right]^{1/2}$$

and for the (V, Ω) case

$$\Delta V = \left[V_0^2 - 2VV_0 \cos \left(\frac{\pi}{2} s_i \Delta \Omega \right) + V^2 \right]^{1/2},$$

the first expression being Edelbaum's famous ΔV equation.

6.4.3 Analytic Split (V, i), (V, Ω) Sequence Transfer

It is now possible to solve a given transfer from (V_0, i_0, Ω_0) to (V_f, i_f, Ω_f) analytically by carrying out the (V, i) portion first meaning that Ω is held constant at Ω_0, its initial value and the transfer taking place from (V_0, i_0) to (V_1, i_1), which is reached at time t_1. Because i_1 is equal to i_f, we have $(\Delta i)_t = i_1 - i_0 = i_f - i_0$ such that from Equation (6.17)

$$(\Delta i)_t = \frac{2}{\pi} \left[\tan^{-1} \left(\frac{ft_1 - V_0 c_{\beta_0}}{V_0 s_{\beta_0}} \right) + \frac{\pi}{2} - \beta_0 \right] \tag{6.81}$$

which can be written as

$$\frac{-1}{\tan \left[\frac{\pi}{2} (\Delta i)_t + \beta_0 \right]} = \frac{ft_1 - V_0 c_{\beta_0}}{V_0 s_{\beta_0}}$$

yielding the time t_1 when the switch to the (V, Ω) leg starts while holding i_1 constant

$$t_1 = \frac{V_0 c_{\beta_0}}{f} - \frac{V_0 s_{\beta_0}/f}{\tan \left[\frac{\pi}{2} (\Delta i)_t + \beta_0 \right]}. \tag{6.82}$$

The velocity V_1, as well as $(\lambda_V)_1$ and β_1 at time t_1, are obtained from

$$V_1 = \left(V_0^2 + f^2 t_1^2 - 2ft_1 V_0 c_{\beta_0} \right)^{1/2} \tag{6.83}$$

$$(\lambda_V)_1 = \frac{V_0 c_{\beta_0} - f t_1}{f V_1} \tag{6.84}$$

$$c_{\beta_1} = \frac{V_0 c_{\beta_0} - f t_1}{V_1} \tag{6.85}$$

λ_i and λ_Ω stay constant during this first leg such that

$$(\lambda_\Omega)_1 = -\frac{V_0 s_{i_0} s_{\beta_0}}{2f} = (\lambda_\Omega)_0$$

$$(\lambda_i)_1 = -\frac{\pi V_0 s_{\beta_0}}{2f} = (\lambda_i)_0 .$$

The value of β_0 is still unknown and must be determined. During the second leg, the (V, Ω) theory is applied instead, with $i_1 = i_0 + (\Delta i)_t = i_f$ staying constant until time t_f and with Ω varying from $\Omega_1 = \Omega_0$ at t_1 until Ω_f is reached at t_f. The total change $(\Delta \Omega)_t = \Omega_f - \Omega_0 = \Omega_f - \Omega_1$ can be used in Equation (6.79) such that

$$(\Delta\Omega)_t = \frac{2}{\pi s_{i_1}} \left\{ \tan^{-1} \left[\frac{f(t_f - t_1) - V_1 c_{\beta_1}}{V_1 s_{\beta_1}} \right] + \frac{\pi}{2} - \beta_1 \right\}$$

leading to

$$f(t_f - t_1) = V_1 s_{\beta_1} \tan \left[(\Delta\Omega)_t \frac{\pi s_{i_1}}{2} - \left(\frac{\pi}{2} - \beta_1 \right) \right] + V_1 c_{\beta_1}. \tag{6.86}$$

Using this expression in the velocity equation

$$V_f = \left[V_1^2 + f^2 (t_f - t_1)^2 - 2f(t_f - t_1) V_1 c_{\beta_1} \right]^{1/2} \tag{6.87}$$

provides the expression from which the angle β_0 can be extracted numerically, that is

$$V_f^2 = V_1^2 + \left\{ V_1 s_{\beta_1} \tan \left[(\Delta\Omega)_t \frac{\pi s_{i_1}}{2} - \left(\frac{\pi}{2} - \beta_1 \right) \right] + V_1 c_{\beta_1} \right\}^2$$

$$- 2V_1 c_{\beta_1} \left\{ V_1 s_{\beta_1} \tan \left[(\Delta\Omega)_t \frac{\pi s_{i_1}}{2} - \left(\frac{\pi}{2} - \beta_1 \right) \right] + V_1 c_{\beta_1} \right\}. \tag{6.88}$$

β_1, t_1 and V_1 are still functions of β_0, so that once β_0 is obtained from Equation (6.88), these variables are then determined and the final remaining unknown t_f is solved for from Equation (6.86).

6.4.4 Analytic split (V, Ω), (V, i) sequence transfer

In this strategy, i is kept fixed at i_0 during the first leg while Ω is changed from Ω_0 to $\Omega_1 = \Omega_f$ at t_1. Then the (V, i) transfer is applied while holding Ω constant at $\Omega_1 = \Omega_f$ while i is changed from $i_1 = i_0$ to i_f at time t_f. As in the previous section, V is continuously changing during both legs.

From Equation (6.79), the total required $(\Delta\Omega)_t = \Omega_1 - \Omega_0$ allows us to solve for t_1 as a function of the unknown β_0, that is

$$(\Delta\Omega)_t = \frac{2}{\pi s_{i_0}} \left[\tan^{-1}\left(\frac{ft_1 - V_0 c_{\beta_0}}{V_0 s_{\beta_0}}\right) + \frac{\pi}{2} - \beta_0 \right]$$

yielding

$$t_1 = \frac{V_0 c_{\beta_0}}{f} - \frac{V_0 s_{\beta_0}/f}{\tan\left[\frac{\pi}{2} s_{i_0} (\Delta\Omega)_t + \beta_0\right]} \tag{6.89}$$

with once again

$$\lambda_{V_1} = \left(V_0^2 + f^2 t_1^2 - 2ft_1 V_0 c_{\beta_0}\right)^{1/2} \tag{6.90}$$

$$V_1 = \frac{V_0 c_{\beta_0} - ft_1}{fV_1} \tag{6.91}$$

$$c_{\beta_1} = \frac{V_0 c_{\beta_0} - ft_1}{V_1}. \tag{6.92}$$

During the second leg of this (V, Ω), (V, i) sequence, the total $(\Delta i)_t$ change, that is, $i_f - i_1 = i_f - i_0$ is used in Equation (6.81)

$$(\Delta i)_t = \frac{2}{\pi} \left\{ \tan^{-1}\left[\frac{f(t_f - t_1) - V_1 c_{\beta_1}}{V_1 s_{\beta_1}}\right] + \frac{\pi}{2} - \beta_1 \right\}$$

yielding

$$f(t_f - t_1) = V_1 s_{\beta_1} \tan\left[(\Delta i)_t \frac{\pi}{2} - \left(\frac{\pi}{2} - \beta_1\right)\right] + V_1 c_{\beta_1} \tag{6.93}$$

which is now used in the following velocity expression

$$V_f = \left[V_1^2 + f^2 (t_f - t_1)^2 - 2f (t_f - t_1) V_1 c_{\beta_1}\right]^{1/2} \tag{6.94}$$

such that the initial value of the thrust angle β_0 is extracted numerically from

$$V_f^2 = V_1^2 + \left\{V_1 s_{\beta_1} \tan\left[(\Delta i)_t \frac{\pi}{2} - \left(\frac{\pi}{2} - \beta_1\right)\right] + V_1 c_{\beta_1}\right\}^2$$
$$- 2V_1 c_{\beta_1} \left\{V_1 s_{\beta_1} \tan\left[(\Delta i)_t \frac{\pi}{2} - \left(\frac{\pi}{2} - \beta_1\right)\right] + V_1 c_{\beta_1}\right\}. \tag{6.95}$$

Once β_0 is thus obtained, the final unknown t_f will be given by Equation (6.93).

6.4.5 Comparison with Precision Integration

The unaveraged system of differential equations valid in circular orbit can be written as follows in terms of the classical elements V, i, Ω, and α where $\alpha = \omega + M$ is the mean angular position equal also to $\theta = \omega + \theta^*$, the angular position defined in terms of the true anomaly θ^* because $M = \theta^*$ in the circular orbit case, as in [8]

$$\dot{V} = -fc_\beta \tag{6.96}$$

$$\dot{i} = fs_\beta c_\alpha / V \tag{6.97}$$

$$\dot{\Omega} = fs_\beta s_\alpha / (Vs_i) \tag{6.98}$$

$$\dot{\alpha} = \frac{V^3}{\mu} - fs_\beta s_\alpha / (V \tan i). \tag{6.99}$$

For minimum-time transfers, the following Hamiltonian leads to the adjoint differential equations

$$H = 1 + \lambda_V \dot{V} + \lambda_i \dot{i} + \lambda_\Omega \dot{\Omega} + \lambda_\alpha \dot{\alpha}$$

$$= 1 + \lambda_V \left(-fc_\beta\right) + \lambda_i \left(\frac{fs_\beta c_\alpha}{V}\right) + \lambda_\Omega \left(\frac{fs_\beta s_\alpha}{Vs_i}\right) + \lambda_\alpha \left(\frac{V^3}{\mu} - \frac{fs_\beta s_\alpha}{V \tan i}\right) \tag{6.100}$$

$$\dot{\lambda}_V = -\frac{\partial H}{\partial V} = \lambda_i \frac{fs_\beta c_\alpha}{V^2} + \lambda_\Omega \frac{fs_\beta s_\alpha}{V^2 s_i} - \lambda_\alpha \left(\frac{3V^2}{\mu}\right) - \lambda_\alpha \left(\frac{fs_\beta s_\alpha}{V^2 \tan i}\right) \tag{6.101}$$

$$\dot{\lambda}_i = -\frac{\partial H}{\partial i} = \lambda_\Omega \frac{fs_\beta s_\alpha c_i}{Vs_i^2} - \lambda_\alpha \frac{fs_\beta s_\alpha}{Vs_i^2} \tag{6.102}$$

$$\dot{\lambda}_\Omega = -\frac{\partial H}{\partial \Omega} = 0 \tag{6.103}$$

$$\dot{\lambda}_\alpha = -\frac{\partial H}{\partial \alpha} = \lambda_i \frac{fs_\beta s_\alpha}{V} - \lambda_\Omega \frac{fs_\beta c_\alpha}{Vs_i} + \lambda_\alpha \frac{fs_\beta c_\alpha}{V \tan i}. \tag{6.104}$$

The optimality condition $\partial H / \partial \beta = 0$ leads to the optimal control

$$\tan \beta = \frac{-\lambda_i \dfrac{c_\alpha}{V} - \lambda_\Omega \dfrac{s_\alpha}{Vs_i} + \lambda_\alpha \dfrac{s_\alpha}{V \tan i}}{\lambda_V} = \frac{s_\beta}{c_\beta}. \tag{6.105}$$

The continuously varying control angle β is used to drive the combined system of state and adjoint Equations (6.96–6.99) and (6.101–6.105) by starting from guessed $(\lambda_V)_0$, $(\lambda_i)_0$, $(\lambda_\Omega)_0$, $(\alpha)_0$, as well as t_f in order to integrate from $V_0, i_0, \Omega_0, (\lambda_\alpha)_0 = 0$ and satisfy the end conditions $V_f, i_f, \Omega_f, (\lambda_\alpha)_f = 0$ and $H_f = 0$ by slowly adjusting the initial guesses. Both initial and final orbital positions are optimized in this search scheme in order to arrive at the overall minimum solution. The example transfer parameters shown in Table 6.1 use an acceleration $f = 3.5 \times 10^{-6}$ km/s^2, with initial and final semi-major axis values of $a_0 = 6{,}563.14$ km and $a_f = 6{,}878$ km. The solutions in Tables 6.1 and 6.2 also involve the Edelbaum and exact transfers with $(\lambda_\alpha)_f = 0.000000061$ s/deg for the exact transfer.

Table 6.1. *Initial and final achieved parameters*

	V_0 (km/s) V_f (km/s)	i_0 (deg) i_f (deg)	Ω_0 (deg) Ω_f (deg)	α_0 (deg) α_f (deg)	$(\lambda_\alpha)_0$ (s/rad) $(\lambda_\alpha)_f$ (s/rad)	t_1 (s) t_f (s)
$(V, i), (V, \Omega)$	7.7931587 7.6126921	10.0 5.000000000	20.0 10.00000000000	NA NA	NA NA	3.054367786×10^5 3.576032781×10^5
$(V, \Omega), (V, i)$	7.7931587 7.6126921	10.0 5.000000000	20.0 10.00000000000	NA NA	NA NA	1.075218944×10^5 4.091171803×10^5
(V, i^*)	7.7931587 7.6126921	$i_0^* = 5.148939835$ $i_f^* = 1.590277 \times 10^{-15}$	20.0 10.00000000000	NA NA	NA NA	NA 3.146467816×10^5
Exact	7.7931587 7.6126921	10.0 4.999983319	20.0 9.999944152	345.4613991 46.85238677	0.0 3.49414×10^{-6}	NA 3.12638781×10^5

Table 6.2. Solutions for the two split-sequence, Edelbaum, and exact transfers

	$(\lambda_V)_0$ (days/km/s)	$(\lambda_i)_0$ (days/deg)	$(\lambda_\Omega)_0$ (days/deg)	β_0 (deg)	H_f	ΔV (km/s)
$(V, i), (V, \Omega)$	0.736837970	−0.688765006	−0.060029826	77.125292270	0.000000000	1.251611474
$(V, \Omega), (V, i)$	0.715748836	−0.571219156	−0.119778934	77.499833680	0.000000000	1.431910131
(V, i^*)	0.769281154	(λ_i^*) −0.687143888	NA	76.54800297	-1.78258×10^{-17}	1.1012637
Exact	0.188059951	0.485445519	0.014422179	−86.739798650	−0.9999785801	1.0942357

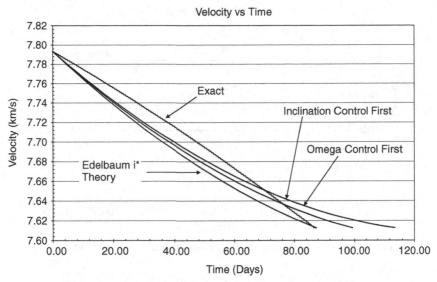

Figure 6.10. Velocity variations for split-sequence, Edelbaum, and exact transfers.

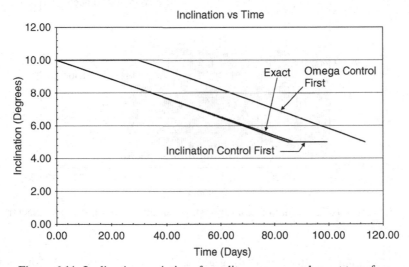

Figure 6.11. Inclination variations for split-sequence and exact transfers.

The velocity variations of the four solutions are shown in Figure 6.10 as a function of time. In this example, the initial inclination being larger than the final value, it is more economical to carry out the (V, i) leg first followed by the (V, Ω) leg instead of the other way around because it is much easier to adjust Ω at the lower inclination. The penalty in using the (V, i), (V, Ω) sequence with respect to the exact solution is about 14% in ΔV, but a much large 30% for the (V, Ω), (V, i) sequence. Figures 6.11 and 6.12 depict the evolutions of the inclination and node variables for the split-sequence transfers as well as the exact solution. Most of the benefit of allowing the orbit size, that is, velocity to vary optimally during each leg of the (V, i), (V, Ω) sequence will be lost if the orbit is raised initially from $a_0 = 6{,}563.14$ km to $a_f = 6{,}878$ km

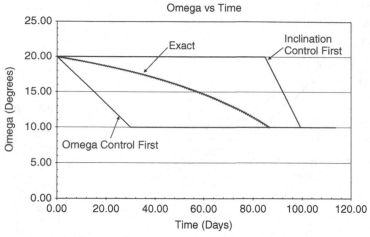

Figure 6.12. Node variations for split-sequence and exact transfers.

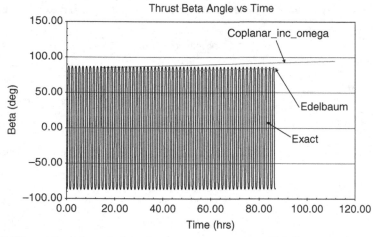

Figure 6.13. Thrust β angle variations for three-step sequence, Edelbaum, and exact transfers.

in a purely coplanar transfer thrusting along the velocity vector, before carrying out the (V,i), (V,Ω) legs. The ΔV needed for this initial coplanar leg is equal to $|V_0 - V_f| = 180.466532$ m/s requiring of time of 51,561.866 s or 14.322740 hrs such that the total transfer time is now $t_f = 4.013070866 \times 10^5$ s or a $\Delta V_{tot} = 1.404574803$ km/s using $f = 3.5 \times 10^{-6}$ km/s^2. This ΔV_{tot} is now about 28% or double the 14% increase over the (V,i), (V,Ω) sequence without the coplanar initial phase, because in the coplanar, (V,i), (V,Ω) transfer, the last two legs (V,i), (V,Ω) are carried out essentially at the constant V_f value such that these legs are almost pure inclination and pure omega changes.

The thrust β angle histories are shown in Figure 6.13 for the exact, Edelbaum and coplanar, (V,i), (V,Ω) cases with the last two cases using analytic averaging that switches the piecewise-constant β angle sign twice during each revolution instead of continuously varying β as in the exact solution.

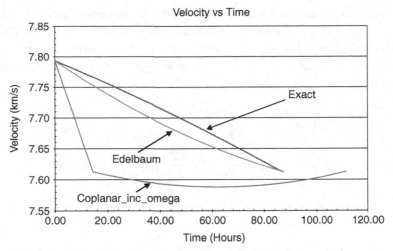

Figure 6.14. Velocity variations for three-step sequence, Edelbaum, and exact transfers.

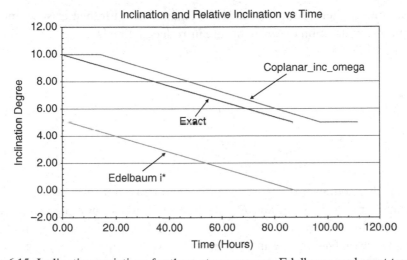

Figure 6.15. Inclination variations for three-step sequence, Edelbaum, and exact transfers.

Figures 6.14 and 6.15 show the variations of the velocity and inclination variables in time for the three cases under discussion with the understanding that in the Edelbaum case, i^* is the relative inclination and not the equatorial one.

6.4.6 Maximization of the Equatorial Inclination in Circular Orbit Using Analytic Averaging under J_2 Influence and Minimum-Time Transfer for Fixed Boundary Conditions

In this problem, we go back to the original averaged equations for V and equatorial i

$$di/dt = \frac{2fs_\beta}{2V} \qquad (6.106)$$

$$dV/dt = -fc_\beta \qquad (6.107)$$

with the averaging done by switching the sign of β at the orbital antinodes such that the thrust acceleration will only induce changes in i and V but not in Ω. This is in a way equivalent to applying impulsive ΔVs at the nodes to carry out inclination and semi major axis changes. The $d\Omega/dt$ equation will give a zero net contribution after averaging in this manner because

$$d\Omega/dt = \frac{f_h s_\alpha}{V s_i} = \frac{f_h s_\theta}{V s_i} \tag{6.108}$$

$$\dot{\Omega} = \frac{1}{2\pi} \int_0^{2\pi} \frac{f_h s_\theta}{V s_i} d\theta = \frac{2}{2\pi} \int_{-\pi/2}^{\pi/2} \frac{f s_\beta}{V s_i} s_\theta d\theta = 0. \tag{6.109}$$

Note that the assumption $\alpha = \theta$ is an excellent one because from the original equation

$$\frac{d\alpha}{dt} = n + \frac{2f_n}{V} - \frac{f_h s_\alpha}{V \tan i} \tag{6.110}$$

$f_n = 0$ and the $f_h s_\alpha$ term contribution is also equal to zero for the thrust components. When the J_2 acceleration components are introduced, that is

$$(f_n)_{J_2} = \frac{3\mu J_2 R^2}{2a^4} \left(1 - 3s_i^2 s_\theta^2\right) \tag{6.111}$$

$$(f_t)_{J_2} = -\frac{3\mu J_2 R^2}{a^4} s_i^2 s_\theta c_\theta \tag{6.112}$$

$$(f_h)_{J_2} = -\frac{3\mu J_2 R^2}{a^4} s_i c_i s_\theta \tag{6.113}$$

with $(f_n)_{J_2}$ directed inward, $(f_t)_{J_2}$ tangential, and $(f_h)_{J_2}$ out-of-plane, and where μ is the Earth gravity constant and R the equatorial radius, then the three terms in Equation (6.110) will be of order $10^{-3}, 10^{-6}$, and 10^{-6} respectively in the more severe LEO case such that $d\alpha/dt \simeq n$ and $\alpha \simeq nt = \theta$. If we now apply the $(f_h)_{J_2}$ component to drive the $\dot{\Omega}$ equation, then

$$\left(\dot{\Omega}\right)_{J_2} = (f_h)_{J_2} s_\theta / V s_i$$

$$\left(\dot{\Omega}\right)_{J_2} = \frac{1}{2\pi} \int_0^{2\pi} \left(\dot{\Omega}\right)_{J_2} d\theta$$

$$\left(\dot{\Omega}\right)_{J_2} = -\frac{3\mu^{1/2} J_2 R^2 c_i}{2a^{7/2}} \tag{6.114}$$

which can also be written as

$$\dot{\Omega} = -\frac{3J_2 R^2 c_i V^7}{2\mu^3} = k_1 c_i V^7 \tag{6.115}$$

because from $V = \mu^{1/2} a^{-1/2}$ we have $a^{7/2} = \mu^{7/2}/V^7$. Thus the thrust-and-J_2-averaged system of equations reduce to Equations (6.106), (6.107), and (6.115) because $(f_h)_{J_2}$ will not contribute to

$$\left(\dot{i}\right)_{J_2} = \frac{1}{2\pi} \int_0^{2\pi} \frac{(f_h)_{J_2} c_\alpha}{V} d\alpha = \frac{1}{2\pi} \int_0^{2\pi} \frac{-3\mu J_2 R^2}{a^4} s_i c_i s_\theta c_\theta d\theta = 0$$

and

$$\left(\dot{V}\right)_{J_2} = -\frac{1}{2\pi} \int_0^{2\pi} (f_t)_{J_2} d\alpha = -\frac{1}{2\pi} \int_0^{2\pi} \frac{-3\mu J_2 R^2}{a^4} s_i^2 s_\theta c_\theta d\theta = 0.$$

Removing the tilde in Equation (6.115) for ease of writing, the Hamiltonian H below leads to three Lagrange multipliers for minimum-time transfer with the present averaging scheme, that is, β sign switching at equatorial antinodes

$$H = 1 + \lambda_i \left(\frac{2fs_\beta}{\pi V}\right) + \lambda_V \left(-fc_\beta\right) + \lambda_\Omega \left(-\frac{3J_2 R^2 c_i V^7}{2\mu^3}\right) \tag{6.116}$$

$$\dot{\lambda}_i = -\partial H/\partial i = -\frac{3J_2 R^2 s_i V^7}{2\mu^3}\lambda_\Omega \tag{6.117}$$

$$\dot{\lambda}_V = -\partial H/\partial V = \frac{2fs_\beta}{\pi V^2}\lambda_i \tag{6.118}$$

$$\dot{\lambda}_\Omega = -\partial H/\partial \Omega = 0. \tag{6.119}$$

Thus $\lambda_\Omega = const = K_1$. The optimality condition $\partial H/\partial \beta = 0$ provides the optimal control law

$$\frac{\partial H}{\partial \beta} = 0 = \lambda_i \frac{2fc_\beta}{\pi V} + \lambda_V fs_\beta \tag{6.120}$$

$$\tan \beta = -\frac{2}{\pi} \frac{\lambda_i}{V\lambda_V} \tag{6.121}$$

H being explicitly independent of t, the transverality condition $H_f = 0$ translates to $H = 0$ throughout the transfer

$$H = 0 = 1 + \lambda_i \left(\frac{2fs_\beta}{\pi V}\right) + \lambda_V \left(-fc_\beta\right) + \lambda_\Omega k_1 c_i V^7 \tag{6.122}$$

and letting $K = K_1 k_1$, Equations (6.70) and (6.72) yield the λ_i and λ_V expressions in terms of the constant λ_Ω such that

$$\lambda_i = -Kc_i V^8 \frac{\pi s_\beta}{2f} \tag{6.123}$$

$$\lambda_V = Kc_i V^7 c_\beta/f. \tag{6.124}$$

In summary, the initial guesses $(\lambda_i)_0$, $(\lambda_V)_0$, $(\lambda_\Omega)_0 = K_1$, and t_f are used to integrate Equations (6.106), (6.107), (6.115) using the control law in Equation (6.121), as well as λ_i and λ_V from Equations (6.123) and (6.124), and they are slowly adjusted until i_f, V_f, Ω_f, and $H_f = 0$ are satisfied.

The maximization of i_f in fixed time, which is the same as the minimization of $-i_f$ using the same averaging scheme described in this section, is carried out by considering the following set of state and adjoint equations

$$\dot{i} = \frac{2fs_\beta}{\pi V} \tag{6.125}$$

$$\dot{V} = -fc_\beta \tag{6.126}$$

$$\dot{\Omega} = k_1 c_i V^7 \tag{6.127}$$

$$\dot{\lambda}_i = k_1 s_i V^7 \lambda_\Omega \tag{6.128}$$

$$\dot{\lambda}_V = \frac{2fs_\beta}{\pi V^2} \lambda_i \tag{6.129}$$

$$\dot{\lambda}_\Omega = 0. \tag{6.130}$$

Here, $i_0, V_0, \Omega_0, V_f, \Omega_f$, and t_f are fixed, and the Hamiltonian reads as

$$H = \lambda_i \frac{2fs_\beta}{\pi V} + \lambda_V \left(-fc_\beta\right) + \lambda_\Omega \left(k_1 c_i V^7\right). \tag{6.131}$$

Letting $\psi_1 = V\left(t_f\right) - V_f = 0$, $\psi_2 = \Omega\left(t_f\right) - \Omega_f = 0$, and $\Phi = -i\left(t_f\right) + \upsilon_1 \left[V\left(t_f\right) - V_f\right] + \upsilon_2 \left[\Omega\left(t_f\right) - \Omega_f\right]$ or $\Phi = \varphi\left(t_f\right) + \upsilon_1 \psi_1 + \upsilon_2 \psi_2$, the boundary conditions on λ_i, λ_V, and λ_Ω at time t_f are given by

$$\lambda_i\left(t_f\right) = \left(\frac{\partial\varphi}{\partial i} + \upsilon^T \frac{\partial\psi}{\partial i}\right)_{t_f} = -1 + \upsilon_1 \frac{\partial\left[V\left(t_f\right) - V_f\right]}{\partial i} + \upsilon_2 \frac{\partial\left[\Omega\left(t_f\right) - \Omega_f\right]}{\partial i} = -1 \tag{6.132}$$

$$\lambda_V\left(t_f\right) = \left(\frac{\partial\varphi}{\partial V} + \upsilon^T \frac{\partial\psi}{\partial V}\right)_{t_f} = \upsilon_1 \tag{6.133}$$

$$\lambda_\Omega\left(t_f\right) = \left(\frac{\partial\varphi}{\partial V} + \upsilon^T \frac{\partial\psi}{\partial\Omega}\right)_{t_f} = \upsilon_2. \tag{6.134}$$

Thus, starting from guesses for $(\lambda_i)_0$, $(\lambda_V)_0$, $(\lambda_\Omega)_0$, υ_1, υ_2, the six differential Equations (6.75)–(6.80) are integrated from i_0, V_0, Ω_0, until the fixed time t_f and the guesses slowly adjusted until the end conditions $\lambda_i\left(t_f\right) = -1$, $\lambda_V\left(t_f\right) = \upsilon_1$, $\lambda_\Omega\left(t_f\right) = \upsilon_2$, $V\left(t_f\right) = V_f$, $\Omega\left(t_f\right) = \Omega_f$, are closely matched to within a small tolerance indicating convergence. Of course the optimal control law in Equation (6.71) is used to steer the thrust vector at each instant of time. The fixed-time minimization of i_f requires $(\lambda_i)_{t_f} = 1$ instead of -1. Tables 6.3 and 6.4 show the achieved final parameters as well as the converged initial guesses for the 10-day-fixed, 20-day-fixed, and the minimum-time transfers, the latter case being for a fixed required $i_f = 70$ deg.

In all cases, both λ_Ω and H remained constant at the values shown, and for the 10-day and 20-day-fixed cases, $\upsilon_2 = (\lambda_\Omega)_f = (\lambda_\Omega)_0$ due to the constancy of λ_Ω. The Hamiltonian being homogeneous in the multipliers, acts like a scaling factor for the multipliers and it stays perfectly constant throughout the transfer because it is

Table 6.3. *Initial and final achieved parameters*

	V_0 (km/s) V_f (km/s)	i_0 (deg) i_f (deg)	Ω_0 (deg) Ω_f (deg)	β_0 (deg) β_f (deg)	t_f (days)	H
10-day fixed	7.5460614 7.5460614	60.0 64.519789	60.0 25.000000	274.32181 97.077658	10.0	-1.247011×10^{-6}
20-day fixed	7.5460614 7.5460614	60.0 82.888829	60.0 25.000000	280.32641 106.94783	20.0	$-4.4430404 \times 10^{-7}$
Min-time	7.5460614 7.5460614	60.0 70.000000	60.0 25.000000	274.53944 99.680491	11.690898	$3.8367853 \times 10^{-11}$

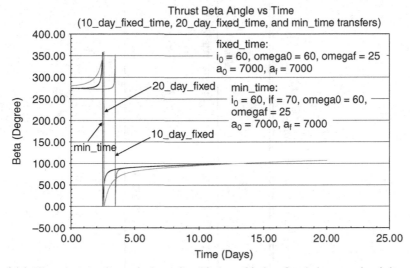

Figure 6.16. Thrust β angle variations for 10-day, 20-day fixed-time, and minimum-time transfers.

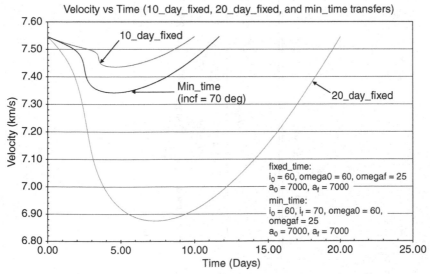

Figure 6.17. Velocity variations for 10-day, 20-day fixed-time, and minimum-time transfers.

Table 6.4. *Optimal solutions for 10-day-fixed, 20-day-fixed, and minimum-time transfer*

	$\begin{array}{c}(\lambda_i)_0\\(\lambda_i)_f\\(\text{s/rad})\end{array}$	$\begin{array}{c}(\lambda_V)_0\\(\text{s/km/s})\end{array}$	$\begin{array}{c}(\lambda_\Omega)_0\\(\text{s/rad})\end{array}$	$\begin{array}{c}v_1 = (\lambda_V)_f\\(\text{s/km/s})\end{array}$	$\begin{array}{c}v_2 = (\lambda_\Omega)_f\\(\text{s/rad})\end{array}$
10-day fixed	$5.424784656 \times 10^{-1}$ $-9.999999999 \times 10^{-1}$	$3.458679474 \times 10^{-3}$	1.518535541	$-1.047474946 \times 10^{-2}$	1.518535541
20-day fixed	$1.892680862 \times 10^{-1}$ $-9.999999974 \times 10^{-1}$	$2.909391473 \times 10^{-3}$	$7.537769312 \times 10^{-1}$	$-2.570884399 \times 10^{-2}$	$7.537769312 \times 10^{-1}$
Min-time	$3.353820437 \times 10^{5}$	$2.246413070 \times 10^{-3}$	$1.303257038 \times 10^{6}$	NA	NA

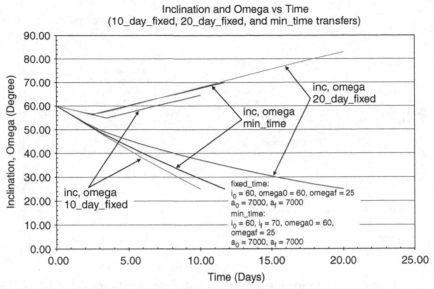

Figure 6.18. Inclination and node variations for 10-day, 20-day fixed-time, and minimum-time transfers.

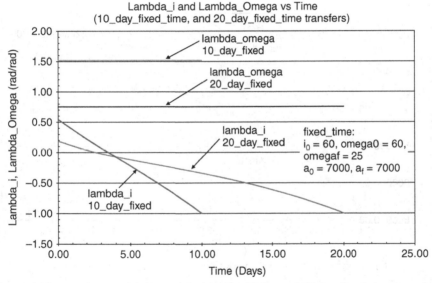

Figure 6.19. λ_i and λ_Ω multipliers variations for 10-day and 20-day fixed-time transfers.

not an explicit function of time. These runs used $J_2 = 1.08263 \times 10^{-3}$ and integration error controls at the 10^{-12} level, with a thrust acceleration of $f = 3.5 \times 10^{-6}$ km/s^2. The numerical results are produced by way of a series of Fortran routines provided in [14], including a shooting, Runge-Kutta driver with adaptive step size control routine, a globally convergent Newton routine using a line search method for Newton step control, and other supporting routines for a robust package fully adequate to solve such problems.

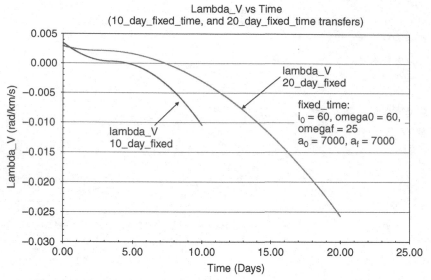

Figure 6.20. λ_V variations for 10-day and 20-day fixed-time transfers.

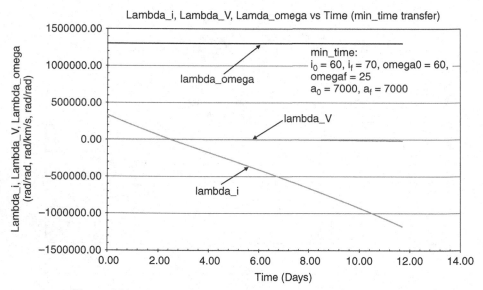

Figure 6.21. λ_i, λ_Ω and λ_V variations for minimum-time transfer.

Figure 6.16 depicts the thrust angle β evolution in time with once again sign switch at the equatorial antinodes to comply with the thrust averaging method used, for the three examples under discussion. Figure 6.17 for the velocity shows how the orbits expand before shrinking back to the 7,000 km size, while Figure 6.18 shows the inclination and omega variations with an initial dip in the inclination corresponding to an increase in orbit size in order to adjust for the required nodal precession due to J_2.

Figures 6.19, 6.20, and 6.21 show the evolutions of the three multipliers λ_i, λ_V, and λ_Ω with perfectly constant λ_Ω for all three transfers and steadily decreasing λ_i

and λ_V values and with $(\lambda_i)_f$ for the 10-day and 20-day-fixed time transfers ending exactly at -1, as also seen in Table 6.4.

REFERENCES

[1] Edelbaum, T. N. (1961) Propulsion Requirements for Controllable Satellites, *ARS Journal*, 1079–1089.

[2] Edelbaum, T. N. (1962) Theory of Maxima and Minima, in *Optimization Techniques with Applications to Aerospace Systems*, ed. G. Leitmann, Academic, New York, 1–32.

[3] Wiesel, W. E., and Alfano, S. (1983) *Optimal Many-Revolution Orbit Transfer*, AAS/AIAA Astrodynamics Specialist Conference, AAS Paper 83-352, Lake Placid, NY.

[4] Cass, J. R. (1983) *Discontinuous Low-Thrust Orbit Transfer*, M. S. Thesis, School of Engineering, Air Force Institute of Technology, AFIT/GA/AA/83D-1, Wright-Patterson AFB, OH.

[5] McCann, J. M. (1988) *Optimal Launch Time for a Discontinuous Low-Thrust Orbit Transfer*, M. S. Thesis, School of Engineering, Air Force Institute of Technology, AFIT/GA/AA/88D-7, Wright-Patterson AFB, OH.

[6] Kéchichian, J. A. (1997) Reformulation of Edelbaum's Low-Thrust Transfer Problem Using Optimal Control Theory, *Journal of Guidance, Control, and Dynamics*, Vol. 20, No. 5, 988–994.

[7] Kéchichian, J. A. (2006) Optimal Altitude-Constrained Low-Thrust Transfer between Inclined Circular Orbits, *The Journal of the Astronautical Sciences*, **54**, No. 3 & 4, 485–503.

[8] Kéchichian, J. A. (2008) *Optimal Low-Thrust Transfer in General Circular Orbit Using Analytic Averaging of the System Dynamics*, F. Landis Markley Astronautics Symposium, AAS Paper 08-272, Cambridge, MD.

[9] Bryson, Jr., A. E., and Ho, Y-C. (1969) *Applied Optimal Control*, Ginn and Company, Waltham, MA.

[10] Hestenes, M. R. (1981) *Optimization Theory*, Krieger, Huntington, NY.

[11] Ewing, G. M. (1985) *Calculus of Variations with Applications*, Dover.

[12] Bryson, Jr., A. E., Denham, W. F., and Dreyfus, S.E. (1963) Optimal Programming Problems with Inequality Constraints I: Necessary Conditions for Extremal Solutions, *AIAA Journal*, **1**, No. 11, 2544–2551.

[13] Speyer, J. L., and Bryson, Jr., A. E. (1968) Optimal Programming Problems with a Bounded State Space, *AIAA Journal*, **6**, No. 8, 1488–1491.

[14] Press, W. H., Teukolsky, S. A., Vetterling, W. T., and Flannery, B. P. (1992) Numerical Recipes in Fortran 77, *The Art of Scientific Computing*, 2nd edition, Cambridge University Press.

Global Optimization and Space Pruning for Spacecraft Trajectory Design

Dario Izzo

European Space Agency, Advanced Concepts Team, Noordwijk, NL

7.1 Introduction

Global optimization algorithms and space pruning methods represent a recent new paradigm for spacecraft trajectory design. They promise an automated and unbiased search of different trajectory options, freeing the final user from the need for caring about implementation details. In this chapter we provide a unified framework for the definition of trajectory problems as pure mathematical optimization problems highlighting their common nature. We then present the detailed definition of two popular typologies, the Multiple Gravity Assist (MGA) and the Multiple Gravity Assist with single Deep Space Manouver (MGA-1DSM). Later we describe in detail the instantiation of four particular problems proposing them as a test set to benchmark the performances of different algorithms and pruning solutions. We take inspiration from real interplanetary trajectories such as Cassini, Rosetta, and the proposed TandEM mission, considering a large search space in terms of possible launch windows and transfer times, but also from rather academic cases such as that of the First Global Trajectory Optimisation Competition (GTOC). We test four popular heuristic paradigms on these problems (differential evolution, particle swarm optimization, simulated annealing with adaptive neighborhood, and genetic algorithm) and note their poor performances both in terms of reliability and solution quality, arguing for the need to use more sophisticated approaches, for example, pruning methods, to allow finding better trajectories. We then introduce the cluster pruning method for the MGA-1DSM problem and we apply it, in combination with the simulated annealing with adaptive neighborhood algorithm, to the TandEM test problem finding a large number of good solutions and a new putative global optima.

Many of the results reported here would not have been possible without the great passion and competence of Tamas Vinko, Marco del Rey Zapatero, and Marek Rucinski, all researching, at different times, different global trajectory optimization aspects. The author also wishes to acknowledge Massimiliano Vasile who, while a research fellow with the Advanced Concepts Team at the European Space Agency, conceived the Ariadna studies on Advanced Global Optimization Tools for Mission Analysis and Design, which ignited the spark of this now incredibly rich research topic.

7.2 Notation

We will make extensive use of a notation that has become quite standard for the engineering community in the past decades. We will denote the components of multi-dimensional vectors with a boldface, $\mathbf{x}$. We will then refer to their components by using a subscript on a nonboldface font, x_i. An important part of the spacecraft trajectory optimization problem definition is the choice of variables that describe the spacecraft state. Here we use an abstract state vector representation $\mathbf{x}$ whenever possible. Otherwise, when explicit equations are useful, a Cartesian choice is used: $\mathbf{x} = [\mathbf{r}, \mathbf{v}, m]$, where $\mathbf{r}$ denotes the spacecraft position in the inertial frame selected, $\mathbf{v}$ its velocity, and m its mass. Other representations can be more suited to particular applications. I_{sp} is the spacecraft propulsion system specific impulse and $g_0 = 9.81$ m/s. The positions and velocity of celestial bodies (planets and asteroid) will be noted with capital letters $\mathbf{R}$ and $\mathbf{V}$. In the case of multiphase problem, we will use a superscript to indicate that a certain quantity is referred to a particular phase, so that $\mathbf{x}^i$ will be the spacecraft state during the i-th phase. We also use the subscripts s and f to indicate the start or final instants of a phase, so that $\mathbf{x}_s^i = \mathbf{x}^i(t_s^i)$ will denote the spacecraft state at the beginning of the phase i.

7.3 Problem Transcription

A spacecraft trajectory optimization problem can, in general, be written in the deceptively familiar form

$$
\mathcal{P} \quad
\begin{aligned}
\text{Optimize:} \quad & \phi(t_s, t_f, \mathbf{x}_s, \mathbf{x}_f) + \int_{t_s}^{t_f} \mathcal{L}(\mathbf{x}(t), \mathbf{u}(t), t)dt \\
\text{Subject to:} \quad & \dot{\mathbf{x}} = \mathbf{a}(\mathbf{x}, t) + \mathbf{u}(t) && \text{dynamic con.} \\
& \mathcal{G}(t_s, t_f, \mathbf{x}_s, \mathbf{x}_f) \leq 0 && \text{boundary con.} \\
& \mathbf{u} \in \mathcal{U}(\mathbf{x}, t) && \text{propulsion / power con.}
\end{aligned}
\quad (7.1)
$$

where $\mathbf{x}(t)$ is the spacecraft state (including at the least its position, velocity, and mass or equivalent quantities), $\mathbf{a}(\mathbf{x}, t)$ represent the external forces acting on the spacecraft (gravitational and non), and $\mathbf{u}(t)$ represent the control acting on the spacecraft and coming from its propulsion system. Note that the set of admissible controls has, in this general form, an explicit state dependence, a characteristic that is not accounted for in the classical optimal control theory. The characteristic that makes problem $\mathcal{P}$ unique in its kind is the extremely complicated nature of the dynamic constraints, which limits the straightforward use of numerical methods to obtain an optimal solution, and the presence of a large number of locally optimal optimal solutions that calls for the use of global optimisation techniques to effectively explore the possible solutions. As a consequence, depending on the particular trajectory problem considered, the problem is often reduced to a simpler form. This process, which we call problem transcription, is central to the spacecraft trajectory optimization process as it heavily influences the performances of the subsequent solution technique, but also the feasibility of the solution obtained, that is, the possibility of transforming it back to an optimal solution of the full problem $\mathcal{P}$. Here we will focus our attention on the case of interplanetary mission design, and in particular on the preliminary phases of the process when the need of a fast exploration of the solution space is more important than a precise description of the trajectory. A comprehensive description of the mission design process may be found in [1]. We will make use of a number of simplifications and hypotheses that are quite commonly applied:

- The only external force considered in writing the term $\mathbf{a}(\mathbf{x}, t)$ is the gravitational attraction $\mathbf{g}(\mathbf{x})$ due to the Sun.
- The sequence of $N + 1$ celestial bodies the spacecraft interacts with is imposed *a priori* and not left to the optimization process to eventually find.
- The spacecraft interaction with the selected celestial bodies happens when the spacecraft state enters the body sphere of influence [2] $\mathbf{x} \in \mathcal{S}_i$, an event that is imposed as a boundary constraint. The interaction with the intermediate bodies (that is, excluding the departure and arrival) is described by a discontinuity in the spacecraft state $\Delta \mathbf{x}^i = \mathbf{x}_s^{i+1} - \mathbf{x}_f^i$, and on the time $\Delta t^i = t_s^{i+1} - t_f^i$, typically a consequence of what is commonly referred to as planetary flyby or a planet rendezvous.

These discontinuities in the state, describing trajectory phases where the spacecraft interacts with a celestial body, are subject to constraints we write in the form: $\Xi^i(\mathbf{x}_s^{i+1}, \mathbf{x}_f^i, t_s^{i+1}, t_f^i) \leq 0$ and we call phase constraints. These hypotheses formally transform problem $\mathcal{P}$ into:

$$
\begin{aligned}
\text{Optimize:} \quad & \phi(t_s^1, t_f^N, \mathbf{x}_s^1, \mathbf{x}_f^N) + \sum_{i=1}^{N} \int_{t_s^i}^{t_f^i} \mathcal{L}^i(\mathbf{x}^i(t), \mathbf{u}^i(t), t)dt \\
\text{Subject to:} \quad & \dot{\mathbf{x}}^i = \mathbf{g}(\mathbf{x}^i) + \mathbf{u}^i(t), \forall i = 1..N & \text{dynamic con.} \\
\mathcal{P}' \qquad & \mathcal{G}(t_s^1, t_f^N, \mathbf{x}_s^1, \mathbf{x}_f^N) \leq 0 & \text{boundary con.} \\
& \Phi^i(t_s^i, t_f^i, \mathbf{x}_s^i, \mathbf{x}_f^i) \leq 0, \forall i = 1..N & \text{phase boundary con.} \\
& \Xi^i(\mathbf{x}_s^{i+1}, \mathbf{x}_f^i, t_s^{i+1}, t_f^i) \leq 0, \forall i = 1..N-1 & \text{phase matching con.} \\
& \mathbf{u}^i \in \mathcal{U}(\mathbf{x}^i, t), \forall i = 1..N & \text{propulsion / power con.}
\end{aligned}
$$

$$(7.2)$$

that is a multiphase optimal control problem where the number of phases N depends on the number of planetary encounters during the overall trajectory. In explicit Cartesian coordinates $\mathbf{x}^i = [\mathbf{r}^i, \mathbf{v}^i, m^i]$, and $\mathbf{u}^i = [\mathbf{0}, \frac{\mathbf{T}^i}{m^i}, 0]$, where T indicates the magnitude of the spacecraft thrust $\mathbf{T}$. In going from $\mathcal{P}$ to $\mathcal{P}'$, we have simplified the problem at the cost of introducing an aprioristic choice on the sequence of planetary encounters, a choice that also needs to be optimized and thus introduces an integer programming element into the interplanetary trajectory design process. The sequence of planetary encounters is enforced in the phase boundary constraints Φ^i. Considering the planetary sphere of influence reduced to one point $\mathcal{S}^i = \{\mathbf{x} \in \mathbb{R}^7, \mathbf{x} = [\mathbf{R}^i(t), ., ., ., .]\}$ (also a common simplification), the boundary constraints Φ^i will include the following conditions that enforce the spacecraft position to be, at the beginning of each phase, equal to the departure planet position, and at the end of each phase equal to the arrival planet position

$$\Phi^i = \begin{cases} \vdots \\ \mathbf{r}_s^i = \mathbf{R}^i(t_s^i) \\ \mathbf{r}_f^i = \mathbf{R}^{i+1}(t_f^i) \\ \vdots \end{cases}, \forall i = 1..N \qquad (7.3)$$

where we have introduced the sequence of $N+1$ planets via their positions $\mathbf{R}^i(t)$, $i = 1..N + 1$. Under the assumption that the $\mathbf{u}^i$ are piecewise continuous functions, problem $\mathcal{P}'$ is further specified to describe low-thrust problems, whereas an impulsive modelling of the thrust describes chemical propulsion problems. In mathematical form, a chemical (impulsive) thrust can be written as $\mathbf{T}^i = \sum_j^M m^i \Delta \mathbf{V}_j^i \delta(t - \tilde{t}_j^i)$. Here $\Delta \mathbf{V}_j^i$ are the velocity increments (impulses) that fully define the thrust law together with the times $\tilde{t}_j^i$. Note that we have used the Dirac delta generalized function $\delta(t - \tilde{t}_j^i)$.

The following sections will focus entirely on chemical propulsion problems, a particular instance of problem $\mathcal{P}'$ where we basically preassign a given analytical shape to the thrust strategy that becomes thus determined unequivocally by a finite number of parameters, effectively reducing the OCP into an NLP (from infinite dimension to a finite number). In the following two sections we introduce two particular instances of chemical propulsion problems: the MGA problem and the MGA-1DSM problem.

7.4 The MGA Problem

The MGA problem definition that follows is a generalization of that given by Izzo et al. [3]. We define the general form of the MGA trajectory optimization problem as the following instance of problem $\mathcal{P}'$

Optimize: $\phi(t_s^1, t_f^N, \mathbf{x}_s^1, \mathbf{x}_f^N)$

Subject to: $\dot{\mathbf{x}}^i = \mathbf{g}(\mathbf{x}^i) + \mathbf{u}^i(t), \forall i = 1..N$ dynamic con.

$\mathcal{G}(t_s^1, t_f^N, \mathbf{x}_s^1, \mathbf{x}_f^N) \leq 0$ boundary con.   (7.4)

$\Phi^i(t_s^i, t_f^i, \mathbf{x}_s^i, \mathbf{x}_f^i) \leq 0, \forall i = 1..N$ phase boundary con.

$\Xi^i(\mathbf{x}_s^{i+1}, \mathbf{x}_f^i, t_s^{i+1}, t_f^i) \leq 0, \forall i = 1..N - 1$ phase matching con.

where:

- Thrusting is possible only at the beginning of each trajectory leg and at arrival and is impulsive: $\mathbf{T}^i = m^i \Delta \mathbf{V}^i \delta(t - t_s^i), \forall i = 1..N - 1, \mathbf{T}^N = m^N \Delta \mathbf{V}^N \delta(t - t_s^N) + \Delta \mathbf{V}^{N+1} \delta(t - t_f^N)$.

- The phase matching constraints include $\Delta m^i = 0$, $\Delta t^i = 0$, $\forall i = 1..N - 1$ and $r_p(\mathbf{v}_f^i, \mathbf{v}_s^{i+1}, t_f^i) \geq \tilde{r}_p^i$, $\Delta \mathbf{V}^i = F_1(\mathbf{v}_f^i, \mathbf{v}_s^{i+1}, t_f^i)$, $\forall i = 2..N - 1$. The expression for the functions F_1 and r_p are given in appendix A and model a planetocentric hyperbolic trajectory phase where the spacecraft is allowed to thrust tangentially at the periplanet (powered flyby).
- The boundary constraints include the launcher performances $m_s^1 = G(\mathbf{v}_s^1, t_s^1)$, the departure thrust impulse definition $\Delta \mathbf{V}^1 = M(\mathbf{v}_s^1, t_s^1)$, and the arrival thrust impulse definition $\Delta \mathbf{V}^{N+1} = N(\mathbf{v}_f^N, t_f^N)$. The expression for G, M, and N are problem dependent and are part of the further problem instantiation.

From an engineering point of view, these assumptions model a spacecraft equipped with chemical propulsion engines, that is, high thrust engines that deliver their acceleration to the spacecraft in a very short time and that are able to thrust only at each planet. As noted before, we no longer have an optimal control problem (OCP) as $\mathbf{u}^i = [0, \frac{\mathbf{T}^i}{m^i}, 0]$ is now fully defined by a finite number of parameters. At this stage, the variables to be optimized, that is, the decision vector, are the initial mass m_s^1 and the start and final epochs of each leg t_s^i, t_f^i. The whole dynamic $\mathbf{x}^i(t)$ during each phase i, and thus all of the constraints can be, as shown in the following, determined by these variables, allowing us to simplify the problem solving explicitly most of the constraints, starting from the dynamic constraints, and thus reducing the problem complexity and dimension.

7.4.1 Spacecraft Position and Velocity

The dynamic constraints relative to $\mathbf{r}^i$ and $\mathbf{v}^i$ may be solved explicitly if we consider them separately grouped with the appropriate phase boundary constraints appearing in Equation (7.3) as follows

$$
\begin{cases}
\dot{\mathbf{r}}^i = \mathbf{v}^i \\
\dot{\mathbf{v}}^i = -\dfrac{\mu}{r^{i3}} \mathbf{r}^i \\
\mathbf{r}_s^i = \mathbf{R}^i(t_s^i), \mathbf{r}_f^i = \mathbf{R}^{i+1}(t_f^i)
\end{cases}
\tag{7.5}
$$

For each phase, the above two-points boundary value problem defines what is commonly known as a Lambert's problem [2]. Such a problem admits $2(1 + 2M_i)$ solutions where $M_i \in [0, \infty)$ is an integer depending on the boundary conditions in a rather complex way. Solutions to the Lambert's problem are commonly divided into posigrade orbits and retrograde orbits according to the direction of their angular momentum vector $\mathbf{h}^i = \mathbf{r}^i \wedge \mathbf{v}^i$ with respect to the ecliptic frame. Also, they are divided into multirevolution and single revolution, according to the number of times the initial position is acquired during the trajectory (one or more). To simplify the problem, we consider only single-revolution and posigrade solutions (a general formal treatment of the MGA problem not using these hypotheses can be found in Izzo et al. [3]). Under these hypotheses, Lambert's problem always admits a unique solution, and we may thus derive the relations $\mathbf{r}^i(t) = \mathbf{L}_r^i(t, t_s^i, t_f^i)$, $\mathbf{v}^i(t) = \mathbf{L}_v^i(t, t_s^i, t_f^i)$.

7.4.2 Spacecraft Mass

Also the remaining dynamic constraint, relative to the mass, can be solved explicitly. From the MGA problem definition we know the thrust law along each phase. Considering last of Equation (7.5) (relative to the spacecraft mass), we have $\dot{m}^i = -\frac{\Delta V_i}{I_{sp}g_0}m^i\delta(t - t_s^i)$. Integrating from t_s^i to t_f^i, we may write explicitly the mass as a function of time using the Heaviside step function H and explicitly using the phase matching constraints $\Delta m^i = 0$

$$
m^i(t) = m_s^1 \exp\left(-\frac{\sum_{i=1}^{N} \Delta V^i H(t - t_s^i) + \Delta V^{N+1} H(t - t_f^N)}{I_{sp}g_0}\right). \tag{7.6}
$$

Eliminating m_s^1 and ΔV_i in the expression above by explicitly using the various constraints F_1, G, M, N in the general MGA problem definition, we may eventually write $m^i(t) = \mathbf{L}_m^i(t, t_s^1, .., t_s^i, t_f^1, .., t_f^i)$.

7.4.3 The Final Form

In the remaining phase, matching constraint we still have, from the definition of the MGA problem, $\Delta t^i = 0 \rightarrow t_f^i = t_s^{i+1}$, which allows us to eliminate $N - 1$ further variables. Choosing as new variable notation $t_0 = t_s^1$ and $t_i = t_f^{i-1}$, $\forall i = 1..N$ we define the decision vector as $\mathbf{p} = [t_0, t_1, .., t_N]$. For each choice of the decision vector, the whole spacecraft state $\mathbf{x}^i(t)$ is known throughout phase i and the general form of the MGA problem described by Equation (7.4) is thus reduced to

$$
\begin{aligned}
\text{Optimize:} \quad & \phi(\mathbf{p}) \\
\text{Subject to:} \quad & \mathcal{G}(\mathbf{p}) \leq 0 \quad && \text{boundary con.} \\
& r_p(t_{i-1}, t_i, t_{i+1}) \geq \tilde{r}_p^i \quad && \text{phase matching con.}
\end{aligned} \tag{7.7}
$$

where no phase boundary constraints are left as they all are explicitly satisfied. The above problem is defined here as the MGA problem and has a dimension $N + 1$.

We have formally presented the procedure to transcribe a trajectory optimization problem into an MGA problem, highlighting the hypotheses that underlie this particular transcription. Further specification of the objective function, of the boundary constraints, and of the decision vector bounds will create different instances of the MGA problem.

7.5 The MGA-1DSM Problem

The MGA-1DSM problem definition that follows generalizes a number of particular problem instances studied in previous works [4, 5, 6]. We define the general form of the MGA-1DSM trajectory optimization problem as the following instance of problem $\mathcal{P}'$

Optimize: $\phi(t_s^1, t_f^N, \mathbf{x}_s^1, \mathbf{x}_f^N)$

Subject to: $\dot{\mathbf{x}}^i = \mathbf{g}(\mathbf{x}^i) + \mathbf{u}^i(t), \forall i = 1..N$ dynamic con.

                $\mathcal{G}(t_s^1, t_f^N, \mathbf{x}_s^1, \mathbf{x}_f^N) \leq 0$ boundary con.   (7.8)

                $\Phi^i(t_s^i, t_f^i, \mathbf{x}_s^i, \mathbf{x}_f^i) \leq 0, \forall i = 1..N$ phase boundary con.

                $\Xi^i(\mathbf{x}_s^{i+1}, \mathbf{x}_f^i, t_s^{i+1}, t_f^i) \leq 0, \forall i = 1..N-1$ phase matching con.

where:

- Thrusting is possible only at departure, at arrival, and once at some point along each phase and is impulsive: $\mathbf{T}^1 = m^1 \Delta \mathbf{V}^0 \delta(t - t_s^1) + \Delta \mathbf{V}^1 \delta(t - \tilde{t}_1)$, $\mathbf{T}^i = m^i \Delta \mathbf{V}^i \delta(t - \tilde{t}^i), \forall i = 2..N-1$, $\mathbf{T}^N = m^N \Delta \mathbf{V}^N \delta(t - \tilde{t}^N) + \Delta \mathbf{V}^{N+1} \delta(t - t_f^N)$.
- The phase matching constraints include $\Delta m^i = 0$, $\Delta t^i = 0$, $\forall i = 1..N-1$, $r_p(\mathbf{v}_f^i, \mathbf{v}_s^{i+1}, t_f^i) \geq \tilde{r}_p^i \ \forall i = 1..N-1$ and $\tilde{v}_{in}^i = \tilde{v}_{out}^i$, $\forall i = 1..N-1$, where $\tilde{v}_{in}^i = |\mathbf{v}_f^i - \mathbf{V}^{i+1}(t_f^i)|$, $\tilde{v}_{out}^i = |\mathbf{v}_s^{i+1} - \mathbf{V}^{i+1}(t_s^{i+1})|$, and the functional relationship r_p is the same as in the MGA problem and is given in appendix.
- The boundary constraints include the launcher performances $m_s^1 = G(\mathbf{v}_s^1, t_s^1)$, the departure thrust impulse definition $\Delta \mathbf{V}^0 = M(\mathbf{v}_s^1, t_s^1)$, and the arrival thrust impulse definition $\Delta \mathbf{V}^{N+1} = N(\mathbf{v}_f^N, t_f^N)$. The expression for G, M, and N are problem dependent and are part of the further problem instantiation.

These hypotheses model a spacecraft equipped with chemical propulsion engines, able to thrust at departure, at arrival, and only once during each trajectory phase and never during planetary flybys. The MGA-1DSM problem removes most of the limitation of the MGA problem and is an accurate problem transcription for many preliminary trajectory design cases. The most important remaining limitation of this problem transcription is in the fixed number of DSM allowed in each phase. As in the MGA problem, the optimal control problem is transformed into an NLP problem as the control $\mathbf{u}^i(t)$ is now fully parameterized by a discrete number of variables. At this stage, the variables to be optimised, that is, the decision vector, are the initial mass m_s^i and, for each phase, the vector $\mathbf{p}^i = [t_s^i, \mathbf{v}_s^i, \tilde{t}^i, t_f^i]$. The whole dynamic during each phase $\mathbf{x}^i(t)$ is, as shown in the following, determined analytically by these variables, allowing us to simplify the problem solving explicitly most of the constraints, starting from the dynamic constraints, and thus reducing the problem complexity and dimension.

7.5.1 Spacecraft Position and Velocity

Let us focus on the i-th phase. Given t_s^i and $\mathbf{v}_s^i$, it is possible to find $\mathbf{r}^i(t)$ and $\mathbf{v}^i(t)$ for $t \leq \tilde{t}^i$ using the known analytical solution to Kepler's problem expressed, for example, in terms of the Lagrange coefficients [2]. The initial position is also known from the phase boundary constraints and is $\mathbf{r}_s^i = \mathbf{R}^i(t_s^i)$. From $\tilde{t}^i$ to t_f^i, we also can derive explicitly the spacecraft position and velocity by considering (as in the MGA case) the dynamic constraints separately grouped with the appropriate phase boundary constraints appearing in Equation (7.3) as follows

$$
\begin{cases}
\dot{\mathbf{r}}^i = \mathbf{v}^i \\
\dot{\mathbf{v}}^i = -\frac{\mu}{r^{i3}}\mathbf{r}^i \\
\mathbf{r}^i(t_s^i) = \mathbf{r}^i(\tilde{t}^i), \mathbf{r}^i(t_f^i) = \mathbf{R}^{i+1}(t_f^i).
\end{cases}
$$

This is again a Lambert's problem that, under the same hypotheses assumed for the MGA case, always admits a unique solution and allows us to derive the relations $\mathbf{r}^i(t) = \mathbf{L}_r^i(t, \mathbf{p}^i)$, $\mathbf{v}^i(t) = \mathbf{L}_v^i(t, \mathbf{p}^i)$. Note that in $\tilde{t}^i$, the spacecraft velocity will be discontinuous as the value evaluated using the Lagrange coefficients will generally differ from that returned by the Lambert's problem solution. Such a discontinuity defines the velocity increment $\Delta \mathbf{V}^i, i = 1..N$ as a function of $\mathbf{p}^i$.

7.5.2 Spacecraft Mass

As was done in the case of the MGA problem, by explicitly using the phase matching constraints $\Delta m^i = 0$ and the equation for the mass, we may derive the simple expression

$$
m^i(t) = m_s^1 \exp\left(-\frac{\Delta V^0 H(t - t_s^1) + \sum_{i=1}^N \Delta V^i H(t - t_s^i) + \Delta V^{N+1} H(t - t_f^N)}{I_{sp}g_0}\right).
$$

$$(7.9)$$

Eliminating m_s^1 and the ΔV^l in the expression above by explicitly using the launcher performance, the departure and arrival thrust impulse definitions, we may eventually write $m^i(t) = \mathbf{L}_m^i(t, \mathbf{p}^1, .., \mathbf{p}^i)$.

7.5.3 The Final Form

All the remaining phase constraints from the definition of the simple MGA-1DSM problem may also be explicitly satisfied using suitable variable changes in the decision vector. In particular we may substitute the variables $\mathbf{v}^i$ for each phase, except the first one, with the periplanet of the planetocentric hyperbola r_p^i and the b-plane orientation β^i using the existing functional relationship

$$
\mathbf{v}_s^{i+1} = F_2(\mathbf{v}_f^i, \mathbf{V}^{i+1}(t_s^{i+1}), r_p^{i+1}, \beta^{i+1})
$$

$$(7.10)$$

that describes explicitly the flyby dynamics satisfying explicitly the phase constraints $\tilde{v}_{in}^i = \tilde{v}_{out}^i, \forall i = 1..N - 1$ (see Appendix A) and allows to transform the nonlinear constraint $r_p(\mathbf{v}_f^i, \mathbf{v}_s^{i+1}, t_f^i) = r_p^i \geq \tilde{r}_p^i$ into a lower bound for the introduced decision vector variable r_p^i. Eventually eliminating further variables using the phase matching constraint $\Delta t^i = 0$, we have a decision vector $\tilde{\mathbf{p}} = [t_s^1, \mathbf{v}_s^1, \tilde{t}^1, t_f^1, r_p^2, \beta^2, \tilde{t}^2, t_f^2, ..]$. To simplify the search space structure, we introduce some variable substitutions. Instead of the heliocentric spacecraft initial velocity $\mathbf{v}_s^1$, we use the variables V_∞, u, v defined as

$$\mathbf{v}_s^1 = V_\infty \left(\cos(\theta)\cos(\phi)\hat{\mathbf{i}} + \sin(\theta)\cos(\phi)\hat{\mathbf{j}} + \sin(\phi)\hat{\mathbf{k}} \right)$$

$$\theta = 2\pi u$$

$$\phi = \arccos(2v - 1) - \pi/2$$

$$\hat{\mathbf{i}} = V^1(t_s^1)/|V^1(t_s^1)|$$

$$\hat{\mathbf{k}} = R^1(t_s^1) \wedge V^1(t_s^1)/|R^1(t_s^1) \wedge V^1(t_s^1)|$$

$$\hat{\mathbf{j}} = \hat{\mathbf{k}} \wedge \hat{\mathbf{i}}.$$

Also, instead of the absolute epochs t_f^i we use the transfer times $T^1 = t_f^1 - t_s^1$, $T^i = t_f^i - t_f^{i-1}$ and instead of $\tilde{t}^i$ we use η^i defined as $\tilde{t}^1 = t_s^1 + T^1 \eta^1$, $\tilde{t}^i = t_f^{i-1} + T^i \eta^i$. The decision vector used will thus be $\tilde{\mathbf{p}}' = [t_s^1, V_\infty, u, v, \eta^1, T^1, r_p^2, \beta^2, \eta^2, T^2, ..]$. This allows us to specify as upper and lower bounds on the decision vector what would otherwise need to be nonlinear constraints. Eventually the general form of the simple MGA-1DSM problem described by Equation (7.8) is reduced to

$$\begin{aligned} \text{Optimize:} \quad & \phi(\tilde{\mathbf{p}}') \\ \text{Subject to:} \quad & \mathcal{G}(\tilde{\mathbf{p}}') \leq 0 \quad \text{boundary con.} \end{aligned} \qquad (7.11)$$

No phase matching constraints or phase boundary constraints are left as we have explicitly satisfied all of them, reducing the search space structure to a hyper-rectangle. The remaining boundary constraints, as detailed later in the description of particular instances of this problem, express, typically, a maximum trajectory length duration or other particular mission requirements. The MGA-1DSM problem, as defined here, includes trajectories with multiple revolutions and with no deep space maneuvers in a particular phase. Thus, this particular transcription creates a continuous optimization problem while being able to describe discrete decision variables such as the number of revolutions or the use of a deep space maneuver during a trajectory phase.

7.6 Benchmark Problems

We now describe the detailed instantiation of some MGA and MGA-1DSM problems that can be used as test problems to study the performance of global optimization solvers or of space pruning techniques. All problems proposed present very large bounds to be representative of the type of trajectory optimization often required in preliminary mission design phases. For each problem we define the flyby sequence, the objective function, the bounds on the decision vector variables, and the constraint expressions. Clearly there are a number of other factors that influence, to some smaller extent, the exact calculation of the objective function, such as the values of the different planetary gravitational constants, the planet ephemerides used (that is, the functions $\mathbf{R}^i, \mathbf{V}^i$), the planets' radii, and so on. To allow the scientific community to share a common implementation of each problem instance, the code in C++ and Matlab of each one of the problems described in detail here is available for download from the European Space Agency Global Trajectory Optimisation Problems (GTOP) database [7]. A preliminary description of some of these test problems is also given by Vinko [8]. We also report the best putative global optima known at the time of writing as taken from the GTOP database where the reader can also find exact numerical details of the solutions here reported.

7.6.1 Cassini1

This problem is a particular instance of the MGA problem as defined in Equation (7.7). It has $N = 5$ phases and hence $N+1 = 6$ celestial bodies are defined in the flyby sequence: Earth, Venus, Venus, Earth, Jupiter, and Saturn. Thus the decision vector is $\mathbf{p} = [t_0, .., t_5]$ and contains the epochs of each planetary encounter. The objective function is defined as $\phi(\mathbf{p}) = -g_0 I_{sp} \log(m_f^N / m_s^1)$, where the ratio between the final and the initial mass m_f^N / m_s^1 is given by Equation (7.6) evaluated at t_f^N. Note that, after taking the logarithm, the objective function is essentially the sum of the various velocity increments: $\phi(\mathbf{p}) = \sum \Delta V_i$. No further constraints are considered except those appearing in the generic MGA problem definition. The Cassini1 problem can be written as

$$
\begin{aligned}
\text{Minimize:} \quad & -g_0 I_{sp} \log(m_f^N / m_s^1) = \sum \Delta V_i \\
\text{Subject to:} \quad & r_p(t_{i-1}, t_i, t_{i+1}) \geq \tilde{r}_p^i, \forall i = 1..N - 2 \quad \text{phase matching con.}
\end{aligned}
\tag{7.12}
$$

The departure thrust impulse is defined as $\Delta V_1(t_0, t_1) = |\mathbf{V}^1(t_0) - \mathbf{v}_s^1|$, the arrival thrust impulse is defined as an orbital insertion as detailed in the Appendix A (r_p^{ins} and e^{ins} are given in Table 7.3). The launcher performance is in this case not relevant, as the value of the initial spacecraft mass m_s^1 does not appear anywhere in the problem definition (the objective function depends only on the velocity increments). Introducing the variable change $T_i = t_i - t_{i-1}, i = 1..5$, the exact values needed to define this test problem are given in Table 7.1. The Cassini1 problem admits a putative global optima at $\phi = 4.93$ km/s and is characterized by a number of local optima with a very strong basin of attraction, particularly noteworthy the one at $\phi = 5.30$ km/s that seems to be very difficult for many optimization techniques to overcome.

7.6.2 GTOC1

This problem is a particular instance of the MGA problem as defined in Equation (7.7). It has $N = 7$ phases and hence $N + 1 = 8$ celestial bodies (the flyby

Table 7.1. *Bounds and other parameters for the problem Cassini1*

Variable	Lower Bound	Upper Bound	units	parameter	value	units
t_0	−1000	0	(MJD2000)	$\tilde{r}_p^1$	6351.8	km
T_1	30	400	days	$\tilde{r}_p^2$	6351.8	km
T_2	100	470	days	$\tilde{r}_p^3$	6778.1	km
T_3	30	400	days	$\tilde{r}_p^4$	671492	km
T_4	400	2000	days	r_p^{ins}	108950	km
T_5	1000	6000	days	e^{ins}	0.98	

Table 7.2. *Bounds and other parameters for the problem GTOC1*

Variable	Lower Bound	Upper Bound	units	parameter	value	units
t_0	3000	10000	(MJD2000)	$\tilde{r}_p^1$	6351.8	km
T_1	14	2000	days	$\tilde{r}_p^2$	6778.1	km
T_2	14	2000	days	$\tilde{r}_p^3$	6351.8	km
T_3	14	2000	days	$\tilde{r}_p^4$	6778.1	km
T_4	14	2000	days	$\tilde{r}_p^5$	600000	km
T_5	100	9000	days	$\tilde{r}_p^6$	70000	km
T_6	366	9000	days	ΔV_{lau}	2.5	km/s
T_7	300	9000	days	m_0	1500	kg
				I_{sp}	2500	sec

sequence) are forced to be encountered by the spacecraft along its trajectory: Earth, Venus, Earth, Venus, Earth, Jupiter, Saturn, TW299. Thus, the decision vector is $\mathbf{p} = [t_0, .., t_7]$ and contains the epochs of each planetary encounter. The objective function is defined as $\phi(\mathbf{p}) = m_f^N |(\mathbf{V}^{N+1} - \mathbf{v}_f^N) \cdot \mathbf{V}^{N+1}|$ and no further constraints are considered except those appearing in the generic MGA problem definition. The GTOC1 problem can be written as

$$\text{Maximize:} \quad \phi(\mathbf{p}) = m_f^N |(\mathbf{V}^{N+1} - \mathbf{v}_f^N) \cdot \mathbf{V}^{N+1}|$$
$$\text{Subject to:} \quad r_p(t_{i-1}, t_i, t_{i+1}) \geq \tilde{r}_p^i, \forall i = 1..N - 2 \quad \text{phase matching con.} \tag{7.13}$$

where the final mass m_f^N is given by Equation (7.6) evaluated in t_f^N. The departure thrust impulse is defined as $\Delta V_1(t_0, t_1) = \max(|\mathbf{V}^1(t_0) - \mathbf{v}_s^1| - \Delta V_{lau}|, 0)$, the arrival thrust impulse is defined as $\Delta V^{N+1} = 0$. The launcher performance is defined as $m_s^1 = m_0$. Introducing the variable change $T_i = t_i - t_{i-1}, i = 1..7$, the exact values needed to define this test problem are given in Table 7.2. The GTOC1 problem admits a putative global optima at $\phi = 1,580,599$ kg km^2 /s^2 and is characterized by a large number of local optima at almost all objective function ranges.

7.6.3 Rosetta

This problem is a particular instance of the MGA-1DSM problem as defined in Equation (7.11). It has $N = 5$ phases and hence $N + 1 = 6$ celestial bodies (the flyby sequence) are forced to be encountered by the spacecraft along its trajectory: Earth, Earth, Mars, Earth, Earth, Jupiter, Saturn, 67P/Churyumov-Gerasimenko.. Thus, the decision vector is $\tilde{\mathbf{p}} = [t_s^1, \mathbf{v}_s^1, \tilde{t}^1, t_f^1, r_p^2, \beta^2, \tilde{t}^2, t_f^2, ..]$. The objective function is defined as $\phi(\mathbf{p}) = -g_0 I_{sp} \log(m_f^N / m_s^1)$, where the ratio between the final and the initial mass m_f^N / m_s^1 is given by Equation (7.9) evaluated in t_f^N. Note that, after taking the logarithm, the objective function is essentially the sum of the various velocity increments: $\phi(\mathbf{p}') = \sum \Delta V_i$. The Rosetta problem can thus be written as

$$\text{Minimize:} \quad -g_0 I_{sp} \log(m_f^N / m_s^1) = \sum \Delta V_i \qquad (7.14)$$

and is an unconstrained global optimization problem. The departure velocity increment is defined as $\Delta V_0 = 0$, the arrival velocity increment is defined as $\Delta V_{N+1}(\mathbf{p}') = |\mathbf{V}^N(t_f^N) - \mathbf{v}^N(t_f^N)|$. The launcher performance is in this case not relevant, as the value of the initial spacecraft mass m_s^1 does not appear anywhere in the problem definition (the objective function depends only on the velocity increments). The exact values needed to define this test problem are given in Table 7.3. The Rosetta problem admits a putative global optima at $\phi = 1.34$ km/s.

Table 7.3. *Lower and upper bounds defining the MGA-1DSM problems Rosetta and TandEM*

Variable	Rosetta LB	Rosetta UB	units	TandEM LB	TandEM UB
t_0	1460	1825	MJD2000	5475	9132
V_{inf}	3	5	km/s	2.5	4.9
u	0	1		0	1
v	0	1		0	1
T_1	300	500	days	20	2500
T_2	150	800	days	20	2500
T_3	150	800	days	20	2500
T_4	300	800	days	20	2500
T_5	700	1850	days		
η^1	0.01	0.9		0.01	0.99
η^2	0.01	0.9		0.01	0.99
η^3	0.01	0.9		0.01	0.99
η^4	0.01	0.9		0.01	0.99
η^5	0.01	0.9			
r_p^1	1.05	9	Planet Radii	1.05	10
r_p^2	1.05	9	Planet Radii	1.05	10
r_p^3	1.05	9	Planet Radii	1.05	10
r_p^4	1.05	9	Planet Radii		
β^1	$-\pi$	π	rad	$-\pi$	π
β^2	$-\pi$	π	rad	$-\pi$	π
β^3	$-\pi$	π	rad	$-\pi$	π
β^4	$-\pi$	π	rad		

7.6.4 Constrained TandEM–Atlas501–EVEEJ

This problem is a particular instance of the MGA-1DSM problem as defined in Equation (7.11). It has $N = 4$ phases and hence $N + 1 = 5$ celestial bodies (the flyby sequence) are forced to be encountered by the spacecraft along its trajectory: Earth, Venus, Earth, Earth, Jupiter. Thus the decision vector is $\tilde{\mathbf{p}} = [t_s^1, \mathbf{v}_s^1, \tilde{t}^1, t_f^1, r_p^2, \beta^2, \tilde{t}^2, t_f^2, ..]$. The objective function is defined as $\phi(\mathbf{p}) = m_f^N$, where the final mass is given by Equation (7.9) evaluated in t_f^N. We also introduce a global constraint on the total trajectory duration. The TandEM-Atlas501-EVEEJ (in the following "TandEM problem" for brevity) can thus be written as

$$
\begin{aligned}
\text{Minimize:} \quad & m_f^N \\
\text{subject to:} \quad & t_f^N - t_s^1 \leq t_{tot}
\end{aligned}
\tag{7.15}
$$

and is a constrained global optimization problem where the total trajectory time is limited to $t_{tot} = 10$ years. The departure thrust impulse is defined as $\Delta V_0 = 0$, the arrival thrust impulse is defined as an orbital insertion as detailed in the Appendix ($r_p^{ins} = 80,330$ km and $e^{ins} = 0.9853$). The launcher performance is that of Atlas-501 obtained as detailed in Appendix A using the table given by NASA Launch Services (NLS) Launcher Performances. Outside the reported declinations (± 28.5 deg), a null mass is considered. The TandEM problem admits a putative global optima at $\phi = 1437.58$ kg.

7.7 Global Optimization

The advantage of transcribing a given trajectory design problem into a well-defined optimization problem stems from the possibility of using computer algorithms to achieve a complete automation of the design. Each problem instance can in fact be coded into a black-box function expressing the functional relationship between the decision vector and a figure of merit expressing the quality of the related trajectory and its constraint violations. Derivative-free global optimization algorithms can then be applied to try finding optimal solutions. This approach to designing spacecraft trajectories is reaching a great maturity and promises a completely unbiased and automated listing of optimal trajectory options. Algorithms such as Differential Evolution, Genetic Algorithms, Particle Swarm Optimization, Simulated Annealing, and, even more recently, Monotonic Basin Hopping, just to quote a few, have all been tried with different degrees of success. It is important to be aware that all of the above algorithms are able to solve some particular instances of the trajectory problems. Simply taking a suitable problem instance, it is possible to achieve good performances in almost all cases. It is crucial to escape the temptation to pick an algorithm, perhaps introduce some modifications, and present a few pseudo-randomly selected trajectory optimizations where its performances seem good. With this respect, the test problems introduced here, in Vinko et al. [8], and in general

Table 7.4. *Performances of off-the-shelf solvers on two MGA test problems over 100 runs*

Problem	Paradigm	DE	PSO	MPSO	SA-AN	SGA
Cassini1 (min = 4.93)						
	Mean	8.57	9.47	7.05	11.67	7.09
	Std	3.29	4.20	2.39	4.236	2.30
	Min	4.93	5.33	5.43	5.12	5.44
	Max	16.71	22.90	15.53	23.44	17.98
GTOC1 (max = 1,580,599)						
	Mean	1,140,759	759,221	602,331	1,179,835	907,781
	Std	146,589	174,463	125,163	139,590	232,290
	Min	836,772	295,143	341,780	846,720	81,820
	Max	1,523,629	1,134,860	876,177	1,511,767	1,416,050

Table 7.5. *Performances of off-the-shelf solvers on two MGA-1DSM test problems over 100 runs*

Problem	Paradigm	DE	PSO	MPSO	SA-AN	SGA
Rosetta (min = 1.34)						
	Mean	7.55	10.36	10.76	4.13	9.95
	Std	1.82	2.72	1.93	0.94	3.29
	Min	3.94	5.34	7.01	2.61	4.35
	Max	13.26	15.79	15.46	6.65	17.52
TandEM (max = 1437.58)						
	Mean	216.32	144.03	108.12	625.26	78.75
	Std	80.35	135.79	50.42	254.62	80.50
	Min	95.28	22.42	33.57	60.43	3.78
	Max	460.87	862.64	321.37	1298.19	498.22

those present in the GTOP database [7] can offer a significant help in understanding the value of any proposed algorithm under the conditions that the bounds, the parameter values, and the underlying models are left untouched. They describe a fair range of quite complex interplanetary spacecraft trajectory optimization problems. Some of them have a rather academic value, like Cassini1 or GTOC1, and some are instead quite close to the type of problems mission designers solve in preliminary phases of the trajectory design process, like Rosetta or the constrained TandEM problem.

In Table 7.4 and Table 7.5, we list the performances that several standard implementation of popular heuristics paradigms achieve on the described test problems. These results can give a feel for the type of performances one can expect by applying a given paradigm but should in no way be considered as a general comparison table between paradigms, as the tables are obtained by choosing a particular algorithmic setting that can, no doubt, be improved by tuning appropriately the various constants or by changing some implementation details. The algorithms tested are all well described in the literature and we here only briefly touch upon them:

- DE: The Differential Evolution paradigm has been found by Myatt et al. [9] to be a good solver for spacecraft trajectory optimization problems and is described in detail in the work by Storn and Price [10]. It is fully defined by the strategy adopted (several are proposed in the original paper) the population size NP, and two parameters, the weighting factor F and the crossover ratio CR.

- PSO: The Particle Swarm Optimization algorithm, in its simplest form, has been proposed by Kennedy and Eberhart [11] and is fully defined by the number of particles NP, the inertia weight ω, by the cognitive component factor η_1, and the social component factor η_2.

- MPSO: The Multiple Particle Swarm Optimization algorithm is a variation to the canonic PSO whereby multiple swarms are performing the search independently except randomly swapping every k iterations the swarm membership. It has been found to provide some advantages over PSO in [12] and is fully defined by k, the number of particles NP, the inertia weight ω, the cognitive component factor η_1, the social component factor η_2, and the number of swarms n.

- SA-AN: The Simulated Annealing with Adaptive Neighborhood paradigm [13] is a variation to the Simulated Annealing, where the sampling neighborhood for each decision vector component is adaptively changed according to the acceptance rate of new solutions. There are many ways of implementing such a paradigm; here we use the algorithm proposed by Corana [13] detailed in the Appendix B using re-annealing and that is fully defined by the starting temperature T_i, the final temperature T_f, and the annealing speed nf_{ann} defining the number of function evaluation allowed for each annealing.

- SGA: The Simple Genetic Algorithm is a basic version of a genetic algorithm [14] that uses roulette wheel selection, exponential crossover, uniform mutation, and an elitist strategy. The free parameters for this algorithm are the population size NP, the mutation probability M of each gene, and the crossover ratio CR defined in the same way as that of the DE algorithm.

In the test results presented here, each algorithm was left free to calculate for enough function evaluations allowing a statistical convergence of the results. That is, the results given in the tables refer to a fixed number of function evaluations $FEVAL$ and would not be significantly statistically different (a pair-wise Welsh test has been carried out with a confidence level of 95%) if the results were compiled after $FEVAL/2$ function evaluations. For all tests, $FEVAL = 1,200,000$ except for Cassini1 where $FEVAL = 80,000$. All population-based algorithms were tested with a population size $NP = 20$. The parameters for DE are $F = 0.7, CR = 0.7$, and the strategy DE/rand/1/exp was used. For PSO we used $\omega = 0.65, \eta_1 = \eta_2 = 2$. For

MPSO we used the same settings as for PSO and $n = 4$ swarms, that is, each swarm had five particles. For the GA we used $M = 0.2$ and $CR = 0.7$. For SA-AN we used a starting temperature $T_i = 10$ for Cassini1 and Rosetta, $T_i = 100,000$ for GTOC1 and $T_i = 1$ for TandEM, a final temperature of $T_f = 0.036$ for Cassini1 $T_f = 138$ for GTOC1, $T_f = 0.073$ for Rosetta, and $T_f = 0.0024$ for TandEM. All SA-AN simulations were allowed $nf_{sa-an} = 10,000$ function evaluations per annealing cycle.

7.7.1 Discussion

The results presented aim at showing that the standard implementations of the global optimization algorithms tested all have, to different degrees, quite poor reliability and performances if applied directly to the difficult interplanetary test problems proposed. The straightforward use of these algorithms seem thus to fail providing the hoped automated and unbiased approach to complex trajectory optimization. Tuning the various algorithms parameters can surely offer a performance improvement that comes, though, at the cost of further objective function evaluations and needs to be performed for each problem instance. From the results reported here and from the experience accumulated by this author in several other experiments, we can claim that DE and SA-AN are the best performing algorithms, DE being particularly efficient in MGA problems and SA-AN, whose performances are reported here for the first time, appearing to outperform all others in MGA-1DSM problems. A further performance improvement can also be obtained by letting heterogeneous versions of these algorithms run in parallel in a so-called island model[1], as recently suggested by Izzo et al. [15]. Other algorithms are available that have not been tested here, and new ones are certain to come in the future. Noteworthy is the case of the Basin Hopping algorithm [16] that has been very recently applied, to the knowledge of this author, for the first time by Bernadetta Addis, Fabio Schoen, and Marco Locatelli to a great number of interplanetary trajectory problem instances (also to the ones described here) locating reliably good solutions and beating consistently all the known global optima [7]. Another approach able to find reliably a large number of good trajectory options in reasonable computational times is that of applying global optimization algorithms on a reduced portion of the search space obtained by pruning out regions according to some predefined criteria. This way the problem instance complexity can be substantially reduced allowing different algorithms to reliably converge in short times to optimal solutions. In the next sections, we will introduce two of such techniques and we will comment on how, while allowing for an automated and efficient search, they introduce the need to perform a number of

---

[1]The island model is one successful paradigm to perform heuristic global optimization in a CPU network by having different sets of solutions being optimized separately in different islands, typically assigned to different CPUs, and letting some solutions stochastically move to new islands following predefined "migration" paths. A careful setup of such a system can bring to an improvement that is superlinear with the respect to the number of CPUs used and eventually to the definition of a new algorithm outperforming those operating on the single islands.

choices on the pruning criteria that need to be carefully made as not to introduce unwanted biases in the optimization process.

7.8 Space Pruning

The term space pruning refers to all techniques that allow reduction of search space focussing the optimization in smaller areas where the optimal solutions are to be found. The output of a typical pruning process is a set of hyperrectangles contained in the original search space where, according to some criteria, good solutions are expected. There are a large number of criteria that can be adopted to define such regions and that are dependent on the particular problem instance considered, here we introduce techniques that are generically applicable to a whole problem class, regardless of the instantiation details.

7.8.1 Pruning the MGA Problem: GASP

For the MGA problem an efficient pruning method is that developed by Myatt et al. [9, 3] and named Gravity Assist Space Pruning (GASP). The details of such a technique are well described in these two references and are thus not reported here. The method is based on the possibility of incrementally dividing the MGA problem in a cascade of two-dimensional problems where grid sampling is computationally efficient. Propagating back and forward pruning criteria defined on the flyby constraint satisfaction and on maximum ΔV^i allowed, it is then possible to reduce the number of sampled points to a fraction of the original space. For the Cassini1 problem, a space reduction of six-order of magnitude is reported. The polynomial complexity of the resulting algorithm has also been demonstrated both with respect to the grid size defined for each two-dimensional problem and to the overall problem dimension. While polynomial complexity as such does not necessarily lead to efficient algorithms, in the case of GASP the low exponents involved produce an incredibly fast pruning algorithm that allow to reliably solve MGA problems locating all good launch windows and the globally optimal solution contained therein. Recent attempts have been made to improve the GASP algorithm to allow a mathematical proof on the global optimality of the found solution [17] or to extend it to low-thrust trajectories [18] and to problems similar to the MGA-1DSM [19, 20, 21]. In these methods, polynomial complexity is retained, but with much larger exponents, or to the price of an excessive problem simplification that make the resulting implementations of a rather limited use.

7.8.2 Pruning the MGA-1DSM Problem: Cluster Pruning

The cluster pruning algorithm we propose here was developed by Marco del Rey Zapatero and by this author during March 2008 at the Advanced Concepts Team and is reported here for the first time. It results in the possibility to achieve full automation of the trajectory optimization process at the cost of employing appropriate computing power. The algorithm is based on the observation that in the MGA-1DSM problem, good solutions are often clustered in a small portion of the solution space rather than being equally distributed all over it. This is quite intuitive for variables such as t_s^1, that is, the launch date, but is also equally valid for the other variables, as we will see. Each cluster of solutions corresponds to a different strategy and can be detected automatically and further explored in more depth as isolated problems by global optimization algorithms. The following pseudoalgorithm illustrates the approach in more detail:

(1) instantiate the problem with bounds LB, UB
(2) while not *convergence-criterion*
(3) perform N optimizations using the algorithm $\mathcal{A}$
(4) refine the N solutions found using a local optimization technique (optional)
(5) identify clusters of good solutions
(6) prune according to the results obtained and define new bounds LB, UB

In step 1, an MGA1-DSM problem is instantiated. In the following main loop, a global optimization algorithm is used to produce N solutions to the problem. The solutions, representing local optima of varying quality, are then optimized locally; this step is not necessary but helps in reducing the total pruning steps. The obtained trajectories are later analyzed to identify regions where the best of them lie and thus to produce reduced bounds. At each iteration, a smaller space is produced, and the N following optimizations produce increasingly better solutions. The baseline performances of the chosen algorithm $\mathcal{A}$ are crucial to the success of the process, as is the method used to produce the new bounds (or the set of new bounds in case multiple clusters are allowed to be selected).

We illustrate the detailed steps of a possible implementation of the algorithm in the particular case of the MGA-1DSM problem TandEM. We select one the best performing algorithms from Table 7.5, that is Simulated Annealing with Adaptive Neighborhood. Performing $N = 100$ optimization on the full problem, we obtain the relatively poor results reported on the same table. We evaluate the p percentile of the different objective functions returned and consider only the decision vectors $\mathbf{x}$ that are above such a value. We set the new bounds to $LB_{new_i} = min(x_i) - (UB_i - LB_i)p/100/2$ and $UB_{new_i} = max(x_i) + (UB_i - LB_i)p/100/2$ only if these are still within the old bounds.

In Table 7.6 we report the results of the $N = 100$ runs of the SA-AN algorithm at each pruning step (we perform four iterations of the cluster pruning algorithm with increasing percentile levels of 80, 90, and 95. At the end of each iteration, we run a local optimizer starting from all the N trajectories found. A total of 400 locally optimal trajectories is thus computed during the process.). In Figure 7.1, as an example, we show the values of u and T_3 for these 400 solutions together with the bounds at each pruning steps. The final best solution can be further improved by performing iteratively local optimization starting from a close neighborhood. We thus find a new putative global optimum at $m_f^N = 1{,}476.01$ kg improving the known

Table 7.6. *Stochastic pruning for the TandEM-Atlas501-6 MGA–1DSM problem*

Problem	Pruning iterations	0	1 ($p = 80$)	2 ($p = 90$)	3 ($p = 95$)
TandEM-Atlas501-6					
(max)	Mean	625.26	838.94	1060.74	1445.45
	Std	254.62	269.31	262.08	38.133
	Min	60.43	249.37	341.2	1370.52
	Max	1298.19	1381.38	1475.22	1475.71

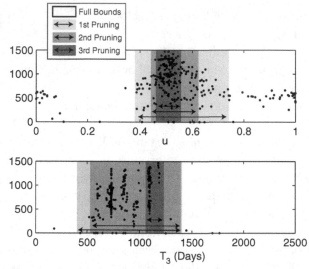

Figure 7.1. Plot of u and T_3 against the final mass for all the 400 computed trajectories. The bounds reduction obtained by pruning is also visualized. Note the Earth-Earth transfer times resonances clearly visible as clusters in the T_3 graph.

best solution reported in the GTOP database [7]. The corresponding trajectory is visualized in Figure 7.2.

At the end of the process, we have not only a putative globally best trajectory, but a large number of other trajectories distributed in those parts of the search space where good solutions are likely to be found and thus a thorough representation of the solution space that is necessary in the preliminary phases of the trajectory design process when requirements change quite often (for example, bounds and constraints) and it is not possible to run a new optimization at each time. It is noteworthy that the global best solution employs only one deep space maneuver, that is, between the two consecutive Earth flybys, but this strategy was not imposed *a priori*; rather it is a result of the optimization process. In the best solution found, no multiple revolutions are present. While these can in principle increase the final mass, the constraint on the total flight duration in this case drives the optimization process toward trajectories that do not make use of multiple revolutions so as not to lose time. Once again, this is not imposed upfront, but it is a result of the automated optimization process. Releasing the constraint on the total flight duration, some tests

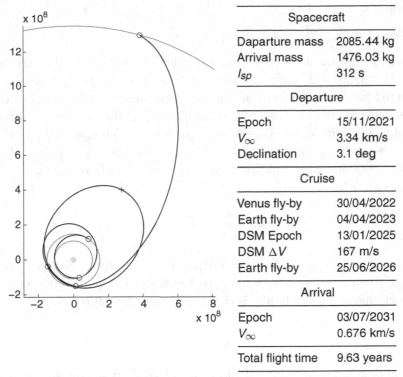

Spacecraft	
Daparture mass	2085.44 kg
Arrival mass	1476.03 kg
I_{sp}	312 s
Departure	
Epoch	15/11/2021
V_∞	3.34 km/s
Declination	3.1 deg
Cruise	
Venus fly-by	30/04/2022
Earth fly-by	04/04/2023
DSM Epoch	13/01/2025
DSM ΔV	167 m/s
Earth fly-by	25/06/2026
Arrival	
Epoch	03/07/2031
V_∞	0.676 km/s
Total flight time	9.63 years

Figure 7.2. Putative globally optimal trajectory found in the TandEM problem using cluster pruning.

revealed that the optimal solutions indeed exploit multiple revolutions to increase the final mass.

7.9 Concluding Remarks

Global optimization meta-heuristics are useful in automatically finding and selecting good trajectory options between the often-many possibilities one has in the preliminary phases of mission design. Their use and efficiency are established for chemical propulsion problems of high complexity (that is, large launch windows and multiple flybys) whenever approaches more sophisticated than the straightforward use of standard algorithms are adopted. It seems likely that future research results will aim at proving the use of these techniques for the automated computation of low-thrust trajectories as well. Preliminary results in this sense are already available and pointing to an increased need of computational resources. Under the assumption that the available computing power will keep increasing at the same pace in the next decade, it thus seems possible to argue that a completely automated trajectory design process, at least in the case of patched two-body problems, may sooner or later be able to replace the current design methods relying substantially on expert knowledge, similar to how we no longer have to perform a full function study to obtain its graph.

Appendix 7A

Definition of r_p (v_{in}, v_{out}, t) and F_1 (v_{in}, v_{out}, t)

The functional expression relating the incoming and outgoing velocities during a given planetary flyby at epoch t is described here. In the following, the spacecraft is allowed to have a tangential and impulsive velocity change at the periplanet so that its trajectory will consist of two hyperbolas patched at the pericenter. Introducing the relative velocities $\tilde{v}_{in} = |v_{in} - V(t)|$, $\tilde{v}_{out} = |v_{out} - V(t)|$, simple astrodynamic calculations show that taking as length unit any L and as velocity unit $\sqrt{\mu_{pla}/L}$ (where μ_{pla} is the gravitational parameter of the planet considered), the angle between $\tilde{v}_{in}$ and $\tilde{v}_{out}$ is given by

$$\alpha_i = \arcsin \frac{a_{in}}{a_{in} + r_p} + \arcsin \frac{a_{out}}{a_{out} + r_p} \qquad (7.16)$$

where $a_{in} = 1/(\tilde{v}_{in} \cdot \tilde{v}_{in})$ and $a_{out} = 1/(\tilde{v}_{out} \cdot \tilde{v}_{out})$. Inverting this equation for r_p, we define the function $r_p = r_p(v_{in}, v_{out}, t)$. The velocity increment necessary at the pericenter, that is, the function F, is given by the simple relation

$$\Delta V = \left| \sqrt{(1/a_{in} + 2/r_p)} - \sqrt{(1/a_{out} + 2/r_p))} \right| = F(v_{in}, v_{out}, t).$$

Note that in a MGA-1DSM problem as $\tilde{v}_{in} = \tilde{v}_{out}$, there is no velocity increment needed ($F = 0$), and the definition of r_p can be obtained by explicit inversion of the first equation.

Definition of the Arrival Thrust Impulse $N(v, t)$ as an Orbit Insertion

In case when at the arrival planet the spacecraft is inserted into an elliptic planeto-centric orbit having an assigned pericenter r_p^{ins} and eccentricity e^{ins}, the tangential velocity increment needed at the pericenter is determined by the spacecraft arrival velocity v at epoch t. Taking as length unit r_p^{ins} and as velocity unit $\sqrt{\mu_{pla}/r_p^{ins}}$ (where μ_{pla} is the gravitational parameter of the planet considered), we have, from simple astrodynamics

$$v_{p-} = \sqrt{\tilde{v}^2 + 2}$$
$$v_{p+} = \sqrt{1 + e^{ins}}$$
$$\Delta V = |v_{p-} - v_{p+}| = N(v, t).$$

Definition of the Launcher Performances $m = G(v, t)$

When we want to model a particular launcher, its performances are often given in terms of a table relating the mass that can be delivered by the launcher to a given value of the hyperbolic escape velocity $\tilde{v}$ and at a certain declination δ (this last being referred to the equatorial reference system). These two quantities may be evaluated from the heliocentric departure velocity $\mathbf{v}$ and the epoch t as

$$\tilde{\mathbf{v}} = \mathbf{v} - \mathbf{V}(t)$$

$$\tilde{\mathbf{v}}_{equ} = \mathbf{R}\tilde{\mathbf{v}}$$

$$\sin\delta = \frac{\tilde{\mathbf{v}}_{equ} \cdot \mathbf{k}}{\tilde{\mathbf{v}}_{equ} \cdot \mathbf{k}}$$

where $\mathbf{R}$ is the rotation matrix that allows a change of reference system to the equatorial planetocentric. Then m can be found by interpolation.

Definition of $v_{out} = F_2(v_{in}, V, r_p, \beta)$

In a patched conic approximation, the spacecraft heliocentric velocity after a flyby can be related to its incoming heliocentric velocity $\mathbf{v}_{in}$, the planet velocity $\mathbf{V}$, the planetocentric hyperbola periplanet r_p, and its plane orientation β. The following equations, implementing standard astrodynamic calculations, describe such a relation

$$\tilde{\mathbf{v}}_{in} = \mathbf{v}_{in} - \mathbf{V}$$

$$e = 1 + r_p/\mu_{pla}\tilde{v}_{in}^2$$

$$\delta = 2\arcsin(1/e)$$

$$\hat{\mathbf{v}}_{out} = \cos(\delta)\hat{\mathbf{i}} + \cos(\beta)\sin(\delta)\hat{\mathbf{j}} + \sin(\beta)\sin(\delta)\hat{\mathbf{k}}$$

$$\tilde{\mathbf{v}}_{out} = \tilde{v}_{in}\hat{\mathbf{v}}_{out}$$

$$\mathbf{v}_{out} = \mathbf{V} + \tilde{\mathbf{v}}_{out}$$

where the unit vectors are $\hat{\mathbf{i}} = \frac{\tilde{\mathbf{v}}_{in}}{\tilde{v}_{in}}$, $\hat{\mathbf{j}} = \frac{\hat{\mathbf{i}}\wedge\mathbf{V}}{|\hat{\mathbf{i}}\wedge\mathbf{V}|}$, and $\hat{\mathbf{k}} = \hat{\mathbf{i}} \wedge \hat{\mathbf{j}}$. Note that in this representation, the incoming and outgoing hyperbolic velocities, $\tilde{v}_{in}$, $\tilde{v}_{out}$, are of equal magnitude (no ΔV is modeled during the hyperbola) and form an angle δ. The second angle, β, determines the position of the outgoing hyperbolic velocity on the cone with axis $\tilde{v}_{in}$ and aperture δ.

Appendix 7B

The particular implementation of the Simulated Annealing algorithm with Adaptive Neighborhood (that we call SA-AN) is given in the following pseudocode describing a single annealing cycle.

1: Select a point $\mathbf{x}_0$
2: for i=1:n_o
3:   for j=1:n_t
4:     for k=1:n_r

5:         for l=1:D
6:             alter x_l component adding a random $\delta \in [-r_l, r_l]$
7:             accept or refuse according to Metropolis criteria
8:         adjust each component of the neighborhood **r** using Corana's method [13]
            (the acceptance rate is evaluated separately for each component)
9:   adjust the temperature using $T = \alpha T$

At the end of the annealing process, the temperature will be $T_f = T_i \alpha^{n_o}$. The total number of function evaluation per annealing cycle has to be set and is $nf_{ann} = n_o n_t n_r D$. We use $n_t = 1$ and $n_r = 20$ so that we adjust the neighborhoods and the temperature simultaneously and we have enough points to evaluate the acceptance rate (here, 20 for each component). The starting neighborhood is set to be equal to $r_l = UB_l - LB_l$. We then perform re-annealing (that is, we restart the algorithm from the point returned by a previous run) up to when we reach the total number of function evaluations *FEVAL*. The algorithm free parameters are T_f, T_i defining the cooling schedule, and nf_{ann} defining the annealing cycle speed.

REFERENCES

[1] Wertz, J., Larson, W., Kirkpatrick, D., and Klungle, D. (1999) *Space Mission Analysis and Design*, Microcosm Press.

[2] Battin, R. (1999) *An Introduction to the Mathematics and Methods of Astrodynamics*, AIAA.

[3] Izzo, D., Becerra, V., Myatt, D., Nasuto, S., and Bishop, J. (2007) Search Space Pruning and Global Optimisation of Multiple Gravity Assist Spacecraft Trajectories, *Journal of Global Optimization*, **38**, No. 2, 283–296.

[4] Vasile, M., and De Pascale, P. (2006) Preliminary Design of Multiple Gravity-Assist Trajectories, *Journal of Spacecraft and Rockets*, **43**, No. 4, 794–805.

[5] Biesbroek, R., and Ancarola, B. (2002) *Optimization of Launcher Performance and Interplanetary Trajectories for Pre-Assessment Studies*. IAF abstracts, 34th COSPAR Scientific Assembly, The Second World Space Congress, held 10-19 October, in Houston, TX, USA., pA-6-07IAF.

[6] Izzo, D. (2006) Advances in Global Optimisation for Space Trajectory Design. *Proceedings of the international symposium on space technology and science*, **25**, p. 563.

[7] Izzo, D., Vinkó, T., and Del Rey Zapatero, M. (2007) GTOP Database: Global Trajectory Optimisation Problems and Solutions. *Web resource*, http://www.esa.int/gsp/ACT/inf/op/globopt.htm

[8] Vinkó, T., Izzo, D., and Bombardelli, C. (2007) *Benchmarking Different Global Optimisation Techniques for Preliminary Space Trajectory Design*. Paper IAC-07-A1.3.01, 58th International Astronautical Congress, Hyderabad, India.

[9] Myatt, D., Becerra, V., Nasuto, S., and Bishop, J. (2004) *Advanced Global Optimisation Tools for Mission Analysis and Design*. Tech. Rep. 03-4101a, European Space Agency, the Advanced Concepts Team, available online at www.esa.int/act

[10] Storn, R., and Price, K. (1997) Differential Evolution – A Simple and Efficient Heuristic for Global Optimization over Continuous Spaces. *Journal of Global Optimization*, **11**, No. 4, 341–359.

[11] Kennedy, J., and Eberhart, R. (1995) Particle Swarm optimization. *Proceedings, IEEE International Conference on Neural Networks*, **4**.

[12] Blackwell, T., and Branke, J. (2004) Multi-Swarm Optimization in Dynamic Environments. *Lecture Notes in Computer Science*, **3005**, 489–500.

[13] Corana, A., Marchesi, M., Martini, C., and Ridella, S. (1987) Minimizing Multimodal Functions of Continuous Variables with the Simulated Annealing Algorithm. *ACM Transactions on Mathematical Software (TOMS)*, **13**, No. 3, 262–280.

[14] Holland, J. (1992) Genetic Algorithms Computer Programs That "Evolve" in Ways That Resemble Natural Selection Can Solve Complex Problems Even Their Creators Do Not Fully Understand. *Scientific American*, **267**, 1992, 66–72.

[15] Izzo, D., Rucinski, M., and Ampatzis, C. (2009) *Parallel Global Optimisation Meta-Heuristics using an Asynchronous Island-model*. IEEE Congress on Evolutionary Computation (IEEE CEC 2009), Trondheim, Norway, May 18-21.

[16] Wales, D., and Doye, J. (1997) Global Optimization by Basin-Hopping and the Lowest Energy Structures of Lennard-Jones Clusters Containing up to 110 Atoms. *Journal of Physical Chemistry*, **101**, No. 28, 5111–5116.

[17] Armellin, R., Di Lizia, P., Topputo, F., and Zazzera, F. (2008) *Gravity Assist Space Pruning Based on Differential Algebra*. New Trends in Astrodynamics and Applications V, Milano, June 30th–July 2nd.

[18] Schutze, O., Vasile, M., Junge, O., Dellnitz, M., and Izzo, D. (2009) Designing Optimal Low-Thrust Gravity-Assist Trajectories Using Space Pruning and a Multi-Objective Approach. *Engineering Optimization*, **41**, 155–181.

[19] Vasile, M., Ceriotti, M., Radice, G., Becerra, V., Nasuto, S., and Anderson, J. (2007) *Global Trajectory Optimisation: Can We Prune the Solution Space When Considering Deep Space Manoeuvres?* Tech. Rep. 06-4101c, European Space Agency, the Advanced Concepts Team, Available online at www.esa.int/act

[20] Zazzera, F., Lavagna, M., Armellin, R., Di Lizia, P., Topputo, F., and Bertz, M. (2007) *Global Trajectory Optimisation: Can We Prune the Solution Space When Considering Deep Space Manoeuvres?* Tech. Rep. 06-4101b, European Space Agency, the Advanced Concepts Team, Available online at www.esa.int/act

[21] Olympio, J., and Marmorat, J. (2007) *Global Trajectory Optimisation: Can We Prune the Solution Space When Considering Deep Space Manoeuvres?* Tech. Rep. 06-4101a, European Space Agency, the Advanced Concepts Team, available online at www.esa.int/act

8 Incremental Techniques for Global Space Trajectory Design

Massimiliano Vasile and Matteo Ceriotti
Department of Aerospace Engineering, University of Glasgow, Glasgow, United Kingdom

8.1 Introduction

Multiple gravity assist (MGA) trajectories represent a particular class of space trajectories in which a spacecraft exploits the encounter with one or more celestial bodies to change its velocity vector. If deep space maneuvers (DSM) are inserted between two planetary encounters, the number of alternative paths can grow exponentially with the number of encounters and the number of DSMs. The systematic scan of all possible trajectories in a given range of launch dates quickly becomes computationally intensive even for moderately short sequences of gravity assist maneuvers and small launch windows. Thus finding the best trajectory for a generic transfer can be a challenge. The search for the best transfer trajectory can be formulated as a global optimization problem. An instance of this global optimization problem can be identified through the combination of a particular trajectory model, a particular sequence of planetary encounters, a number of DSMs per arc, a particular range for the parameters defining the trajectory model, and a particular optimality criterion. Thus a different trajectory model would correspond to a different instance of the problem even for the same destination planet and sequence of planetary encounters. Different models as well as different sequences and ranges of the parameters can make the problem easily solvable or NP-hard. However, the physical nature of this class of transfers allows every instance to be decomposed into subproblems of smaller dimension and smaller complexity. Each subproblem can be approached incrementally, adding one segment of the trajectory at the time. At each incremental step, a portion of the search space can be pruned out.

In a work by Myatt et al. [1] it was demonstrated that if the trajectory model does not contain deep space maneuvers and a powered swing-by model is adopted for the gravity assist maneuver, then an algorithm with polynomial complexity exists that can prune the solution space efficiently. In this particular instance of the MGA problem, each subproblem is a bi-impulsive planet-to-planet transfer and can be approached independently of the other subproblems. In the work of Myatt, the

---

The authors are grateful to Dr. Rüdiger Jehn of the European Space Operations Centre for the reference solution for the EVVMeMe test case.

two-dimensional search space associated with each subproblem was explored with a simple grid sampling. Unfortunately, when deep space maneuvers are inserted along a planet-to-planet transfer leg and an unpowered swing-by model for gravity assist maneuver is considered [2], a grid sampling approach becomes problematic due to the higher dimensionality. The main issue associated to the use of grid sampling is that if a coarse grid and an aggressive pruning are used, many optimal solutions are lost; on the other hand, if a sufficiently fine grid is used, the computational time becomes unacceptable even for a limited number of planetary swing-bys. As it will be explained in the remainder of this chapter, this is due to a dependency problem peculiar to this particular way of modeling MGA trajectories.

Systematic approaches for trajectory models with DSMs have been already proposed in the past, leading to the effective generation of optimal solutions even for complex MGA transfers [3–6]. Some approaches made use of reduced models [7] or graphical tools like the Tisserand Graph [5, 8]. More recently some authors tackled the problem with stochastic based approaches of different nature. Vasile et al. [9, 10] proposed a combination of branch and bound and evolutionary search, Olds et al. [11] showed the effectiveness of Differential Evolution, Rosa Sentinella [12] proposed a combination of various evolutionary algorithms, Gurfil et al. [13] proposed a memetic genetic algorithm, and Vasile et al. [14] a hybridization of Differential Evolution and Monotonic Basin Hopping.

In this work, we present an incremental approach in which, after decomposition of the problem into subproblems, the grid sampling is substituted with a global search through a stochastic method. At each sublevel, the search for the global optimum is substituted with the search for a feasible set. Then the feasible set is preserved and the rest of the search space is pruned out. It will be shown how the proposed incremental search performs an effective pruning of the search space, providing interesting results with a lower computational cost compared to a nonincremental approach. In particular we compare the proposed incremental process to the direct application of known stochastic methods for global optimization, such as a simple Multi-Start (MS), Differential Evolution [15] (DE) and Monotonic Basin Hopping [16] (MBH) to the whole problem.

8.2 Modeling MGA Trajectories

In this section, we present two particular MGA models that lead to problems of different complexity and prone to different solution approaches. Although the incremental approach presented in the following section could, in principle, be applied to both models, we will show its application only to the most complicated of the two. ·

8.2.1 The Linked Conic Approximation

A multiple gravity assist trajectory can be defined as a sequence of transfer arcs and swing-bys of gravitational bodies, starting from a departure one and ending at a target one (or a target orbit). A change in the velocity vector along the trajectory can be obtained either by firing the engine of the spacecraft or by exploiting the gravity of a celestial body. On the scale of the solar system and for fast swing-bys, both the propelled maneuvers and the gravity assist maneuvers can be generally considered instantaneous. Thus an acceptable approximation is to consider the heliocentric position of the spacecraft fixed during each maneuver, either propelled or gravity assisted. In other words, each maneuver has the effect of introducing a discontinuity in the velocity vector, but not in the position vector. This particular model of a multiple gravity assist trajectory is called linked-conic approximation, since it is made of conic arcs (the transfer arcs) linked together by impulsive changes in the velocity vector (given by the swing-bys or by a DSM). For each instant of time, the position and velocity of the celestial bodies are given by analytical ephemerides with respect to a heliocentric, ecliptic, inertial reference frame. Therefore, given a sequence of celestial bodies and times of encounter, the position of each gravity assist maneuver is fully determined. For the cases presented in this chapter, all the celestial bodies are planets. At the departure planet, the velocity of the spacecraft is the sum of the launch velocity and the heliocentric velocity of the planet and is normally limited by the launch capabilities.

8.2.1.1 Gravity Assist Models

Following the assumption of the linked conic approximation, the effect of the gravity of a planet is to instantaneously change the velocity vector of the spacecraft. The incoming velocity vector and the outgoing velocity vector relative to the planet have the same modulus but different directions; therefore, the heliocentric outgoing velocity results to be different from the heliocentric incoming one. In the linked conic model, the spacecraft is assumed to follow a hyperbolic trajectory with respect to the swing-by planet. The angular difference between the incoming relative velocity $\tilde{\mathbf{v}}_i$ and the outgoing one $\tilde{\mathbf{v}}_o$ depends on the modulus of the incoming velocity, on the gravity parameter of the celestial body μ_P, and on the pericenter radius $\tilde{r}_p$. Both the relative incoming and outgoing velocities belong to the plane of the hyperbola. Thus the linked conic approximation of a gravity maneuver can be modeled with the following set of algebraic equations

$$\mathbf{r}_i = \mathbf{r}_o = \mathbf{r}_P \tag{8.1}$$

$$\tilde{v}_i = \tilde{v}_o \tag{8.2}$$

$$\langle \tilde{\mathbf{v}}_o, \tilde{\mathbf{v}}_i \rangle = -\cos(2\beta)\,\tilde{v}_i^2 \tag{8.3}$$

$$\beta = \arccos\left(\frac{\mu_P}{\tilde{v}_i^2 \tilde{r}_p + \mu_P}\right) \tag{8.4}$$

where $\mathbf{r}_i$, $\mathbf{r}_o$ and $\mathbf{r}_P$ are respectively the incoming heliocentric position, the outgoing heliocentric position, and the heliocentric position of the planet. If the incoming and outgoing velocities are available, for example, as free parameters of the global optimization problem, Equations (8.2) and (8.3) represent a set of nonlinear constraints. Thus a feasible trajectory must satisfy Equations (8.2) and (8.3).

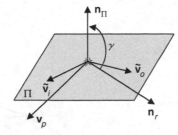

Figure 8.1. Geometry of the swing-by model.

If only the incoming velocity vector is available from the transfer leg prior to the swing-by, then the outgoing vector can be computed provided that the attitude of the plane of the hyperbola is defined. The attitude of the plane of the hyperbola Π can be defined through its normal vector $\mathbf{n}_\Pi$, and the angle γ between the normal vector and a reference plane (see Figure 8.1). There are different possible choices for the definition of the attitude angle γ; the one proposed in [2] is to define γ as the angle between the vector $\mathbf{n}_\Pi$ and the reference vector $\mathbf{n}_r$, that is normal to the plane containing the incoming relative velocity and the velocity of the planet $\mathbf{v}_P$.

If the outgoing velocity is computed from the incoming velocity, all trajectories are feasible with respect to the swing-by constraints, but the problem dimension is higher because a new free parameter needs to be introduced. Furthermore, the outgoing velocity now depends on the incoming one, therefore the leg after the swing-by cannot be generated without knowing the incoming velocity. Expressing the outgoing velocity as a function of the incoming one rather than satisfying constraints in Equations (8.2) and (8.3) has important consequences on the complexity of the solution algorithm.

8.2.2 Velocity Formulation

Assuming a two-body dynamic model, an orbit arc can be characterized in two ways: by assigning a value to the initial position and velocity vector $\mathbf{r}$, $\mathbf{v}$ and to the transfer time T, or by assigning a value to the initial and final position vectors $\mathbf{r}_1$, $\mathbf{r}_2$ and to the transfer time T. The number of free parameters is the same for both cases, but while in the former case an initial value problem needs to be solved, in the latter case the solution of a boundary value problem (or Lambert's problem) is required. We can call the former characterization *velocity formulation* of the conic arc, while the latter is the *position formulation* of the conic arc.

Assume now that a trajectory is made of two contiguous arcs separated by a discontinuity in the velocity vector. If the velocity formulation is adopted, the velocity and position vectors at the end of the first arc have to be computed before generating the second arc. On the other hand, if the position formulation is used, the second arc can be generated independently of the first arc.

From the definition of velocity and position formulation, we can call the computation of the outgoing vector from the incoming vector in the gravity assist model the velocity formulation of the GA model, while solving constraints in Equations (8.2)

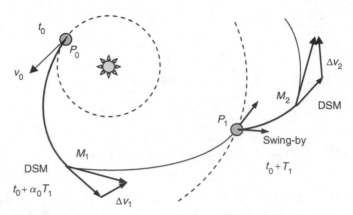

Figure 8.2. Schematic representation of a multiple gravity assist trajectory.

and (8.3) is the position formulation of the GA model. More in general, trajectory models for which each stage of the trajectory depends on the full state vector (position and velocity) at the end of the previous stage will be called velocity formulations.

If the velocity formulation for the GA model is adopted, a complete MGA trajectory can be modeled as in Figure 8.2. Given a sequence of N_P planets, there exist $N_P - 1$ legs, each of them beginning and ending with an encounter with a planet. Each leg i is made of two conic arcs: the first, propagated analytically forward in time, ends where the second, the solution of a Lambert's problem [17], begins. The two arcs have a discontinuity in the absolute heliocentric velocity at their matching point M_i. Each DSM is computed as the vector difference between the velocities along the two conic arcs at the matching point. Given the time of flight (TOF) T_i and the variable $\alpha_i \in [0, 1]$ relative to each leg i, the matching point is at time $t_{DSM,i} = t_{f,i-1} + \alpha_i T_i$, where $t_{f,i-1}$ is the final time of the leg $i - 1$. The relative velocity vector at the departure planet $\mathbf{v}_0$ can be a design parameter and is expressed as

$$\mathbf{v}_0 = v_0 \left[\sin \bar{\delta} \cos \bar{\theta}, \sin \bar{\delta} \sin \bar{\theta}, \cos \bar{\delta} \right]^T \tag{8.5}$$

with the angles $\bar{\delta}$ and $\bar{\theta}$ respectively representing the declination and the right ascension with respect to a local reference frame, with the x axis aligned with the velocity vector of the planet, the z axis normal to orbital plane of the planet, and the y axis completing the coordinate frame. This choice allows easily constraining the escape velocity and asymptote direction while adding the possibility of having a deep space maneuver in the first arc after the launch. This is often the case when the escape velocity must be fixed due to the launcher capability or to the requirement of a resonant swing-by of the Earth (Earth-Earth transfers).

In order to have a uniform distribution of random points on the surface of the sphere defining all the possible launch directions, the following transformation [18] can be applied

$$\theta = \frac{\bar{\theta}}{2\pi}; \; \delta = \frac{\cos \left(\bar{\delta} + \pi / 2 \right) + 1}{2}. \tag{8.6}$$

It results that the sphere surface is uniformly sampled when a uniform distribution of points θ, $\delta \in [0, 1]$ is chosen. Once the heliocentric velocity at the beginning of leg i, which can be the result of a swing-by maneuver or the asymptotic velocity after launch, is computed, the trajectory is analytically propagated until time $t_{DSM,i}$. The second arc of leg i is then solved through a Lambert's algorithm, from M_i, the Cartesian position of the deep space maneuver, to P_i, the position of the target planet of phase i, for a time of flight $(1 - \alpha_i) T_i$. Two subsequent legs are then joined together using the swing-by model. Given the number of legs of the trajectory $N_L = N_P - 1$, the complete solution vector for this model is

$$\mathbf{x} = [v_0, \theta, \delta, t_0, \alpha_1, T_1, \gamma_1, r_{p,1}, \alpha_2, T_2, \ldots, \gamma_i, r_{p,i}, \alpha_{i+1}, T_{i+1}, \ldots, \gamma_{N_L-1}, r_{p,N_L-1}, \alpha_{N_L}, T_{N_L}] \tag{8.7}$$

where t_0 is the departure date. Now the design of a multiple gravity assist transfer can be transcribed into a general nonlinear programming problem, with simple box constraints, of the form

$$\min_{\mathbf{x} \in D} f(\mathbf{x}). \tag{8.8}$$

One of the appealing aspects of this formulation is its solvability through a general global search method for box constrained problems. Depending on the kind of problem under study, the objective function can be defined in different ways. Here we choose to focus on the problem of minimizing the total Δv of the mission, therefore defining

$$f(\mathbf{x}) = v_0 + \sum_{i=1}^{N_L} \Delta v_i + \Delta v_f \tag{8.9}$$

where Δv_i is the velocity change due to the DSM in the ith leg, and Δv_f is the maneuver needed to inject the spacecraft into the final orbit.

The trajectory model in Figure 8.2 solves explicitly the gravity assist constraints. To do that, it requires the incoming velocity before computing the outgoing velocity. Furthermore, the computation of the transfer leg from planet P_i to DSM at position M_{i+1} requires the velocity at the beginning of the transfer arc. As a consequence, the transfer leg from M_{i+1} to P_{i+1} is dependent on the full state vector at the end of the preceding arc. Velocity formulations suffer from a dependency problem: since each stage of the trajectory cannot be computed without all the preceding stages, if we discretize the state vector at the end of each stage, the number of possible paths, corresponding to each discrete value of the state vector, grows exponentially with the number of stages.

8.2.3 Position Formulation

The alternative to the velocity formulation is the position formulation (see Figure 8.3). In the position formulation, we assign the position and time of each

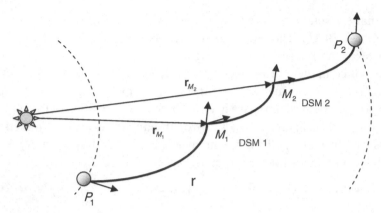

Figure 8.3. Schematic representation of the position formulation of the MGA problem with DSMs.

event along the trajectory: DSMs and swing-bys. Each leg is then computed as a solution of a Lambert's problem. Velocities are computed *a posteriori* as a result of the solution of the Lambert's problems. The position formulation allows us to compute each arc independently of the other arcs. Now let us assume, without the loss of generality, that the problem is planar and that the distance of each DSM from the center of the coordinate system is constant; then the position M_i of each DSM can be identified by a single variable: the angle θ_{M_i}. The time at which the DSM happens requires another variable, t_{M_i}. If we consider that each angle θ_{M_i} can assume only k values, and the time t_{M_i}, only h values, then the number of possible distinct arcs leading to a DSM is equal to hk. The position of the planets is determined through the ephemerides, thus only one parameter, the epoch, has to be specified. Once again it is assumed that the time can be discretized in h elements. This means that there are

$$h \cdot hk = h^2 k$$

possible distinct arcs to go from planet P_1 (at a given time t_1) to M_1 (at a given time and position). These arcs can be computed independently of the rest of the trajectory once $t_1, t_{M_1}, \theta_{M_1}$ are given. The same holds for connecting the last DSM to the arrival planet.

An arc connecting two consecutive deep space maneuvers is determined when the time and position of the two DSMs is fixed. Thus, the total number of independent arcs is

$$hk \cdot hk = h^2 k^2.$$

Once again, these arcs can be computed independently of the other parts of the trajectory.

For the trajectory given as an example, with 2 DSMs, the total number of independent legs is

$$h^2 k + h^2 k^2 + h^2 k = 2h^2 k + h^2 k^2$$

and in general, if N deep space maneuvers are considered

$$2h^2k + (N-1)h^2k^2.$$

Therefore, the position formulation does not suffer from the dependency problem, and the growth is polynomial. On the other hand, the position formulation requires the solution of a set of nonlinear constraints for each gravity maneuver in order to link the incoming and the outgoing velocities. As an alternative, a powered swing-by model can be used to match the incoming velocity to the outgoing velocity, but it should be noted that performing a corrective maneuver at the pericenter of the swing-by hyperbola may increase the efficiency of the transfer. As a result, the same transfer trajectory may result in being more efficient when computed with the position formulation than when computed with the velocity formulation.

8.3 The Incremental Approach

The generic Δv_i in Equation (8.9) can be computed once the trajectory is completed up to leg i. This means that only the part of the solution vector $\mathbf{x}$ concerning legs 1 to i is needed, and the value of f is independent of the variables associated to legs $i+1$ to N_L. This allows decomposing the problem into subproblems, or *levels*, and solving it incrementally adding one level at a time. Let us call $D_{L,i}$ the dimensional slice of the global domain D that is composed only of the variables related to level i. For the model used here, the variables and domains for levels from 1 to i are listed in Table 8.1. Let us also define $D_i = \prod_{k=1}^{i} D_{L,k}$, such that the trajectory up to level i is defined on the domain D_i.

Let us introduce a partial objective function, for each level, of the form:

$$f_i(\mathbf{x}_i) = f_{i-1}(\mathbf{x}_{i-1}) + \phi_i(\mathbf{x}_i), \quad \mathbf{x}_i \in D_i, \quad i = 1 \ldots N_L \tag{8.10}$$

such that

$$f(\mathbf{x}) = f_{N_L}(\mathbf{x}_{N_L}) = f_{N_L-1}(\mathbf{x}_{N_L-1}) + \phi_{N_L}(\mathbf{x}_{N_L}) \tag{8.11}$$

where $\phi_i(\mathbf{x}_i)$ is a function that is specific to a given level i. It is important to stress that the function f_i associated with level i depends only on the part of the solution vector related to the legs from 1 to i. Then, according to Bellman's principle of optimality, if all the trajectory legs from 1 to i are optimal, f_i is a lower bound for f_j when $j > i$, and for the whole objective function f. Although this is generally true, it does not help us define a proper partial objective function since a minimum, local or global, of the partial objective function is not necessarily a minimum for the whole objective function. On the other hand, if $\mathbf{x}^*$ is the global minimum of the whole objective function f, we can identify at each level i a set $\bar{D}_i$ such that $\mathbf{x}_i^* \in \bar{D}_i$, where $\mathbf{x}_i^*$ is the partial vector containing the components of $\mathbf{x}^*$ up to level i. We call the set $\bar{D}_i \subseteq D_i$ the feasible set at level i and we define it as

$$\bar{D}_i = \{\mathbf{x}_i \in D_i | \Phi_i(\mathbf{x}_i)\} \tag{8.12}$$

Table 8.1. *Levels and related variables*

Level	Variables	Domain
1	$t_0, \theta, \delta, \alpha_1, T_1$	$D_{L,1}$
2	$\gamma_1, r_{p,1}, \alpha_2, T_2$	$D_{L,2}$
...	...	...
i	$\gamma_{i-1}, r_{p,i-1}, \alpha_i, T_i$	$D_{L,i}$

where $\Phi_i(\mathbf{x}_i)$ is a property common to all the solutions in the set. The interest is therefore to converge to a set of solutions with property $\Phi_i(\mathbf{x}_i)$. For example, if the property is to have a partial objective function with a value below a given threshold $\bar{f}_i$ then the property $\Phi_i(\mathbf{x}_i)$ can become the Boolean condition

$$\Phi_i(\mathbf{x}_i) = f_i \leq \bar{f}_i. \tag{8.13}$$

Note that Equation (8.13) is consistent with Bellman's principle; furthermore, conceptually, this approach can be equally applied to the position formulation without modification if we assume that the feasible set is made of those solutions that satisfy the constraints and have f_i below a given threshold. We call the property $\Phi_i(\mathbf{x}_i)$ a pruning criterion since we can prune out the portion of the solution space that does not belong to $\bar{D}_i$ and consider for level $i+1$ the new solution space

$$\bar{D}_i \times D_{L,i+1}. \tag{8.14}$$

The overall process is called *incremental pruning* and requires the definition of a pruning criterion at each level i. Furthermore we need to introduce a special partial objective function $\phi_i(\mathbf{x}_i)$ such that, although Equation (8.11) is not satisfied, all the solutions in $\bar{D}_i$ are minimal for $\phi_i(\mathbf{x}_i)$. What makes this approach interesting is that the evaluation of a partial objective function can be remarkably less expensive than the evaluation of the complete function f, and the associated search space is easier to explore. Thus it is possible to analyze level 1 using f_1 on $D_1 \equiv D_{L,1}$ and ideally remove (or prune) from the search space all the sets of values for which the partial objective function is above the threshold. The result is a pruned partial domain $\bar{D}_1 \subseteq D_1$. Then the process continues with level 2, considering f_2, on $\bar{D}_1 \times D_{L,2}$. Note that this partial domain has a smaller volume than $D_1 \times D_{L,2}$, as there are sets of points in D_1 that have already been discarded during the pruning of level 1. The reduction of the search space at level i makes the search at level $i+1$ more effective. At the last level, the complete objective function f is then minimized on the remaining part of the search domain that was not pruned at previous levels, which is $\bar{D}_{N_L}$.

Note that care must be taken in defining the different pruning criteria and the partial objective functions: for example, if the search for the feasible sets was exhaustive, the threshold $\bar{f}_i$ could be chosen arbitrarily low, but for general search methods the threshold needs to be relaxed to avoid overpruning the search space. Furthermore, as it will be shown later in the EEM test case, for some particular kinds of transfer, the partial objective functions $\phi_i(\mathbf{x}_i)$ are not related to f. In fact, for some cases, it is possible to exploit the knowledge of the physics of the problem to create a partial objective function in which $\phi_i(\mathbf{x}_i)$ does not contribute to the value of the objective function of the whole problem but is specifically devized to prune the search space at level i.

8.3.1 Solution of the Subproblems

In order to find $\bar{D}_i$ at each level i, we can use a global search to identify the regions for which the partial objective function f_i is below a given threshold. We use two approaches: a gradient-based approach derived from MBH and an evolutionary-based approach. Given a local minimum $\mathbf{x}_l$ and a neighborhood of this local minimum $N(\mathbf{x}_l, \rho_l) \subseteq D_i$, with radius ρ_l, MBH selects a random point $\mathbf{x}_c \in N(\mathbf{x}_l, \rho_l)$ and runs a local optimization. If the new local minimum $\bar{\mathbf{x}}_l$ obtained starting from the candidate point $\mathbf{x}_c$ is better (that is, lower value of the objective function) than $\mathbf{x}_l$, then $\mathbf{x}_l = \bar{\mathbf{x}}_l$. MBH saves only the local best and therefore would be unusable to explore the feasible set once one point is identified. Therefore, MBH was modified to store all candidate local optimal points $\mathbf{x}_c$ or $\bar{\mathbf{x}}_l$ that satisfy the conditions $\Phi_i(\mathbf{x}_i)$. The modified MBH can be found in Algorithm 8.1.

The evolutionary-based search is derived from the Multi-Agent Collaborative Search (MACS) described in [10]. The basic idea underneath the MACS is to assign the task of looking for a set of solutions to a population P of agents that perform a combination of local and global searches. An agent is identified by a solution vector $\mathbf{x}_j$, a sub-region of the search space $N(\mathbf{x}_j, \rho_j) \subseteq D_i$, and a local search operator, or

Algorithm 8.1. *Modified MBH*

1:	Select $\mathbf{x}$ in D_i, initialize $n_{eval} = 0$, $n_{trials} = 0$
2:	Run local optimizer from $\mathbf{x}$ to local minimum $\mathbf{x}_l$
3:	Select a candidate point $\mathbf{x}_c \in N(\mathbf{x}_l, \rho_l)$; update n_{eval};$n_{trials} \leftarrow n_{trials} + 1$
4:	**If** $n_{trials} > n_{tmax}$
5:	goto Step 1
6:	**End If**
7:	**If** $\Phi_i(\mathbf{x}_i)$ **Then**
8:	$A_f \leftarrow A_f \cup \{\mathbf{x}_c\}$; goto Step 3
9:	**End If**
10:	Run local optimizer from $\mathbf{x}_c \rightarrow \bar{\mathbf{x}}_l$, update n_{eval}
11:	**If** $f_i(\bar{\mathbf{x}}_l) \leq f_i(\mathbf{x}_l)$ **Then**
12:	$\mathbf{x}_l \leftarrow \bar{\mathbf{x}}_l$
13:	**If** $\Phi_i(\mathbf{x}_i)$
14:	$A_f \leftarrow A_f \cup \{\mathbf{x}_l\}$
15:	**End If**
16:	**If** $f_i(\bar{\mathbf{x}}_l) < f_i(\mathbf{x}_l)$
17:	$n_{trials} = 0$
18:	**End If**
19:	**End If**
20:	**Termination** Unless $n_{eval} \geq n_{max}$, goto Step 3

Algorithm 8.2. *Modified MACS*

1:     Initialize a population P_0 in D_i, with n_{pop} agents, initialize $n_{eval} = 0$, $g = 0$

2:     $\forall \mathbf{x}_j \in P_g$, **If** $\Phi_i(\mathbf{x}_i)$ **Then**

3:           $A_f \leftarrow A_f \cup \{\mathbf{x}_j\}$

4:     **End If**

5:     Apply communication operator: $com : P_g \rightarrow P_c; n_{eval} \leftarrow n_{eval} + n_{pop}$

6:     Apply greedy selection operator:

         $\forall \mathbf{x}_j \in P_g$, $\mathbf{x}_j \leftarrow \mathbf{x}_{j,c} \in P_c$ if $f(\mathbf{x}_{j,c}) \leq f(\mathbf{x}_j)$, for $j = 1, \ldots, n_{pop}$

7:     Rank the population P_g

8:     Apply local search operator to the best $n_{popratio}$ agents in P_g:

         for $\mathbf{x}_j \in P_g$, with $j = 1, \ldots, n_{popratio}$, generate n_{trials} candidate points $\mathbf{x}_{j,l} \in N(\mathbf{x}_j, \rho_j)$;

         $n_{eval} \leftarrow n_{eval} + n_{popratio} n_{trials}$

9:     Apply greedy selection operator:

         $\mathbf{x}_j \leftarrow \mathbf{x}_{j,l}$ for $j = 1, \ldots, n_{popratio}$ and $l = 1, \ldots, n_{trials}$ if $f_i(\mathbf{x}_{j,l}) \leq f_i(\mathbf{x}_j)$

10:    **If** $\Phi_i(\mathbf{x}_i)$**Then**

11:         $A_f \leftarrow A_f \cup \{\mathbf{x}_{j,l}\} A_f = A_f + \{\mathbf{x}_{j,l}\}$

12:    **End If**

13:    $P_{g+1} \leftarrow P_g; g \leftarrow g + 1$

14:    **If** $\forall \mathbf{x}_j, \mathbf{x}_k \in P_g, \max\limits_{j,k} \|\mathbf{x}_j - \mathbf{x}_k\| < tol_{conv}$ **Then**

15:         restart the worse of the two

16:    **End If**

17:    **Termination** Unless $n_{eval} \geq n_{max}$, goto Step 5

---

individualistic behavior function $\beta(\mathbf{x}_j, n_{trials})$ that generates n_{trials} samples $\mathbf{x}_{p,l}$ such that $\mathbf{x}_{p,l} \in N(\mathbf{x}_j, \rho_j)$. At every generation g, a communication operator recombines pair-wise the agents in the current population P_g and generates a new candidate population P_c made of solution points $\mathbf{x}_{j,c}$. A greedy selection operator selects only the solutions in P_c that improve the solutions in P_g and updates P_g. Before applying the communication operator, the points in P_g that satisfy condition $\Phi_i(\mathbf{x}_i)$ are stored in an archive A_f. After the communication operator, the local search operator is applied to the best $n_{popratio} < n_{pop}$ agents. All the sample points $\mathbf{x}_{p,l} \in N(\mathbf{x}_j, \rho_j)$ satisfying condition $\Phi_i(\mathbf{x}_i)$ are stored in the archive A_f. The overall MACS is presented in Algorithm 8.2. Furthermore, in order to facilitate the local exploration in a neighborhood of a feasible solution without any modification to the two search algorithms, we assigned the value -2 to the objective function of all the feasible solutions.

The modified MBH was complemented with a restart of the process after a number $n_{t,max}$ of local trials. This restart is fundamental to avoid stagnation and coverage of only one portion of the feasible set when the feasible set is disconnected. For MACS we adopted a similar idea but the domain D_i is partitioned in subdomains at every restart and MACS is restarted within a subdomain. The domain D_i is partitioned by dividing in two parts one coordinate belonging to a subset of the coordinates at level i (for example, only the second and the third coordinate are divided while the others remain unchanged). At each restart, the coordinate with the longest edge is cut in two parts generating two new subdomains. For each subdomain, we evaluate the two criteria vector

$$\boldsymbol{\psi}_{D_q} = \left[-V_{D_q}, \frac{n_{D_q}}{n_{D_i}} \right] \tag{8.15}$$

where V_{D_q} is the volume of the subdomain D_q and n_{D_q}/n_{D_i} is the ratio between the feasible solutions n_{D_q} in $D_q \subseteq D_i$, such that $D_i = \bigcup_q D_q$, and the total number of feasible solutions n_{D_i} in D_i. For each subdomain, we evaluate its Pareto optimality with respect to the criteria vector in Equation (8.15) by computing the dominance index

$$I_d(D_q) = |\{j|D_j \succ D_q\}| \tag{8.16}$$

where $|.|$ is the cardinality of the set and a subdomain dominates another when both criteria are better, that is, $D_j \succ D_q \Rightarrow -V_{D_j} < -V_{D_q} \wedge n_{D_j}/n_{D_i} < n_{D_q}/n_{D_i}$. Furthermore we count the number of subdivisions $m_d(D_q)$ that do not produce any increase in the number of feasible solutions. For example, assume that the number of feasible solutions in D_i is 100 and that we subdivide D_i in D_q and D_{q+1}; then the subdivision index for each of the subdomains is increased by one unit. After selecting one of the two, we run MACS; if the number of feasible solutions is higher than 100, then m_d is set to 0. In order to select a subdomain where we want to restart MACS, we use the cumulative quality index $I_q(D_q) = I_d(D_q) + m_d(D_q)$. If more subdomains have the same quality index we pick one randomly among them.

8.3.2 Clustering of the Solutions

Once a set of feasible solutions belonging to $\bar{D}_i$ is found, the solutions are grouped with a box clustering procedure. Assuming that we have d dimensions, each coordinate c is subdivided into $m(c)$ bins $b_{q,c} = [b_l, b_u]_{q,c}$ with $q = 1, \ldots, m(c)$, then for the rth solution $\mathbf{x}_r$ in the feasible set we consider the box

$$B_r = b_{q_1,1} \times b_{q_2,2} \times \cdots \times b_{q_d,d} |\mathbf{x} \in B_r \tag{8.17}$$

We add a new box every time a feasible solution is not included in any of the existing boxes. When a level $i+1$ is added to the incremental pruning, the union of all the boxes at the previous level $\widehat{D}_i = \bigcup_r B_r$ is used instead of the actual feasible set $\bar{D}_i$; however, it should be noted that if the search for feasible solutions is exhaustive, $\widehat{D}_i \supseteq \bar{D}_i$.

   If the search engine finds n_{sol} solutions, the algorithmic complexity of this clustering procedure is $n_{sol} \cdot \sum_{c=1}^d m(c)$, which in the case of an equal number of bins per coordinate becomes $n_{sol} \cdot m \cdot d$ and therefore grows linearly with the number of coordinates. Furthermore, in order to reduce the number of boxes that are carried from one level to the next, we also perform a clustering of the boxes: if two or more adjacent boxes can be exactly covered with a single box, the covering box is substituted for the original set of adjacent boxes. The overall incremental process is summarized in Algorithm 8.3. Note line 6 in Algorithm 8.3. The back-pruning procedure removes from levels 1 to $i-1$ the portion of the feasible sets that contain solutions that are not feasible at level i.

8.3.3 Exploration of the Pruned Space

The result of the incremental approach proposed in this chapter is not a specific solution but a set of boxes containing families of solutions. The identification of

Algorithm 8.3. *Incremental Pruning*

1:   Set feasible set D to the whole domain, Start from level 1: $i \leftarrow 1$
2:   **While** $i < n_{levels}$, **Do**
3:          Run search for feasible solutions at level i on feasible set $\bar{D}_{i-1} \times D_{L,i}$
4:          Cluster the feasible solutions
5:          Prune the search space at level i and define new feasible set $\bar{D}_{L,i}$ at level i
6:          Apply back-pruning and redefine feasible sets $\bar{D}_{L,j}$ at levels $j = 1, \ldots, i-1$
7:          Define the new feasible set $\bar{D}_i = \prod\limits_{j=1}^{i-1} \bar{D}_{L,j} \times \bar{D}_{L,i}$
8:   $i \leftarrow i + 1$; **End Do**

a specific solution within the set requires a further exploration step. The objective function within each box is not necessarily convex, and the associated subdomain can still contain many local minima. Thus in the following, the exploration of each box is performed with one of the three generic global optimization methods: Monotonic Basin Hopping, Multi-Start or Differential Evolution.

8.3.3.1 Multi-Start

The Multi-Start (MS) technique is an extremely basic approach to global optimization. In its simplest form, it performs a global sampling of the solution space and then starts a local search from each sampled point. Several variants of this basic principle exist but their description is beyond the scope of this chapter. In the examples presented here, the initial sampling of the search space is performed with Latin Hypercube [19] to obtain a good spreading of the sampled points along each coordinate.

8.3.3.2 Monotonic Basin Hopping

Monotonic Basin Hoping (MBH) was first applied to special global optimization problems, the molecular conformation ones [20], and later extended to general global optimization problems [21]. In its basic version it is quite similar to MS. It is also based on multiple local searches and the only difference is represented by the distribution of the starting points for local searches: while in MS these are randomly generated over the whole feasible region, in MBH they are generated in a neighborhood $N(\mathbf{x}_l)$ of the current local minimizer $\mathbf{x}_l$. Algorithm 8.1 represents, in summary, the working principle of MBH, except for the memorization of the solutions satisfying property Φ_i in the archive A_f. Because of its restart of the local search within $N(\mathbf{x}_l)$, MBH is particularly effective on functions that present a single funnel structure [16] or are globally convex. For multifunnel structures or globally nonconvex cases, MBH needs to be restarted to avoid stagnation at the bottom of each funnel. In the test cases presented in this chapter, both for MS and MBH, the local optimizer is the MATLAB function *fmincon* that implements an SQP method with BFGS update of the Hessian [22].

8.3.3.3 Differential Evolution

Differential Evolution (DE) is a population-based evolutionary algorithm proposed for the first time by Ken Price and Rainer Storn in 1995. Given a population P_k, the basic DE heuristic generates a new candidate point in the search space by updating the position of each individual $\mathbf{x}_{i,k} \in P_k$ with the following iteration

$$\mathbf{x}_{i,c} = \mathbf{x}_{i,k} + \mathbf{e}(C_r)[(\mathbf{x}_{i_1,k} - \mathbf{x}_{i,k}) + F(\mathbf{x}_{i_2,k} - \mathbf{x}_{i_3,k})] \qquad (8.18)$$

where F is the so-called mutation or perturbation parameter, $\mathbf{e}$ is a vector whose components are either 0 or 1 with probability C_r, and the product between $\mathbf{e}$ and the content of the square bracket has to be intended component wise. The individual $\mathbf{x}_{i_1,k}$ can be the best of the population or a random one. In the former case, that will be called strategy *best* in the following, fast convergence is favored, while in the latter case, exploration is favored. The other two individuals in (8.18) instead are chosen randomly, and the mutation parameter F plays an important role in determining the speed of convergence of the algorithm: for example, when strategy *best* is chosen, a small value of F increases convergence while a high value of F favors exploration. The perturbation parameter and the crossover parameter C_r are generally chosen in the interval [0 1]. Other variants of the basic iteration (8.18) exist in the literature; the interested reader can refer to [23] for further information.

Each candidate point becomes part of the new population P_{k+1} according to the greedy selection heuristic

$$\mathbf{x}_{i,k+1} \leftarrow \mathbf{x}_{i,c} \Leftrightarrow f(\mathbf{x}_{i,c}) < f(\mathbf{x}_{i,k}). \qquad (8.19)$$

Note that although (8.19) represents an elitism condition for the selection of the new population, that is, it always preserves the best individual, iteration (8.18) does not generally allow sampling arbitrary subsets of finite measure with nonzero probability. Therefore, the global convergence theorem for evolutionary algorithms proposed by Rudolph [24] does not apply and the asymptotic convergence to the global minimum is not guaranteed.

8.3.4 Discussion

Before proceeding to the next section, it is worthwhile to examine some of the characteristics of the proposed incremental approach. One key assumption of the incremental approach is that a complete solution to the MGA problem, that is, a complete trajectory, can be built by adding individual trajectory legs, starting from the departure to the arrival or vice versa. Therefore, although the global minimum of each subproblem does not represent the global minimum of the whole problem, we can build the solution space of the whole problem by incrementally adding up the search spaces associated with each subproblem in such a way that the resulting total search space contains the global minimum.

The objective functions that are used to prune the search space associated with each subproblem do not directly depend on the chosen objective function for the whole problem. Therefore, the incremental approach is independent of the objective function of the whole problem but is strongly dependent on the characteristics of the trajectory model. In particular, for the trajectory model presented in this paper, the partial objective function (and pruning criterion) associated with each subproblem cannot be evaluated without considering all the previous levels. This represents a fundamental difference with respect to what is done in [1]. In fact, a trajectory model in which gravity maneuvers are modeled as powered swing-bys does not need to build the whole solution incrementally (or as a cascade of subproblems) but each subproblem can be tackled in parallel with the others. Furthermore, in the proposed incremental approach, the search space is built up incrementally, therefore the number of dimensions of each subproblem increases as a new level is added to the list. On the other hand, the number of dimensions of each subproblem in [1] remains constant throughout the whole pruning process.

8.4 Testing Procedure and Performance Indicators

When a new optimization approach is proposed, it is good practice to test its performance against existing approaches on a known benchmark. In order for the tests to be significant, the testing procedure and the performance indicators need to be rigorously defined. The tests will compare the performance of a generic global optimizer, when applied to the search of the solution of a given problem, over the whole search space (called *all-at-once* approach in the following) against the performance of the same optimizer operating on the reduced search space after pruning. In fact, a key advantage of the proposed incremental pruning approach is to increase the probability to find sets of good solutions without increasing the computational cost. Furthermore, since the incremental approach makes use of stochastic-based techniques to identify the feasible set, some tests will demonstrate the reproducibility of the result of the incremental pruning itself. We will start from the definition of a general testing procedure for global optimization algorithms and then we will define some specific performance indicators for a generic optimization algorithm and for the incremental approach.

If we call A a generic solution algorithm and p a generic problem, we can define the procedure in Algorithm 8.4.

Algorithm 8.4. *Convergence test*

1:	Set the max number of function evaluations for A equal to N
2:	Apply A to p for n times
3:	**For** $i \in [1, \ldots, n]$, **Do**
4:	$\quad \varphi(N, i) = \min f(A(N), p, i)$
5:	**End For**
6:	Compute $\varphi_{min}(N) = \min_{i \in [1, \ldots, n]} \varphi(N, i)$; $\varphi_{max}(N) = \max_{i \in [1, \ldots, n]} \varphi(N, i)$

Now and in the following we say that an algorithm A is globally convergent when for a number of function evaluations N that goes to infinity the two functions φ_{min} and φ_{max} converge to the same value, which is the global minimum value denoted as f_{global}. An algorithm A is simply convergent, instead, if for N that goes to infinity the two functions φ_{min} and φ_{max} converge to the same value, which is not necessarily a global or a local minimum for f.

If we fix a tolerance value tol_f, we could consider the following random variable as a possible quality measure of a globally convergent algorithm

$$N^* = \min \left\{ N : \varphi_{max}\left(N'\right) - f_{global} \leq tol_f, \forall N' \geq N \right\}. \tag{8.20}$$

The larger (the expected value of) N^* is, the slower is the global convergence of A. Figure 8.4 a) and b) show the convergence profile for the bi-impulsive transfer from the Earth to asteroid Apophis obtained with 50 repeated independent runs of a Multi-Start algorithm: a number of samples were generated in the solution space with a Latin hypercube sampling procedure and a local optimization that was run from each sample. Slightly more than 1,000 initial samples are required to have a 100% convergence to the global minimum. However, the procedure in Algorithm 8.4 can be impractical since, although finite, the number N^* could be very large. In practice, what we would like is not to choose N large enough so that a success is always guaranteed, but rather, for a fixed N value, we would like to maximize the probability of hitting a global minimizer. Now let us define the following quantities

$$\delta_f\left(\mathbf{x}\right) = \left|f_{global} - f\left(\mathbf{x}\right)\right|; \quad \delta_x\left(\mathbf{x}\right) = \left\|\mathbf{x}_{global} - \mathbf{x}\right\|. \tag{8.21}$$

In case there is more than one global minimum point, $\delta_x\left(\mathbf{x}\right)$ denotes the minimum distance between $\mathbf{x}$ and all global minima. Moreover, in case the global minimum point $\mathbf{x}_{global}$ is not known, we can substitute it with the best-known point $\mathbf{x}_{best}$. We can now define a new procedure, summarized in Algorithm 8.5.

A key point is setting properly the value of n. In fact, a value of n too small would correspond to an insufficient number of samples to have proper statistics. The

Algorithm 8.5. *Convergence to the global optimum*

1:	Set the max number of function evaluations for A equal to N
2:	Apply A to p for n times
3:	Set $j_s = 0$
4:	**For** $i \in [1, \ldots, n]$, **Do**
5:	$\varphi\left(N, i\right) = \min f\left(A\left(N\right), p, i\right)$
6:	$\mathbf{x} = \arg \varphi\left(n, i\right)$
7:	Compute $\delta_f\left(\mathbf{x}\right)$ and $\delta_x\left(\mathbf{x}\right)$
8:	**If** $\delta_f < tol_f \wedge \delta_x < tol_x$ **Then**
9:	$j_s = j_s + 1$
10:	**End If**
11:	**End For**

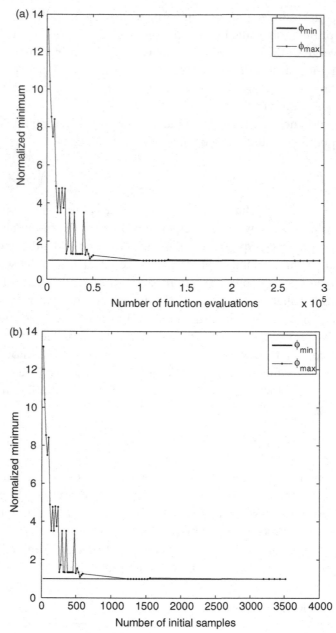

Figure 8.4. Convergence profile for a bi-impulsive Earth–Apophis transfer: (a) convergence as a function of the number of function evaluations; (b) convergence as a function of the number of initial samples for a Multi-Start algorithm.

number n is problem dependent and is related to the complexity of the problem and to the heuristics implemented in the solution algorithm. A proper value for n should give a little or null fluctuations on the value of j_s/n, that is, by increasing n, the value of j_s/n should remain constant or should have a small variation. Note that the values of the tolerance parameter tol_f and tol_x depend on the problem at hand.

Algorithm 8.5 is applicable to general problems either presenting a single solution with value function f_{global} (or f_{best}) or presenting multiple solutions with value f_{global} (or f_{best}). On the other hand, in the following we are not interested in distinguishing between solutions with equal f and different $\mathbf{x}$, therefore we will use a reduced version of Algorithm 8.5 in which the condition $\delta_x(\mathbf{x}) \le tol_x$ is not considered.

Finally, we remark that the two procedures described in Algorithm 8.4 and Algorithm 8.5 only consider the computational cost to evaluate f but not the intrinsic computational cost of A. The intrinsic cost of A is related to its complexity and to the number of pieces of information A is handling. For instance, for a simple grid search, such intrinsic cost is represented by the cost of sweeping through all the N points on the grid at which the objective function is evaluated. The intrinsic cost varies from algorithm to algorithm, but here we are assuming that the computational effort of the algorithms is dominated by the function evaluation cost and therefore, we do not take intrinsic costs into account. Note that if the algorithm A is deterministic then we can set $n = 1$. Indeed, each time A is applied to p, it always returns the same value. Then for deterministic algorithms, given a value N, a reasonable performance index is simply $J_d(N) = \varphi(N, 1)$, that is, the best value returned by the algorithm. Instead, for stochastic based algorithms, different performance indexes can be defined.

Commonly used indexes are the best, the mean, and the variance of all the results returned by the n runs, or the probability of success of a single run. However, the use of best value, mean, and variance present some difficulties. In fact, the distribution of the best values is not Gaussian. Therefore, the distance between the best and the mean values, or the value of the variance in general, does not give an exact indication of the repeatability of the result. Moreover, it changes during the process, therefore we cannot define *a priori* the required number of runs to produce a correct estimation of mean and variance. In addition to that, the minimum number of samples that are required to have sufficient statistics is not well defined for space problems. Note that the use of the best value could be misleading since statistically, even a simple random sampling can converge to the global optimum. On the other hand, an algorithm converging, on average, to a good value with a small variance does not guarantee that it will be able to find the best possible solution. For example, given the integer numbers between 0 and 10, let us assume that we want to find the minimum one, which is, 0, and that an algorithm returns 50% of the times the value 0 and 50% of the times the value 10. The mean would be 5 and the variance about 28, which would lead to the conclusion that the value 0 can be found with probability 0.014 under the assumption of Gaussian distribution. This conclusion is clearly wrong. Assume now that an algorithm returns solutions with mean value equal to 5 and variance equal to 30, and another algorithm returns solutions with mean equal to 3 and variance equal to 10. In this case, which of the two algorithms is better performing is not well defined because the algorithm with a higher mean value has a higher variance and thus also a higher probability to generate solutions better than the average one. Then, if the distribution that describes the statistical phenomenon is not known, these two numbers are not sufficient to claim that one algorithm is better than the other. Note that, because of this evidence, statistical tests like the *t-test*, which start from the assumption of a Gauss distribution, are not applicable or provide unreliable results.

An alternative index that can be used to assess the effectiveness of a stochastic algorithm is the success rate p_s, which is related to j_s in Algorithm 8.5 by $p_s = j_s/n$. Considering the success as the referring index for a comparative assessment implies two main advantages. First, it gives an immediate and unique indication of the algorithm effectiveness, addressing all the issues highlighted above, and second, the success rate can be represented with a binomial probability density function (pdf) independent of the number of function evaluations, the problem, and the type of optimization algorithm. This latter characteristic implies that the test can be designed fixing a priori the number of runs n on the basis of the error we can accept on the estimation of the success rate. We propose here a statistical theory developed by Minisci et al. [25]. It is assumed that the sample proportion p_s of successes (the success rate for a given n in our case) can be approximated with a normal distribution, that is, $p_s \sim N_p \{\theta_p, \theta_p(1 - \theta_p)/n\}$, where θ_p is the unknown true proportion of successes, and that the probability of p_s to be at distance d_{err} from θ_p, $P_r\left[|p_s - \theta_p| \le d_{err}|\theta_p\right]$ is at least $1 - \alpha_b$ (see [26]). This leads to the expression

$$n \ge \theta_p(1 - \theta_p)\chi^2_{(1),\alpha_b} \Big/ d^2_{err} \tag{8.22}$$

and to the conservative rule

$$n \ge 0.25\chi^2_{(1),\alpha_b} \Big/ d^2_{err} \tag{8.23}$$

obtained if $\theta_p = 0.5$. For our tests we required an error $d_{err} \le 0.06$ with a 92% confidence ($\alpha_b = 0.08$), which, according to Equation (8.23), yields $n \ge 94$. This was extended to $n_{runs} = 100$ for all the tests in this chapter in order to have a higher confidence in the result. In order to have a feeling of the speed of convergence, stochastic-based methods were applied to the solution of the whole problem for an increasing number of function evaluations.

For the incremental approach, we defined a number of performance indicators that aim at establishing if the reduction of the search space operated during the incremental search is reliable and efficient. It is important to remind here that the aim of the incremental approach is not to generate optimal solutions but to generate a set of subdomains $\bar{D}_j \subseteq D$ bounding sets of locally optimal solutions. Therefore, the following indicators aim at measuring the ability of the incremental approach to repeatedly generate a tight enclosure of good solutions. Ideally, a good pruning would always yield few small boxes enclosing the global optimum together with all the solutions satisfying $\delta_f < tol_f$. The performance indicators for the incremental pruning are:

- **Percentage of inclusion of the best solutions.** Through the all-at-once approach, a number of solutions satisfying the condition $\delta_f < tol_f$ will be identified. Those solutions are considered to be neighbors of the best one in the criteria space.

The incremental approach is expected to identify, for every run, at least one box containing one or more of the solutions neighboring the best one, that is, one or more solution below the threshold. This indicator gives a measure of the ability of the incremental approach to identify good region of the search space without discarding areas containing potentially good solutions.

- **Percentage of pruned space.** This indicator measures the effectiveness of the reduction of the search space.
- **Average number of blocks.** After the incremental pruning has completed the reduction of the search space, each block coming from the pruning can be further explored to find locally optimal solutions. A small number of blocks is therefore desirable, although multiple blocks can correspond to multiple equivalent launch opportunities. Thus this indicator has to be used together with the percentage of inclusion of the best solutions. In the following we will also considered the standard deviation on the number of generated blocks to provide an indication of the dispersion of the results.
- **Coverage.** This indicator measures the ability of the incremental approach to perform a repeatable pruning of the search space. Given a block $\bar{D}_{j,k}$ for run number k, we compute the number of times that $\bar{D}_{j,k}$ is covered partially or completely in all the other runs

$$I_j(k) = \left| \{i : \forall k, p, k \neq p \wedge \bar{D}_{j,k} \cap \bar{D}_{i,p} \neq \emptyset \} \right| \tag{8.24}$$

with $|\cdot|$ denoting the cardinality of the set. The other coverage index is the actual percentage of a block $\bar{D}_{j,k}$ at run k that is covered by another block $\bar{D}_{i,p}$ at run p

$$\varsigma_j(k) = \frac{1}{n_{runs}} \sum_{p=1}^{n_{runs}} \sum_{i=1}^{n_b(p)} Vol\left(\bar{D}_{j,k} \cap \bar{D}_{i,p}\right) / Vol\left(\bar{D}_{i,p}\right) \tag{8.25}$$

where $n_b(p)$ is the number of blocks resulting from run p, and Vol $(\cdot)$ is the volume of a given block.

8.5 Case Studies

The incremental algorithm was tested on the optimization of two MGA trajectories: the first one is an Earth to Mars transfer, with a single gravity assist of the Earth (EEM sequence); the second is a transfer to Mercury, exploiting two swing-bys of Venus and one of Mercury (EVVMeMe sequence). These two test cases are representative of the class of MGA transfers with resonant swing-bys and well illustrate the complexity of this kind of problems. The incremental approach was compared to the direct solution of the whole problem (all-at-once approach) with two stochastic global optimization methods, DE and MBH, and two deterministic global optimizers, DIRECT [27] (DIvided RECTangles) and Multilevel Coordinate Search [28] (MCS).

8.5.1 EEM Test Case

This single gravity assist test case consists of an Earth-Earth-Mars transfer. Although it is quite simple, the aim of this test is twofold: it demonstrates the effectiveness of the pruning approach, and it is useful to define a particular class of problem-dependent functions $\phi_i(\mathbf{x}_i)$ and pruning criteria $\Phi_i(\mathbf{x}_i)$. The Earth gravity assist is used to increase the kinetic energy of the spacecraft with respect to the Sun when the launch capabilities are limited. In order to gain the required Δv, the spacecraft has to reach the Earth with a relative velocity vector different from the one at departure. This is achieved with a DSM along the Earth-Earth transfer leg. Thus the optimal design of the first leg is essential in order to exploit the encounter of the Earth properly and gain the energy to reach Mars. The departure velocity vector depends on the launch capabilities; therefore, its modulus was set to 2 km/s for this test case, while the nondimensional declination δ and right ascension θ were left free. In the incremental approach, the whole problem is decomposed into two levels: level 1 consists of the Earth-Earth transfer, while level 2 computes the Earth swing-by and the Earth-Mars transfer leg. Since v_0 is constant, the solution vector has only five decision variables at level 1, and v_0 can be removed from all the objective functions. Table 8.2 presents the bounds for the variables of the problem. The global objective function f is the sum of the Δv's of the two deep space maneuvers, plus the difference between the spacecraft velocity and Mars velocity at arrival. The table also reports the number of bins and the number of function evaluations for the pruning performed with MACS and MBH. MBH was run with a perturbation radius ρ_l equal to 10% of the range of the variable at level D_i. MACS was run with a population of 20 agents with $n_{popratio}$ of 10, and a tol_{conv} equal to 10^{-4} of the range of the variables at level D_i. For the restart of MBH, we set the maximum number of trials to 30, while for MACS we partitioned only the second and fifth coordinate since they represent the most critical ones for the EE leg.

The interesting aspect of this problem is that the choice of the partial objective function f_1 for the incremental approach at level 1 is tricky. In fact, the cheapest

Table 8.2. *Search space for the EEM case*

	Lower bound	Upper bound	N. fun. eval. MACS	N. fun. eval. MBH	N. bins	Level
t_0, d, MJD2000	3650	9128.75 (3650 + 15 years)	40000	40000	1	1
θ	0.2	1.2			10	
δ	0	1			5	
α_1	0.01	0.99			5	
T_1,d	50	1000			30	
γ_1, rad	$-\pi$	π	10000–40000	10000–40000	N/A	2
$r_{p,1}$, planet radii	1	5			N/A	
α_2	0.01	0.99			N/A	
T_2, d	50	1000			N/A	

way to perform an Earth-Earth transfer is to move from the Earth orbit as little as possible (or not move at all). Therefore, if the sum of the DSM and v_0 is chosen as objective to minimize, the optimizer returns solutions with no maneuver. These solutions, however, arrive at the Earth with a relative velocity that is not suitable to exploit the swing-by properly. Furthermore, it is known from the physics of the problem that the zero-maneuver solution is a local minimizer even for the whole EEM transfer. Since the gravity assist maneuver requires an accurate timing to reach the swing-by planet with the right incoming conditions, its effect is to narrow down the basin of attraction of each minima that do not correspond to a zero-maneuver solution. In fact, a gravity assist maneuver is more sensitive to a small variation of the variables than a direct transfer. Consequently, the gradient of the objective function in a neighborhood of a solution with Earth swing-by is higher than the gradient in a neighborhood of a solution without Earth swing-by, and the basin of attraction is expected to be narrower. Now a zero-maneuver solution for the EE leg physically corresponds simply to a delayed departure from Earth, with no gravity assist. All the zero-maneuver solutions, therefore, have a much wider basin of attraction. This can be easily verified by applying a general stochastic global optimizer to the whole EEM problem. The optimizer will return the zero-maneuver solutions with higher probability if no special condition is imposed on the departure velocity at the Earth.

In order to minimize the Δv on the EM leg, the incoming velocity vector at the Earth should be such as to have an outgoing relative velocity vector aligned with the velocity vector of the Earth (optimum increase in the kinetic energy) at a time $t_0 + T_1$ such that the EM transfer is close to a Hohmann transfer. A suitable criterion to optimize the first leg can be found by studying the characteristics of the relative velocity vector at the end of the Earth-Earth transfer. Figure 8.5 represents the in-plane components (radial v_r and transversal v_θ) of the normalized incoming relative velocity vector for the best solutions found minimizing the total EEM Δv with

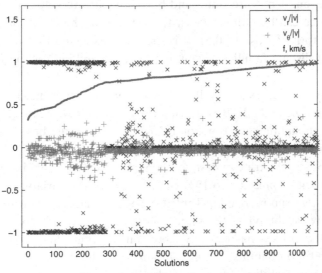

Figure 8.5. Normalized in-plane components of the incoming relative velocity vector before the Earth swing-by, for the best solutions found, and corresponding objective value.

the all-at-once approach. On the same plot the objective function for the complete problem is also represented.

For the best solutions (from 1 to about 300), the direction of the relative velocity is almost completely radial, while for the rest of the solutions the radial component or the entire velocity drops to zero. Therefore, the following partial objective function can be chosen for all the levels in which there is the need for an outgoing velocity parallel to the velocity of the planet either to brake or to accelerate

$$f_i = \omega_i \frac{v_\theta^2 + v_h^2}{v_r^2} + \sum_{k=1}^{i} \Delta v_k. \qquad (8.26)$$

This function tries to minimize the DSM while maximizing the radial component v_r of the relative velocity before the subsequent swing-by, with respect to the other components v_θ, v_h. The weighing factor ω_i is set to 1 km/s in the following. Although this criterion is derived for a specific case, it has general validity and applies to two classes of MGA transfers: aphelion-rising gravity maneuvers and perihelion-lowering gravity maneuvers, as it will be shown in the next test case.

The threshold for level 1 is set to 1 km/s to leave some flexibility in the search for optimal resonant transfers. At level 2 the objective function is the total Δv, therefore it is not pruned.

A first test was run, applying two deterministic-based global optimizers, DIRECT and MCS, and three stochastic-based global optimizers, DE, MS, and MBH to the entire problem (all-at-once approach) to assess the performance of these global optimizers for an increasing number of function evaluations. The best known solution for this problem has a total Δv of 2.908 km/s (see Figure 8.7); therefore, we set tol_f to 0.05 km/s. Figure 8.6 shows the distribution of the values of the variables at level 1 for 200 solutions with an objective function below 2.958 km/s. It is interesting to note that the solutions belong to different launch windows (different departure time) and have a departure velocity with respect to the Earth, which can be either against the velocity of the Earth (θ close to 1) or perpendicular to it (θ close to 0.5). It is therefore expected that both MACS and MBH will find distinct families of solutions when searching for the feasible set at level 1.

For the all-at-once test, we followed the procedure presented in the previous section, that is, DE and MBH were run 100 times and we recorded the number of solutions with an objective function below or equal to the best-known solution. The DE algorithm was run with perturbation parameter $F = 0.8$, crossover parameter $C_r = 0.75$, search strategy *best*, and a population size of 90 individuals, while MBH was run with a ρ_l equal to 10% of the range of the variables. The results of the all-at-once test, summarized in Table 8.3, suggest that deterministic approaches, though they predictably yield the same solution at every run, do not provide satisfactory results. All stochastic approaches, on the other hand, are able to find, with almost 100% probability, better solutions than the deterministic ones. Therefore, though the probability of finding the best-known solution remains small for all

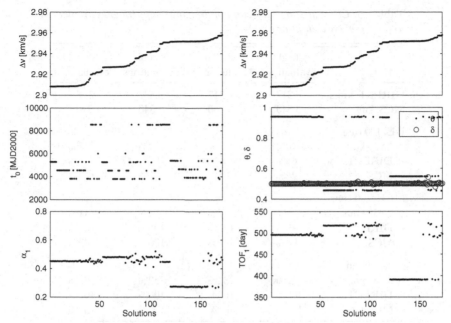

Figure 8.6. Parameters of the first level for the local minima of the complete problem below 2.958 km/s. It is clearly visible that there exist solutions with objective value very close to the global minimum but having substantially different solution vectors.

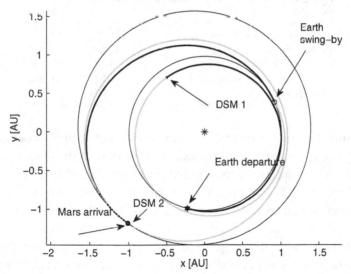

Figure 8.7. Projection on the ecliptic plane of the best solution found by the incremental algorithm. The total Δv is 2.908 km/s.

stochastic methods except MS, their use is advisable. Table 8.3, however, suggests that sophisticated global search methods, such as DE and MBH, are not the right choice; in particular, DE is the worst performing algorithm. The reason is the fast convergence of DE with the selected settings. As theoretically demonstrated in [14] and [29], DE can converge to a fixed point in D, which is not necessarily a local or

Table 8.3. *Comparison of different optimization approaches applied to the EEM case all-at-once*

Solver	20,000 evaluations	40,000 evaluations	80,000 evaluations	160,000 evaluations
DIRECT, km/s	4.317	4.317	3.822	3.809
MCS, km/s	3.840	3.840	3.840	3.812
DE, 100 runs				
< 2.958 km/s	0%	7%	27%	27%
< DIRECT	68%	99%	100%	100%
< MCS	24%	85%	100%	100%
MBH, 100 runs				
< 2.958 km/s	1%	5%	18%	41%
< DIRECT	99%	100%	100%	100%
< MCS	96%	100%	100%	100%
MS, 100 runs				
< 2.958 km/s	22%	32%	52%	67%
< DIRECT	100%	100%	100%	100%
< MCS	100%	100%	100%	100%

Table 8.4. *Incremental approach: performance on the EEM case over 100 runs*

Performance index	MACS	MBH
Inclusion of best solution	100%	100%
Pruned space (mean value)	90.43%	93.51%
Number of blocks, average	8.64	15.9
Number of blocks, standard deviation	1.64	0.082
Average coverage	88.58%	72.21%

global optimum. Once DE has converged, an increase in the number of function evaluations does not improve the performance. Furthermore, if the objective function is globally non-convex, that is, it presents multiple similar funnel structures, MBH may not be effective and DE could quickly converge but within a single funnel, mainly due to the selection heuristic. A simple Multi-Start algorithm instead can yield better performance provided that the local optimization algorithm converges quickly.

The incremental algorithm applied to the first level yields the results in Table 8.4. The best solutions found with the all-at-once approach are always included in at least one of the boxes, which proves the reliability of the pruning, confirmed also by the value of the coverage indicators. At the same time, the percentage of pruned space is over 90% for both MACS and MBH. Therefore, it is expected that when stochastic optimizers such as DE, MBH, and MS operate on the reduced search space, the percentage of times they find solutions with a value lower than 2.958 km/s increases significantly compared to the results in Table 8.3. The performance of the incremental pruning based on MBH is not as good as the one of the incremental algorithms based on MACS, mainly because it generates about twice the number of boxes and with a lower average coverage.

Table 8.5. *Performance of DE and MBH on the box containing the reference solution over 100 runs*

Solver	10,000 evaluations	20,000 evaluations	40,000 evaluations
DE 100 runs < 2.958 km/s	52%	48%	56%
MBH 100 runs < 2.958 km/s	41%	52%	55%
MS 100 runs < 2.958 km/s	84%	97%	100%

Table 8.5 reports the performance of DE, MBH, and MS on the reduced search space pruned at level 1. Differential Evolution was run with a reduced population of 45 individuals and with the same setting and strategy of the all-at-once case. If we look at the percentage of times the best solution is included in at least one box, the average number of boxes, and the percentage of success for 10,000 evaluations on the box containing the best solution, we can conclude that the overall probability of identifying the best solution, after pruning, has considerably increased. On the other hand, the total number of function evaluations, accounting for both the incremental pruning and the search in all the boxes, has decreased. Note that even if the number of function evaluations was the same, the total cost would be lower due to the lower cost of the evaluation at level 1. In fact, on an Intel Pentium 4 with 3 GHz processing capacity running a MATLAB coded algorithm under Microsoft Windows, the cost for a single function evaluation at level 1 is 2.33 ms while the cost at level 2 is 3.68 ms.

Assuming 40,000 function evaluations for the pruning at level 1 and 10,000 function evaluations at level 2, for an average of 9 boxes, the total computational time for the incremental approach is 424,400s against 478,400s for the all-at-once approach with equal number of function evaluations. On the other hand, all the tested optimizers display a lower probability of success in the all-at-once case, even for a higher number of function evaluations.

Figure 8.8 shows a section of the domain D_1 along the first, second, and fifth coordinate. The figure clearly shows the feasible set, identified by the clusters of solutions found by MACS and the resulting boxes. The distribution of the feasible solutions is in agreement with what was found in Figure 8.6, in particular the two clusters around $\theta = 1$ and $\theta = 0.5$, corresponding to the two optimal directions of launch, and the clusters around $T_1 = 500$ d and $T_1 = 300$ d. Also note that at level 1, as expected, all the departure dates are equivalent and cannot be distinguished.

8.5.2 EVVMeMe Transfer

For this test, the search was performed over the interval $t_0 \in [3457, 5457]$ d, MJD2000. In this interval of launch dates, there exists a particularly good solution for the EVVMeMe that was considered as a possible chemical option for the ESA Bepi-Colombo mission [30]. This trajectory will be used as reference solution in the

Table 8.6. *Bounds for the EVVMeMe test case*

	Lower bound	Upper bound	N. fun. eval. MACS	N. fun. Eval. MBH	N. Bins	Level
t_0, MJD2000	3457	5457	5000	15000	40	1
T_1, d	90	180			10	
γ_1, rad	$-\pi/2$	$\pi/2$	70000	150000	3	2
$r_{p,1}$, planet radii	1.01	2			1	
α_2	0.01	0.6			4	
T_2, d	448	673			5	
γ_2, rad	$-\pi/2$	$\pi/2$	140000	185000	3	3
$r_{p,2}$, planet radii	1.01	2			1	
α_3	0.01	0.9			4	
T_3, d	90	220			5	
γ_3, rad	$-\pi/2$	$\pi/2$	10000–80000	10000–80000	N/D	4
$r_{p,3}$, planet radii	1.01	1.1			N/D	
α_4	0.01	0.5			N/D	
T_4, d	260	352			N/D	

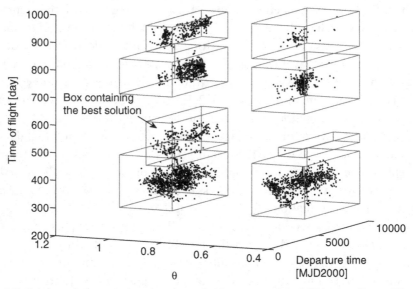

Figure 8.8. Section of the search space along the first, second, and fifth coordinate. The figure shows the clusters of feasible solutions and the corresponding bounding boxes.

following. The upper and lower bounds for the other variables are reported in Table 8.6 and define a search space D with 14 dimensions. At first, we tested DE, MBH, DIRECT, and MCS for an increasing number of function evaluations from 200,000 up to 1,600,000. DIRECT could not reach the highest number of function evaluations and was excluded from the comparison. DE and MBH were run 100 times, according to the testing procedure proposed above, and we recorded the

Table 8.7. *Comparison of global optimization methods applied to the EVVMeMe case*

Solver	200,000 evaluations	400,000 evaluations	800,000 evaluations	1,600,000 evaluations
MCS, km/s	14.35	13.05	13.05	12.01
DE, 100 runs				
< ESA (9.467 km/s)	16%	16%	21%	16%
< MCS	100%	81%	86%	61%
MBH, 100 runs				
< ESA (9.467 km/s)	4%	3%	7%	16%
< MCS	89%	78%	82%	100%

number of solutions with an objective function below or equal to the reference solution. DE was run with $F = 0.8$, $C_r = 0.75$, search strategy, *best*, and a population of 140 individuals while MBH was run with a ρ_l equal to 10% of the range of the variables. The results are reported in Table 8.7. On such a complex search space, a deterministic search like MCS cannot do better than 12 km/s. DE and MBH instead can find solutions that are better than 9.467 km/s, the ESA reference one. The probability of success, however, is limited to maximum 21% if DE and MBH are run on the whole search space. Furthermore, note that there is no statistical difference between the result of DE at 200,000, 400,000, 800,000, and 1,600,000 function evaluations, suggesting that DE converges too quickly. On the other hand, the aim of this preliminary set of runs is not to test DE but to have a standard of comparison for the application of the incremental pruning of the search space. All the solutions found with DE and MBH are represented in Figure 8.9 a) and b) respectively, together with the reference solution, in the plane containing the departure date, in MJD2000, and the total Δv, in km/s. DE and MBH identify four main launch windows, with two of them containing solutions that are better than the reference one. Note that we compare solutions according to the total Δv only, while the reference solution was designed to fulfill other requirements also. Figure 8.10 shows the reference solution together with three other solutions with lower overall Δv, all projected in the ecliptic plane.

The pruning was performed on all the variables except the swing-by variable r_p because in the specified range it has a very low impact. The number of resonant orbits, that is, the number of revolutions per leg, was preassigned. In particular, we use the following resonance strategy: $[0, 2, 0, 1, 2]$ where each number represents the number of full revolutions of the spacecraft around the Sun after the deep space maneuver and before meeting the next planet. For example, the EV leg is not performing any complete revolution around the Sun (after the DSM), while the VV leg performs two complete revolutions. The boundaries on the TOF for each leg were computed as a function of the number of revolutions. In particular, assuming circular the orbits of the departure and destination planets of each leg, the TOF is the period of the Hohmann transfer between the two, times the number of revolutions, while the

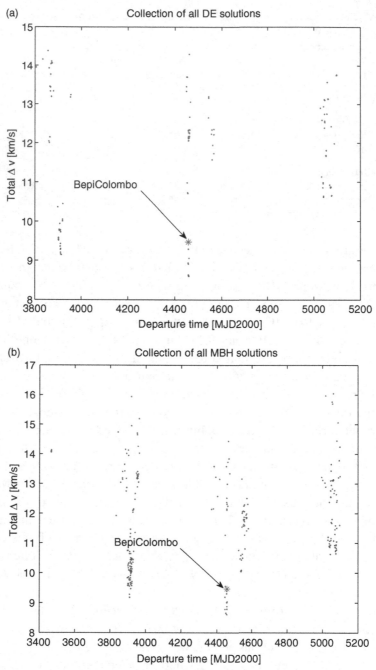

Figure 8.9. Solutions found by DE (a) and MBH (b) over all 100 runs with all-at-once approach.

upper bound is the period of the same Hohmann transfer times the number of full revolutions plus one. Note that, in principle, the trajectory model would not require the specification of the number of revolutions as the arc propagated up to the deep space maneuver, can be as long as required. The resulting trajectory would have

(a)   Departure time = 4457.2 MJD2000, f_{obj} = 9.467 km/s

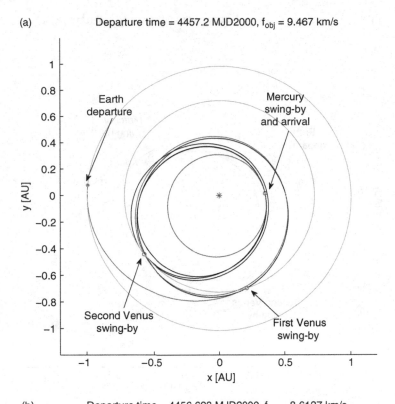

(b)   Departure time = 4456.693 MJD2000, f_{obj} = 8.6127 km/s

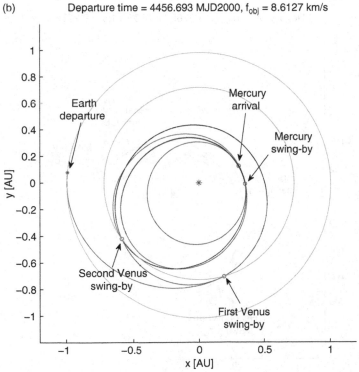

Figure 8.10. Projection on the ecliptic plane of the reference solution and of three improved solutions found with the all-at-once approach: a) the reference solution; b) an improved version for the same launch window; c) a modified version; d) an improved solution for a different launch window.

(c)   Departure time = 4455.8974 MJD2000, f_{obj} = 9.0111 km/s

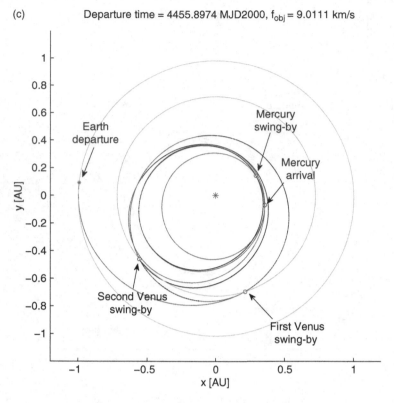

(d)   Departure time = 3911.2789 MJD2000, f_{obj} = 9.2252 km/s

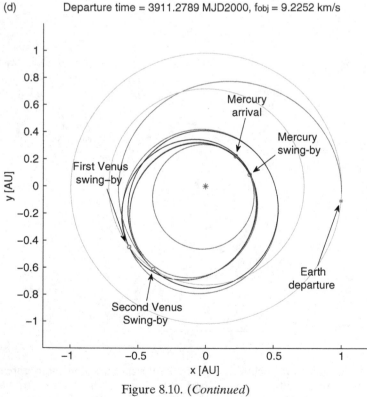

Figure 8.10. (*Continued*)

Table 8.8. *Incremental approach: performance on the EVVMeMe case over 100 runs*

Performance index	MACS	MBH
Inclusion of best solution	100%	94%
Inclusion of ESA solution	91%	89%
Pruned space (mean value)	97.81%	97.8%
Number of blocks, average	53.1	45.21
Number of blocks, standard deviation	19.48	15.21
Average Coverage	80.1%	81.8%

each deep space maneuver after the sequence of resonant orbits. This would make it not comparable to the ESA solution that has each deep space maneuver before the sequence of resonant orbits. Because, in the velocity formulation model, the arc following the deep space maneuver is the solution of a Lambert's problem, then the number of revolutions has to be specified explicitly.

The number of bins and number of function evaluations for each level of the incremental pruning are shown in Table 8.6. The aim of the test is to show how a space reduction of all the levels from 1 to 3 can improve the search with DE and MBH at level 4. For this reason, level 4 was not pruned. In general we can consider that the last level is the least significant for the pruning. MBH was run with a perturbation radius ρ_l equal to 10% of the range of the variable at level D_i. MACS was run with a population of 20 agents with $n_{popratio} = 10$, and a tol_{conv} equal to 10^{-4} of the range of the variables at level D_i. For the restart of MBH, we set the maximum number of trials to 30, while for MACS we use no partitioning at level 1 and we partitioned only γ and the time of flight at each subsequent level.

Special partial pruning criteria were used for the legs ending with Venus and Mercury. In particular, for the arrival conditions at Venus, we used the objective function in Equation (8.26), with $\omega_i = 1$ km/s, while for the arrival conditions at Mercury (last level), the objective function was

$$f_3 = v_\infty + \sum_{k=1}^{3} \Delta v_k \qquad (8.27)$$

where v_∞ is the spacecraft velocity relative to Mercury at arrival. The pruning functions were selected based on the required effect of the gravity assist maneuvers. In particular, Venus gravity maneuvers are expected to maximize the change in the perihelion while the maneuvers at Mercury, combined with the DSM, are supposed to minimize the relative velocity at Mercury. The pruning thresholds are 4, 4.5, and 7 km/s for level 1, 2, and 3 respectively. The first threshold is set according to launch capabilities, the second accepts a DSM no larger than 0.5 km/s for the VV resonant flyby, while the third threshold estimates the arrival v_∞ at Mercury to be lower than 6 km/s. Table 8.8 reports the performance of the incremental pruning on the EVVMeMe case, with pruning up to level 3 included. Note that the reference solution and the best solutions (all the ones better than the reference) are included in at least one of the boxes in the majority of the cases, in particular when the MACS is used to

Table 8.9. *Performance of DE and MBH on the box containing the ESOC solution over 100 runs*

Solver	10,000 evaluations	20,000 evaluations	40,000 evaluations	80,000 evaluations
DE 100 runs < ESA (9.467 km/s)	89%	91%	99%	99%
MBH 100 runs < ESA (9.467 km/s)	32%	48%	72%	88%

Table 8.10. *Average time to evaluate the partial objective functions, for each level*

Level 1	Level 2	Level 3	Level 4
1.69 ms	5.2 ms	5.7 ms	6.94 ms

look for the feasible set. The higher reliability of MACS compared to the modified MBH is marginal and not statistically significant. On the other hand, MACS tends to generate a higher number of boxes. Note that the significance of the variance in the table has to be taken with care because the process is not necessarily Gaussian. The only thing that can be said is that already at level 3, the number of boxes is high despite the fact that many of them are clustered. The last row of Table 8.8 reports the coverage metric. The reproducibility of the pruning is quite good for both algorithms. Table 8.9 shows the percentage of success of DE and MBH on the pruned space, in particular on the most promising box. As can be seen, the increase in performance is considerable, reaching almost 100%, fully justifying the space reduction obtained with the incremental approach. Finally, Table 8.10 reports the time of evaluation of the partial objective function at each level. Again, the lower cost of the partial objectives at lower levels leads to an overall gain in computational speed during the incremental pruning since a good deal of the function evaluations are used when exploring the lower levels. In this particular case, 215,000 function evaluations were used to prune levels from 1 to 3 with MACS and 350,000 with MBH, but the cost of each function evaluation was between one-third and one-tenth of the cost at level 4.

8.6 Conclusions

In this chapter, we presented a simple approach to the design of multiple gravity assist trajectories. The approach decomposes the whole trajectory into subproblems of lower dimension and complexity and proceeds to reduce the search space incrementally, adding one leg at a time. This incremental space reduction, tested on a number of cases, demonstrated to significantly increase the chance of finding good solutions at a relatively low computational cost: the probability of success of all tested optimizers on the pruned space is from two to six times higher and with a similar overall number of function evaluations. Furthermore, due to reduced computational cost of the evaluation of the partial objective functions, the total computational time of the incremental approach is lower even for the same number of function evaluations. Therefore, we obtained an increase in the reliability of the optimization process (that is, higher success rate) with a reduction of its computational cost. An additional advantage of the proposed incremental approach is the generation of a set of feasible solutions rather than a single optimal one. The feasible set corresponds to families of possible mission opportunities for different launch dates. Thus the decision maker, or mission designer, is presented with multiple options together with an enclosure of their neighboring solution space. The amount of available information is, therefore, higher and would allow an easier identification of baseline and backup designs together with their robustness. In fact, the size of the neighborhood, or subset of solutions below a given threshold, can be seen as a measure of their sensitivity to small changes in the design parameters.

The critical aspects for an efficient implementation of the incremental solution of MGA trajectories are: the definition of specific partial objective functions for each incremental level, and the definition of appropriate pruning criteria. Furthermore, the search for the feasible set can substantially change the performance. The two methods presented in this chapter, MACS and MBH, performed substantially in the same way for all the test cases. The main difference is that, while MACS can be used with discontinuous and noisy functions, MBH needs a smooth and continuous function. In both cases, the incremental algorithm remains unchanged, therefore it is recommended to start with a simple search procedure, even a simple Multi-Start algorithm, if the partial objective functions are smooth and differentiable.

Although the incremental procedure is generally applicable to all the trajectory models presented in this chapter, it was tested on the one built with the velocity formulation because this model leads to an exponential growth of the possible alternative path with the number of swing-bys. Furthermore, we presented a forward incremental procedure but, following the same principle, the procedure can be performed backward or forward and backward at the same time.

REFERENCES

[1] Myatt, D. R., Becerra, V. M., Nasuto, S. J., and Bishop, J. M. (2004) *Advanced Global Optimisation for Mission Analysis and Design*, European Space Agency, Advanced Concepts Team, Ariadna Final Report 03-4101a.

[2] De Pascale, P., Vasile, M., and Casotto, S. (2004) Preliminary Analysis of Low-Thrust Gravity Assist Trajectories by an Inverse Method and a Global Optimization Technique, *Proceedings of the 18th International Symposium on Space Flight Dynamics*, European Space Agency, Noordwijk, 2200 AG, Netherlands, Munich, Germany, 493–498.

[3] Petropoulos, A. E., Longuski, J. M., and Bonfiglio, E. P. (2000) Trajectories to Jupiter Via Gravity Assists from Venus, Earth, and Mars, *Journal of Spacecraft and Rockets*, **37**, No. 6, 776–783.

[4] Sims, J. A., Staugler, A. J., and Longuski, J. M. (1997) Trajectory Options to Pluto Via Gravity Assists from Venus, Mars, and Jupiter, *Journal of Spacecraft and Rockets*, **34**, No. 3, 347–353.

[5] Strange, N. J., and Longuski, J. M. (2002) Graphical Method for Gravity-Assist Trajectory Design, *Journal of Spacecraft and Rockets*, **39**, No. 1, 9–16.

[6] Longuski, J. M., and Williams, S. N. (1991) Automated Design of Gravity-Assist Trajectories to Mars and Outer Planets, *Celestial Mechanics and Dynamical Astronomy*, **52**, No. 3, 207–220.

[7] Pessina, S. M., Campagnola, S., and Vasile, M. (2003) Preliminary Analysis of Interplanetary Trajectories with Aerogravity and Gravity Assist Manoeuvres, *Proceedings of the 54th International Astronautical Congress*, **1**, International Astronautical Federation, IAF, Paris, 75015, France, Bremen, Germany, 671–681.

[8] Labunsky, A. V., Papkov, O. V., and Sukhanov, K. G. (1998) *Multiple Gravity Assist Interplanetary Trajectories*, Earth Space Institute Book Series, Gordon and Breach Science Publishers.

[9] Vasile, M., and De Pascale, P. (2006) Preliminary Design of Multiple Gravity-Assist Trajectories, *Journal of Spacecraft and Rockets*, **43**, No. 4, 794–805.

[10] Vasile, M., and Locatelli, M. (2008) A Hybrid Multiagent Approach for Global Trajectory Optimization, *Journal of Global Optimization*.

[11] Olds, A. D., Kluever, C. A., and Cupples, M. L. (2007) Interplanetary Mission Design Using Differential Evolution, *Journal of Spacecraft and Rockets*, **44**, No. 5, 1060–1070.

[12] Sentinella, M. R. (2007) Comparison and Integrated Use of Differential Evolution and Genetic Algorithms for Space Trajectory Optimisation, *Proceedings of the IEEE Congress on Evolutionary Computation*, Institute of Electronics and Electric Engineering Computer Society, Singapore, 973–978.

[13] Pisarevskya, D. M., and Gurfil, P. (2009) A Memetic Algorithm for Optimizing High-Inclination Multiple Gravity-Assist Orbits, *Proceedings of the IEEE Congress on Evolutionary Computation, CEC 2009*, Trondheim, Norway.

[14] Vasile, M., Minisci, E. A., and Locatelli, M. (2009) A Dynamical System Perspective on Evolutionary Heuristics Applied to Space Trajectory Optimization Problems, *Proceedings of the IEEE Congress on Evolutionary Computation, CEC 2009*, Trondheim, Norway.

[15] Price, K. V., Storn, R. M., and Lampinen, J. A. (2005) *Differential Evolution: A Practical Approach to Global Optimization*, Natural Computing Series, Springer, Berlin.

[16] Leary, R. H. (2000) Global Optimization on Funneling Landscapes, *Journal of Global Optimization*, **18**, No. 4, 367–383.

[17] Battin, R. H. (1999) *An Introduction to the Mathematics and Methods of Astrodynamics*, Revised edition, Aiaa Education Series, AIAA, New York.

[18] Weisstein, E. W. (2007) *MathWorld – A Wolfram Web Resource* Sphere Point Picking, http://mathworld.wolfram.com/SpherePointPicking.html

[19] Van Dam, E. R., Husslage, B., Den Hertog, D., and Melissen, H. (2007) Maximin Latin Hypercube Designs in Two Dimensions, *Operations Research*, **55**, No. 1, 158–169.

[20] Wales, D. J., and Doye, J. P. K. (1997) Global Optimization by Basin-Hopping and the Lowest Energy Structures of Lennard-Jones Clusters Containing up to 110 Atoms, *The Journal of Physical Chemistry A*, **101**, No. 28, 5111–5116.

[21] Addis, B., Locatelli, M., and Schoen, F. (2005) Local Optima Smoothing for Global Optimization, *Optimization Methods and Software*, **20**, No. 4–5, 417–437.

[22] Powell, M. J. D. (1978) A Fast Algorithm for Nonlinearly Constrained Optimization Calculations, in *Numerical Analysis*, edited by G.A. Watson, Lecture Notes in Mathematics, Springer, Berlin, 144–157.

[23] Storn, R., and Price, K. (1997) Differential Evolution – a Simple and Efficient Heuristic for Global Optimization over Continuous Spaces, *Journal of Global Optimization*, **11**, 341–359.

[24] Rudolph, G. (1996) Convergence of Evolutionary Algorithms in General Search Spaces, *Proceedings of the IEEE International Conference on Evolutionary Computation, ICEC'96*, IEEE, Nagoya, Japan, 50–54.

[25] Minisci, E. A., and Avanzini, G. (2009) Orbit Transfer Manoeuvres as a Test Benchmark for Comparison Metrics of Evolutionary Algorithms, *Proceedings of the IEEE Congress on Evolutionary Computation, CEC 2009*, Trondheim, Norway.

[26] Adcock, C. J. (1997) Sample Size Determination: A Review, *The Statistician*, **46**, No. 2, 261–283.

[27] Jones, D. R., Perttunen, C. D., and Stuckman, B. E. (1993) Lipschitzian Optimization without the Lipschitz Constant, *Journal of Optimization Theory and Applications*, **79**, No. 1, 157–181.

[28] Huyer, W., and Neumaier, A. (1999) Global Optimization by Multilevel Coordinate Search, *Journal of Global Optimization*, **14**, No. 4, 331–355.

[29] Vasile, M., Minisci, E., and Locatelli, M. (2008) On Testing Global Optimization Algorithms for Space Trajectory Design, *Proceedings of the AIAA/AAS Astrodynamics Specialist Conference and Exhibit*, Honolulu, Hawaii.

[30] Garcia Yárnoz, D., De Pascale, P., Jehn, R., Campagnola, S., Corral, C., et al. (2006) *Bepicolombo Mercury Cornerstone Consolidated Report on Mission Analysis*, ESA-ESOC Mission Analysis Office, MAO Working Paper No. 466.

Optimal Low-Thrust Trajectories Using Stable Manifolds

Christopher Martin
Dept. of Aerospace Engineering, University of Illinois at Urbana-Champaign, Urbana, IL

Bruce A. Conway
Dept. of Aerospace Engineering, University of Illinois at Urbana-Champaign, Urbana, IL

9.1 Introduction

The three-body system has been of interest to mathematicians and scientists for well over a century dating back to Poincaré [1]. Much of the interest in recent years has focused on using the interesting dynamics present around libration points to create trajectories that can travel vast distances around the solar system for almost no fuel expenditure traversing the so-called "Interplanetary Super Highway" (IPS) [2]. It has also been proposed to use Lagrange points as staging bases for more ambitious missions [3].

Lagrange points are equilibrium points of the three-body system that describes the motion of a massless particle in the presence of two massive primaries in a reference frame that rotates with the primaries. There are five Lagrange points (labeled $L_1, \ldots, L_5$), the three collinear points along the line of the two primaries, and the two equilateral points that form an equilateral triangle with the two primaries. It is the collinear points that are of the most interest and in particular the L_1 point between the two primaries and the L_2 point on the far side of the smaller primary. Since Poincaré, there has been much work on finding periodic solutions to the three-body problem. Early work was confined to analytic studies that are restricted to approximations as there exists no closed form analytical solution to the three-body system equations of motion.

Several recent missions have targeted periodic orbits for their advantageous properties. For example, a periodic orbit about the Earth-Sun L_2 Lagrange point minimizes solar radiation, allowing the extremely low temperatures required for missions such as Herschel [4] and Planck [5] to be reached and maintained. The Genesis mission used a periodic orbit about the Earth-Sun the L_2 Lagrange point in order to maximize the incidence of solar wind [6].

The high cost of putting mass into orbit has encouraged the use of low-thrust propulsion, taking advantage of its high specific impulse to realize savings in propellant mass. However, low-thrust engines impart very small thrust. For example, the recently launched Dawn mission is powered by Deep Space 1 heritage xenon-ion thrusters producing only 90mN of thrust [7, 8]. Therefore, the engine may need to be operating for a significant portion of the trajectory to impart the Δv required. If a low-thrust engine is the sole method of propulsion, there will be a significant period of spiralling about the target or departure planet to impart the necessary energy change for escape or capture. Complicated fuel-efficient trajectories can be obtained by combining low-thrust trajectories with terminal or initial coasts along the stable and unstable manifolds of the periodic orbits. These stable and unstable manifolds of the periodic orbits are regions of phase space that are asymptotic to the periodic orbits as $t \to \infty$ or $t \to -\infty$ respectively. The "Interplanetary Super Highway" consists of trajectories created by patching together these manifold–low-thrust–manifold trajectories.

A prerequisite for determining these combined low-thrust/manifold trajectories is a method for accurately calculating individual periodic orbits. Richardson [9] constructed 1st and 3rd order analytic approximations by linearizing the equations of motion about the L_1 and L_2 points. Higher order approximations were computed by Gomez and Mancote [10]. Various others methods have been employed; generating functions [11], Fourier analysis [12], and multiple shooting [13].

In this work, the periodic orbit generation problem is instead formulated as an optimal control problem. In an optimal control problem, the dynamic system is described by a system of differential equations; the goal is then to minimize a cost functional subject to path constraints on the states and controls. The optimal control for this dynamical system can be computed using direct or indirect methods. Indirect methods introduce adjoint variables and use the calculus of variations or the maximum principle to determine necessary conditions that must be satisfied by an optimal solution [14]. The determination of the controls from the necessary conditions generally results in a two point boundary value problem (TPBVP). Indirect methods allow very accurate computation of the optimal solution; however, the region of convergence can be small and therefore an accurate initial guess of the states, adjoints, and controls is required. Direct methods, in which the optimal control problem is converted into a nonlinear programming problem (NLP), also require an initial guess but are more robust. That is, such a method will converge to a solution satisfying the system governing equations and terminal conditions from a poor initial guess of the optimal solution. They do not explicitly use the necessary conditions and therefore do not require the addition of adjoint variables.

This chapter will describe how to create a fuel-minimizing trajectory from one planetary body to another using periodic orbits about Lagrange points and their stable and unstable manifolds. This chapter will first describe the Circular Restricted Three Body Problem (CR3BP) dynamics, the location of equilibrium points, periodic orbits and their generation, and determination of the stable and unstable manifolds. Then it will discuss how to link a low-thrust departure trajectory into a manifold in an optimal way. Finally it will describe how to depart the manifold on a trajectory to another body and optimally combine with a low-thrust transfer into orbit about that body. These elements will be illustrated in an example of an Earth-Moon low-thrust trajectory that uses the periodic orbit about the L_1 point and its manifolds.

9.2 System Dynamics

The differential equations of the CR3BP describe the motion of a point mass P_3 with mass m_3 under the gravitational influence of two massive primaries P_1 and P_2 with masses m_1 and m_2 respectively, where $m_1 > m_2 \gg m_3 \approx 0$. Therefore, P_3 exerts negligible influence on the primaries. The motion is considered in a noninertial frame that moves with the two primaries as they rotate about the system barycenter at constant radius, as shown in Figure 9.1.

It is convenient to normalize the system. The constant separation of the two primaries P_1 and P_2 is chosen to be the length unit, the combined mass of the two primaries $m_1 + m_2$ becomes the mass unit, and the time unit is then selected to make the orbital period of the two primaries about the system barycenter equal to 2π time units. To further simplify, the universal gravitational constant G becomes unity, therefore the mean motion n of the primaries is also equal to 1. The system can now be solely described by a single parameter, the mass ratio μ

$$\mu = \frac{m_2}{m_1 + m_2}. \tag{9.1}$$

By convention, $m_1 \geq m_2$, therefore $\mu \in [0, 0.5]$. The normalized masses of the primaries are then $P_1 = (1 - \mu)$ and $P_2 = \mu$ and they orbit the system barycenter at radii μ and $(1 - \mu)$ length units respectively.

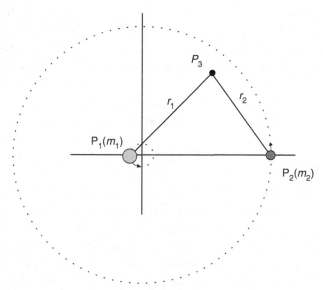

Figure 9.1. Circular restricted three body system geometry.

The circular restricted three body system equations of motion are

$$\dot{x} = v_x$$
$$\dot{y} = v_y$$
$$\dot{z} = v_z$$
$$\dot{v}_x = 2v_y + x - \frac{(1-\mu)(x+\mu)}{r_1^3} - \frac{\mu(x-1+\mu)}{r_2^3} \tag{9.2}$$
$$\dot{v}_y = -2v_x + y - \frac{(1-\mu)y}{r_1^3} - \frac{\mu y}{r_2^3}$$
$$\dot{v}_z = \frac{-(1-\mu)z}{r_1^3} - \frac{\mu z}{r_2^3}$$

where,

$$r_1 = \sqrt{(x+\mu)^2 + y^2 + z^2}$$
$$r_2 = \sqrt{(x-1+\mu)^2 + y^2 + z^2}.$$

The idealized low-thrust engine used in this study provides a constant thrust acceleration of magnitude T_a. The thrust pointing angles u_1 and u_2 are defined with respect to the instantaneous velocity vector $\mathbf{v}$ which makes an angle ϕ with the $x-y$ plane. The projection of $\mathbf{v}$ to the $x-y$ plane makes an angle γ with the x-axis as shown in Figure 9.2. The equations of motion (in first order form) then become

$$\dot{x} = v_x$$
$$\dot{y} = v_y$$

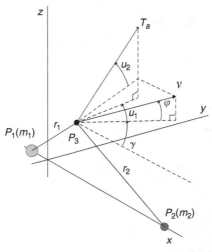

Figure 9.2. Locating the low-thrust spacecraft P_3 in the three body system.

$$\dot{z} = v_z$$

$$\dot{v}_x = 2v_y + x - \frac{(1-\mu)(x+\mu)}{r_1^3} - \frac{\mu(x-1+\mu)}{r_2^3} + T_a \cos(\gamma + u_1)\cos(\phi + u_2)$$

$$\dot{v}_y = -2v_x + y - \frac{(1-\mu)y}{r_1^3} - \frac{\mu y}{r_2^3} + T_a \cos(\phi + u_2)\sin(\gamma + u_1) \qquad (9.3)$$

$$\dot{v}_z = -\frac{(1-\mu)z}{r_1^3} - \frac{\mu z}{r_2^3} + T_a \sin(\phi + u_2)$$

where,

$$\gamma = \tan^{-1}\left(\frac{v_y}{v_x}\right) \phi = \tan^{-1}\left(\frac{v_z}{\sqrt{v_x^2 + v_y^2}}\right).$$

Since only low-thrust engines are used in these cases, significant spiraling about the departure and arrival planets with be required. It is awkward to describe these spirals in cartesian coordinates, so two alternative coordinate systems will be used: orbit elements $(a, e, i, \omega, \Omega, f)$ and equinoctial elements $(a, P_1, P_2, Q_1, Q_2, L)$ [15]. In both cases, the thrust acceleration is described in the local body-fixed axes and modeled as a perturbation. The attraction of the third body is also included as a perturbation. The perturbations R, T, N are accelerations in the body-fixed radial, tangential, and normal directions respectively.

9.2.1 Orbit Elements

The familiar orbital elements $(a, e, i, \Omega, \omega, f)$ are used for the escape spiral from Earth. Since the formula for $\dot{f}$ explicitly depends on time, true anomaly (f) is replaced by two new variables (χ, ξ) related to the mean anomaly [16]

$$M = \xi - \chi. \qquad (9.4)$$

The variational equations become

$$\dot{a} = \frac{2a^{\frac{3}{2}}}{\sqrt{\mu_e (1 - e^2)}}\left(Re\sin f + T(1 + e\cos f)\right)$$

$$\dot{e} = \sqrt{\frac{a(1 - e^2)}{\mu_e}}\left(R\sin f + T(\cos f + \cos E)\right)$$

$$\dot{i} = \sqrt{\frac{a(1 - e^2)}{\mu_e}}\frac{\cos\theta}{1 + e\cos f}N$$

$$\dot{\Omega} = \sqrt{\frac{a(1 - e^2)}{\mu_e}}\frac{\sin\theta}{\sin i(1 + e\cos f)}N \qquad (9.5)$$

$$\dot{\omega} = -\dot{\Omega}\cos i + \sqrt{\frac{a(1 - e^2)}{\mu_e}}\frac{-R\cos f + T\sin f(2 + e\cos f)}{e(1 + e\cos f)}$$

$$\dot{\chi} = \sqrt{\frac{a}{\mu_e}} \frac{(1-e^2)}{e(1+e\cos f)} \left(R\left(2e - \cos f - e\cos^2 f\right) + T\sin f\left(2 + e\cos f\right) \right)$$

$$\dot{\xi} = \sqrt{\frac{\mu_e}{a^3}}.$$

The perturbing accelerations (R, T, N) have two components, the first is due to the thrust from the engine ($\mathbf{F}_{lt}$) and the second is due to the attraction of the Moon ($\mathbf{F}_m$). To compute the attraction due to the Moon, we first need the position vector (in the body fixed frame) of the Moon relative to the spacecraft in terms of the orbit elements.

$$\mathbf{r}_m = r_m \begin{bmatrix} \cos\theta\cos(\lambda_m - \Omega) + \cos i\sin\theta\sin(\lambda_m - \Omega) \\ -\sin\theta\cos(\lambda_m - \Omega) + \cos i\cos\theta\sin(\lambda_m - \Omega) \\ -\sin i\sin(\lambda_m - \Omega) \end{bmatrix}$$

where λ_m is the angle between the Moon and the reference axis in the plane of the Moon's orbit. The acceleration due to the Moon is then given by

$$\mathbf{F}_m = \mu_m \frac{\mathbf{r}_m - \mathbf{r}}{|\mathbf{r}_m - \mathbf{r}|^3} - \mu_m \frac{\mathbf{r}_m}{|\mathbf{r}_m|^3}$$

The acceleration due to the low-thrust engine is simply

$$\mathbf{F}_{lt} = T_A \begin{bmatrix} \sin(\gamma + u_1)\cos u_2 \\ \cos(\gamma + u_1)\cos u_2 \\ \sin u_2 \end{bmatrix}.$$

The total perturbing acceleration is then

$$\begin{bmatrix} R \\ T \\ N \end{bmatrix} = (\mathbf{F}_m + \mathbf{F}_{lt}).$$

9.2.2 Equinoctial Elements

For the example to follow in Section 9.5 the desired final orbit is a zero inclination circular orbit around the moon. The orbit element equations have a singularity at $i = 0$, therefore the equinoctial elements $(a, P_1, P_2, Q_1, Q_2, l)$ are used for the Halo orbit to Moon phase of the trajectory. The variational equations for the equinoctial elements are [15]

$$\dot{a} = \frac{2a^2}{h}\left((P_2\sin L - P_1\cos L)R + \frac{pT}{r}\right)$$

$$\dot{P_1} = \frac{r}{h}\left(-\left(\frac{p}{r}\cos L\right)R + \left(P_1 + \left(1 + \frac{p}{r}\right)\sin L\right)T - P_2(Q_1\cos L - Q_2\sin L)N\right)$$

$$\dot{P_2} = \frac{r}{h}\left(-\left(\frac{p}{r}\sin L\right)R + \left(P_2 + \left(1 + \frac{p}{r}\right)\cos L\right)T - P_1(Q_1\cos L - Q_2\sin L)N\right)$$

$$\dot{Q}_1 = \left(\frac{r}{2h} \left(1 + Q_1^2 + Q_2^2 \right) \sin L \right) N \qquad\qquad (9.6)$$

$$\dot{Q}_2 = \left(\frac{r}{2h} \left(1 + Q_1^2 + Q_2^2 \right) \cos L \right) N$$

$$\dot{l} = n - \frac{r}{h} \left(\left(\frac{a}{a+b} \right) \frac{p}{r} \left(P_1 \sin L + P_2 \cos L \right) + \frac{2b}{a} \right) R$$

$$\qquad - \frac{r}{h} \left(\left(\frac{a}{a+b} \right) \left(1 + \frac{p}{r} \right) \left(P_1 \cos L + P_2 \sin L \right) + \frac{2b}{a} \right) T$$

$$\qquad - \frac{r}{h} \left(Q_1 \cos L - Q_2 \sin L \right) N$$

where

$$K = 1 - P_1 \cos K + P_2 \sin K$$

$$r = a \left(1 - P_1 \sin K - P_2 - \sin K \right)$$

$$b = \left| a \sqrt{1 - P_1^2 - P_2^2} \right|$$

$$n = \sqrt{\frac{\mu}{a^3}}$$

$$\sin L = \frac{a}{r} \left(\left(1 - \left(\frac{a}{a+b} \right) P_2^2 \right) \sin K + \left(\frac{a}{a+b} \right) P_1 P_2 \cos K - P_1 \right)$$

$$\cos L = \frac{a}{r} \left(\left(1 - \left(\frac{a}{a+b} \right) P_1^2 \right) \cos K + \left(\frac{a}{a+b} \right) P_1 P_2 \sin K - P_2 \right)$$

$$p = r \left(1 + P_1 \sin L + P_2 \cos L \right)$$

$$h = nab.$$

In this case, where the Moon is the primary body, it is necessary to consider third body perturbations due to the Earth. The position vector of the Earth relative to the spacecraft (in local body axes) in terms of the equinoctial elements is

$$\mathbf{r}_e = \frac{r_e}{1 + Q_1^2 + Q_2^2} \begin{bmatrix} \frac{r}{r_e} - \cos(L - \lambda_e) + (r + \cos(L - \lambda_e)) Q_1^2 - 2 \sin(L + \lambda_e) Q_1 Q_2 + \cdots \\ (r - \cos(L + \lambda_e)) Q_2^2 \\ \sin(L - \lambda_e) - \sin(L + \lambda_e) Q_1^2 - 2 \cos(L + \lambda_e) Q_1 Q_2 + \sin(L + \lambda_e) Q_2^2 \\ 2 (Q_2 \sin \lambda_e - Q_1 \cos \lambda_e) \end{bmatrix}$$

where λ_e is the angle between the Earth and the reference axis in the plane of the Moon's orbit about the Earth. The acceleration due to the Earth is then given by

$$\mathbf{F}_e = \mu_e \frac{\mathbf{r}_e - \mathbf{r}}{|\mathbf{r}_e - \mathbf{r}|^3} - \mu_e \frac{\mathbf{r}_e}{|\mathbf{r}_e|^3}.$$

The acceleration due to the low-thrust engine is simply

$$\mathbf{F}_{lt} = T_A \begin{bmatrix} \sin(\gamma + u_1) \cos u_2 \\ \cos(\gamma + u_1) \cos u_2 \\ \sin u_2 \end{bmatrix}.$$

The total perturbation acceleration is then

$$\begin{bmatrix} R \\ T \\ N \end{bmatrix} = (\mathbf{F}_m + \mathbf{F}_{lt}).$$

9.2.3 Equilibrium Points and Periodic Orbits

The system equations for the CR3BP are autonomous, hence there exists the possibility of stationary or equilibrium points in the phase space. For a stationary point (x^*, y^*, z^*), Equations (9.2) become

$$-x^* = -\frac{(1-\mu)(x^*+\mu)}{r_1^3} - \frac{\mu(x^*-1+\mu)}{r_2^3} \tag{9.7}$$

$$-y^* = -\frac{(1-\mu)y^*}{r_1^3} - \frac{\mu y^*}{r_2^3} \tag{9.8}$$

$$0 = -\frac{(1-\mu)z^*}{r_1^3} - \frac{\mu z^*}{r_2^3}. \tag{9.9}$$

Clearly from Equation (9.9), any equilibrium points must lie in the $x-y$ plane, as $z^* = 0$ is the only solution to Equation (9.9). Consider solutions on the x-axis, with $y^* = z^* = 0$. Equation (9.7) becomes

$$x^* - \frac{(1-\mu)(x+\mu)}{|x^*+\mu|^3} - \frac{\mu(x^*-1+\mu)}{|x^*-1+\mu|^3} = 0. \tag{9.10}$$

This is equivalent to a quintic polynomial in x^* with three real solutions. These solutions yield the locations of the three collinear Lagrange points L_1, L_2, and L_3.

Now let $r_1 = r_2 = 1$; Equations (9.7) and (9.8) are both satisfied, hence there exist equilibria at the two points in the $x-y$ plane unit distance from both primaries, the equilateral Lagrange points L_4 and L_5, where $L_4 = \left(\frac{1}{2} - \mu, \frac{\sqrt{3}}{2}\right)$ and $L_5 = \left(\frac{1}{2} - \mu, -\frac{\sqrt{3}}{2}\right)$ as shown in Figure 9.3.

The equilateral Lagrange points (L_4, L_5) are stable but the collinear points (L_1, L_2, L_3) are not. However, there exist periodic orbits about the unstable collinear points.

9.2.4 Periodic Orbits

Periodic solutions to the CR3BP have been sought since Poincaré. Since computational power was not readily available, early work focused on analytic approximations. Richardson [9] constructed first and third order analytic approximations by linearizing the equations of motion about the L_1 and L_2 points. Gomez and Marcote [10] extended the analysis to higher orders. More accurate solutions have been obtained through computational means; however, all numerical methods have their

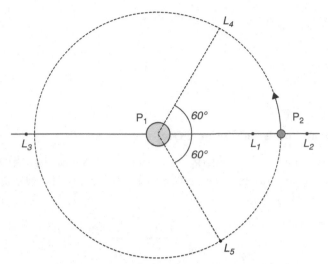

Figure 9.3. Locations of the Lagrange points.

disadvantages. Shooting methods, such as those used by Howell and Pernicka [13], are highly sensitive to the quality of the initial guess. Generating functions as used by Scheeres and Guibout [11] have a limited (spatial) range of applicability. In this work, an alternative and more robust method for generating periodic orbits is used.

9.2.5 Stable and Unstable Manifolds

It will be necessary for the Earth-Moon transfer via a periodic orbit about L_1 to compute the periodic orbit's stable and unstable manifolds (W_s, W_u). The manifolds are tangent to the eigenvectors of the monodromy matrix, which is computed by evaluating the variation matrix (9.11) at time equal to one period of the periodic orbit.

$$\left(\frac{\partial x_i}{\partial x_{0j}}\right) = \begin{pmatrix} \dfrac{\partial x}{\partial x_0} & \dfrac{\partial x}{\partial y_0} & \dfrac{\partial x}{\partial z_0} & \dfrac{\partial x}{\partial v_{x0}} & \dfrac{\partial x}{\partial v_{y0}} & \dfrac{\partial x}{\partial v_{z0}} \\[2mm] \dfrac{\partial y}{\partial x_0} & \dfrac{\partial y}{\partial y_0} & \dfrac{\partial y}{\partial z_0} & \dfrac{\partial y}{\partial v_{x0}} & \dfrac{\partial y}{\partial v_{y0}} & \dfrac{\partial y}{\partial v_{z0}} \\[2mm] \dfrac{\partial z}{\partial x_0} & \dfrac{\partial z}{\partial y_0} & \dfrac{\partial z}{\partial z_0} & \dfrac{\partial z}{\partial v_{x0}} & \dfrac{\partial z}{\partial v_{y0}} & \dfrac{\partial z}{\partial v_{z0}} \\[2mm] \dfrac{\partial v_x}{\partial x_0} & \dfrac{\partial v_x}{\partial y_0} & \dfrac{\partial v_x}{\partial z_0} & \dfrac{\partial v_x}{\partial v_{x0}} & \dfrac{\partial v_x}{\partial v_{y0}} & \dfrac{\partial v_x}{\partial v_{z0}} \\[2mm] \dfrac{\partial v_y}{\partial x_0} & \dfrac{\partial v_y}{\partial y_0} & \dfrac{\partial v_y}{\partial z_0} & \dfrac{\partial v_y}{\partial v_{x0}} & \dfrac{\partial v_y}{\partial v_{y0}} & \dfrac{\partial v_y}{\partial v_{z0}} \\[2mm] \dfrac{\partial v_z}{\partial x_0} & \dfrac{\partial v_z}{\partial y_0} & \dfrac{\partial v_z}{\partial z_0} & \dfrac{\partial v_z}{\partial v_{x0}} & \dfrac{\partial v_z}{\partial v_{y0}} & \dfrac{\partial v_z}{\partial v_{z0}} \end{pmatrix} . \tag{9.11}$$

The transition matrix satisfies the differential equation

$$\frac{d}{dt}\left(\frac{\partial x_i}{\partial x_{0j}}\right) = \sum_{k=1}^{6} \frac{\partial \dot{x}_i}{\partial x_k} \frac{x_k}{x_{0j}} \tag{9.12}$$

which yields

$$
\left(\frac{\partial \dot{x}_i}{\partial x_k}\right) = \begin{bmatrix} 0 & 0 & 0 & 1 & 0 & 0 \\ 0 & 0 & 0 & 0 & 1 & 0 \\ 0 & 0 & 0 & 0 & 0 & 1 \\ \Omega_{xx} & \Omega_{xy} & \Omega_{xz} & 0 & 2 & 0 \\ \Omega_{yx} & \Omega_{yy} & \Omega_{yz} & -2 & 0 & 0 \\ \Omega_{zx} & \Omega_{zy} & \Omega_{zz} & 0 & 0 & 0 \end{bmatrix}
\tag{9.13}
$$

where

$$
\Omega(x, y, z) = \frac{x^2 + y^2}{2} + \frac{1 - \mu}{r_1} + \frac{\mu}{r_2} + \frac{\mu (1 - \mu)}{2}.
\tag{9.14}
$$

Combined with Equation (9.2), there are then 42 simultaneous differential equations that must be solved to yield the variation matrix. The 6 eigenvalues $(\lambda_1, ... \lambda_6)$ of the monodromy matrix are in three pairs [17]

$$
\lambda_1 > 1, \lambda_2 < 1, \lambda_1 \lambda_2 = 1
$$
$$
\lambda_3 = \lambda_4 = 1
$$
$$
\lambda_5 = \lambda_6^*
\tag{9.15}
$$
$$
|\lambda_5| = |\lambda_6| = 1.
$$

The unstable and stable manifolds are approximated near the periodic orbits by the eigenvectors corresponding to λ_1 and λ_2 respectively (V^{W_u}, V^{W_s}). The manifolds are then computed by numerically integrating forward (unstable) or backward (stable) in time from initial conditions given by displacing the spacecraft a small distance d from the periodic orbit along the eigenvector (or its opposite). The displacement distance d must be small enough to retain the validity of the linear approximation but also large enough to enable the spacecraft to get far enough away from the periodic orbit in a reasonable time of flight. A value of 5 km was used for this study. The manifolds are computed by combining the trajectories obtained from many initial points on the periodic orbit $(\bar{x}(t_i))$

$$
\mathbf{X}_s = \mathbf{x}(t_i) \pm d.\mathbf{V}^{W_s}
\tag{9.16}
$$
$$
\mathbf{X}_u = \mathbf{x}(t_i) \pm d.\mathbf{V}^{W_u}
\tag{9.17}
$$

9.3 Basics of Trajectory Optimization

Consider the optimal control problem with system governing equations

$$
\dot{\mathbf{x}}(t) = \mathbf{f}(\mathbf{x}(t), \mathbf{u}(t), t), t_0 \leq t \leq t_f
\tag{9.18}
$$

and objective function

$$
J = \Phi(\mathbf{x}(t_f), t_f) + \int_{t_0}^{t_f} L(\mathbf{x}(t), \mathbf{u}(t), t) dt
\tag{9.19}
$$

subject to the path constraints

$$\mathbf{g}(\mathbf{x}, \mathbf{u}, t) \leq 0, t_0 \leq t \leq t_f \tag{9.20}$$

and terminal constraints

$$\Psi(\mathbf{x}(t_f), t_f) = 0. \tag{9.21}$$

The method used to solve optimal control problems in this work is the method of direct collocation with nonlinear programming [18]. In this method, the trajectory is approximated by a piecewise polynomial defined by the values of the state and control variable at discrete nodes. The trajectory of each state and control are discretized; an example for the state variable history is shown in Figure 9.4.

The trajectory is now wholly defined by the state variables, $\mathbf{x}$, and control variables, $\mathbf{u}$, defined at $(N + 1)$ nodes and possibly at some interior points. For each interval, the time history of each state is approximated over the interval by a polynomial. Herman and Conway [18] found the accuracy and computational efficiency of direct collocation using fifth degree polynomial trajectory approximations superior to approximations using polynomials of lower degree. Six conditions are required to uniquely define the approximating quintic polynomial over the segment. The values of the states at the segment boundary nodes $(i, i+1)$ are defined, and the derivatives can be calculated from the system equation $\dot{\mathbf{x}} = \mathbf{f}(\mathbf{x}, \mathbf{u}, t)$ providing four conditions $(\mathbf{x}_i, \mathbf{f}_i, \mathbf{x}_{i+1}, \mathbf{f}_{i+1})$. That is, the approximating polynomial must evaluate to $\mathbf{x}_i$ at the left side node and its slope must correspond to $\mathbf{f}_i = \mathbf{f}(\mathbf{x}_i)$ there. At the right side node, the polynomial must evaluate to $\mathbf{x}_{i+1}$ and its slope must be $\mathbf{f}_{i+1}$. The remaining two conditions are defined by adding a node at the center of the segment with corresponding state $\mathbf{x}_c$ and time rate of change $\dot{\mathbf{x}}_c = \mathbf{f}(\mathbf{x}_c, \mathbf{u}_c, t_c)$, where $\mathbf{u}_c$, the control at the center point, and $\mathbf{x}_c$ are free parameters.

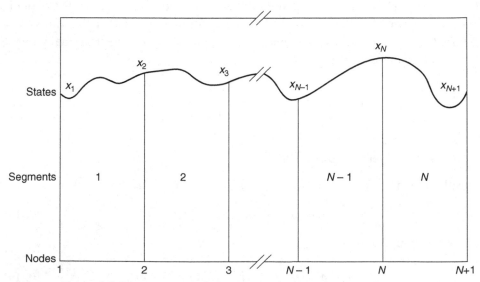

Figure 9.4. Illustration of the discretization of the continuous system.

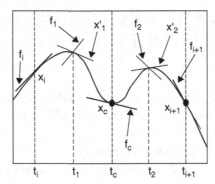

Figure 9.5. Illustration of the Gauss-Lobatto quintic polynomial.

These six conditions determine a quintic polynomial satisfying the governing equations at the left and right nodes and at the center of the segment as shown in Figure 9.5. An additional constraint is imposed that the derivative of the quintic polynomial be equal to the derivative calculated from the system Equation (9.2) at two interior collocation points.

The collocation points (t_1, t_2) are selected to minimize the error in the polynomial estimation of the state [18]. For a fifth degree Gauss-Lobatto polynomial, the collocation points are located at

$$t_1 = t_c - \sqrt{\frac{3}{7}\frac{1}{2}}\Delta t_i$$

$$t_2 = t_c + \sqrt{\frac{3}{7}\frac{1}{2}}\Delta t_i \tag{9.22}$$

where Δt_i is the width of the ith time segment, i.e. $\Delta t_i = t_{i+1} - t_i$.

The states at the collocation points in segment i are obtained from the polynomial evaluated at times t_1 and t_3

$$\mathbf{x}_1 = \frac{1}{686} \left\{ \begin{array}{l} \left(39\sqrt{21} + 231\right)\mathbf{x}_i + 224 + \left(-39\sqrt{21} + 231\right)\mathbf{x}_{i+1} \\ +\Delta t_i \left[\left(3\sqrt{21} + 21\right)\mathbf{f}_i - 16\sqrt{21}\mathbf{f}_c + \left(3\sqrt{21} - 21\right)\mathbf{f}_{i+1}\right] \end{array} \right\} \tag{9.23}$$

$$\mathbf{x}_2 = \frac{1}{686} \left\{ \begin{array}{l} \left(-39\sqrt{21} + 231\right)\mathbf{x}_i + 224 + \left(39\sqrt{21} + 231\right)\mathbf{x}_{i+1} \\ +\Delta t_i \left[\left(-3\sqrt{21} + 21\right)\mathbf{f}_i + 16\sqrt{21}\mathbf{f}_c + \left(-3\sqrt{21} - 21\right)\mathbf{f}_{i+1}\right] \end{array} \right\} \tag{9.24}$$

where $\mathbf{f}_i = \mathbf{f}(\mathbf{x}_i, \mathbf{u}_i, t_i)$, $\mathbf{f}_c = \mathbf{f}(\mathbf{x}_c, \mathbf{u}_c, t_c)$ and $\mathbf{f}_{i+1} = \mathbf{f}(\mathbf{x}_{i+1}, \mathbf{u}_{i+1}, t_{i+1})$.

The system constraints are then

$$\mathbf{C}_{5,1}(\mathbf{x}_i, \mathbf{x}_{i+1}) = \frac{1}{360} \left\{ \begin{array}{l} \left(32\sqrt{21} + 180\right)\mathbf{x}_i - 64\sqrt{21}\mathbf{x}_c + \left(32\sqrt{21} - 180\right)\mathbf{x}_{i+1} \\ \Delta t_i \left[\left(9 + \sqrt{21}\right)\mathbf{f}_i + 98\mathbf{f}_1 + 64\mathbf{f}_c + \left(9 - \sqrt{21}\right)\mathbf{f}_{i+1}\right] \end{array} \right\} = 0 \tag{9.25}$$

$$\mathbf{C}_{5,2}(\mathbf{x}_i, \mathbf{x}_{i+1}) = \frac{1}{360} \left\{ \begin{array}{l} \left(-32\sqrt{21} + 180\right)\mathbf{x}_i + 64\sqrt{21}\mathbf{x}_c + \left(-32\sqrt{21} - 180\right)\mathbf{x}_{i+1} \\ \Delta t_i \left[\left(9 - \sqrt{21}\right)\mathbf{f}_i + 64\mathbf{f}_c + 98\mathbf{f}_2 + \left(9 + \sqrt{21}\right)\mathbf{f}_{i+1}\right] \end{array} \right\} = 0$$

$$(9.26)$$

where $\mathbf{f}_1 = \mathbf{f}(\mathbf{x}_1, \mathbf{u}_1, t_1)$, $\mathbf{f}_2 = \mathbf{f}(\mathbf{x}_2, \mathbf{u}_2, t_2)$. Note that the values of the controls at collocation points t_1 and t_2 must be specified. This then adds two more control parameters per segment per control.

The state parameters $\mathbf{x}$, control parameters $\mathbf{u}$, and the event variables $\mathbf{E}$ are collected into a single vector $\mathbf{P}$. Event variables are extra parameters required to describe the trajectory such as time of flight, engine burn times, departure date, etc.

$$\mathbf{P} = [\mathbf{Z}, \mathbf{E}] \tag{9.27}$$

where

$$\mathbf{Z} = [\mathbf{x}_1, \ldots \mathbf{x}_{N+1}, \mathbf{u}_1, \ldots, \mathbf{u}_{N+1}].$$

The nonlinear programming (NLP) problem is then to minimize $\Phi(P)$ subject to

$$\mathbf{b}_L \le \left\{ \begin{array}{c} \mathbf{P} \\ A\mathbf{P} \\ C(\mathbf{P}) \end{array} \right\} \le \mathbf{b}_U \tag{9.28}$$

where $A\mathbf{P}$ is a vector of linear constraints defined by the matrix A, and $C(\mathbf{P})$ is a vector of nonlinear constraints virtually all of which are the implicit integration constraints (9.25) and (9.26). $\mathbf{b}_L$ and $\mathbf{b}_U$ are the lower and upper bounds, respectively, on the states and constraints.

The fifth degree Gauss-Lobatto transcription was used to solve the periodic orbit generation problem. This problem happens to have no control variables, that is, $\mathbf{u} = \mathbf{0}$. For determining optimal low-thrust trajectories, that is, for case where $\mathbf{u} \ne \mathbf{0}$ and in fact $\mathbf{u}$ must be specified at many points on the trajectory, a different transcription method, Runge-Kutta parallel-shooting (Section 3.2) is a better choice. This is the transcription used for the low-thrust Earth-Moon trajectory.

The NLP problem solvers used SNOPT [19, 20] or *fmincon* from MATLAB requiring an initial guess for the state vector $\mathbf{P}$.

9.4 Generation of Periodic Orbit Constructed as an Optimization Problem

Periodic orbits are trajectories such that $\mathbf{x}(t_0) = \mathbf{x}(t_0 + NT)$ for some integer N and orbital period T. Therefore, the search for periodic solutions to the circular restricted three body problem of specified amplitude or period can be described as an optimal control problem [21] with the objective of minimizing the function

$$J = |\mathbf{x}(t_o) - \mathbf{x}(t_0 + NT)| \tag{9.29}$$

subject to the system Equations (9.2). The problem can be transposed into a problem to be solved with direct collocation. The native MATLAB NLP solver *fmincon* was used in this study.

Periodic orbits were sought about the interior Lagrange point L_1 in the Earth-Moon system, although the method would apply equally well to other Lagrange points and/or different three body systems. In normalized units, the Earth-Moon system is fully defined by the mass ratio $\mu = m_{\text{moon}}/(m_{\text{earth}} + m_{\text{moon}}) = 0.0122$. The full three-dimensional solution to Equations (9.2) was sought. The time histories of the states were discretized into N segments (as shown in Figure 9.4). The values of state variables at the boundaries and at the centers of each segment and the final time t_f then constitute the NLP parameter vector ($\mathbf{P}$). This vector is typically hundreds or thousands of elements in length.

The fifth degree Gauss-Lobatto collocation constraints (Equations [9.25–9.26]) were employed to ensure the trajectory satisfies the system equations of motion. To ensure that the resulting trajectory has the required amplitude, the position $r_1 = [x_1, y_1, z_1]$ at the initial node is specified. Since any periodic orbit must return to its initial position after one orbit period, the position state at the final node is also fixed, that is

$$r_{N+1} = [x_{N+1}, y_{N+1}, z_{N+1}] = [x_1, y_1, z_1] = r_1. \tag{9.30}$$

For a periodic solution, the velocities at the initial and final nodes must also be equal. This is achieved by minimizing the objective (penalty) function

$$\Phi(\mathbf{x}) = \sqrt{(v_{x_i} - v_{x_f})^2 + (v_{y_i} - v_{y_f})^2 + (v_{z_i} - v_{z_f})^2}. \tag{9.31}$$

This function has a global minimum $\Phi(x) = 0$ when $v_{x_i} = v_{x_f}$, $v_{y_i} = v_{y_f}$ and $v_{z_i} = v_{z_f}$, (that is, when the trajectory is a periodic orbit). Thus the pure penalty function of Equation (9.29) is not required. Solutions were obtained with the equality constraint (9.30) for the positions and the penalty function constraint (9.31) for the velocities.

The NLP problem solver requires an initial guess of the optimal solution. For this case, Richardson's first order analytical approximation [9] is used, in which

$$x = A_x \cos(\lambda t + \Phi)$$
$$y = k A_x \sin(\lambda t + \Phi) \tag{9.32}$$
$$z = A_z \sin(\lambda t + \psi)$$

where Φ and ψ are phase angles and

$$k = \frac{2\lambda}{\lambda^2 + 1 - c_2}. \tag{9.33}$$

The linearized frequency λ is found from the solution to

$$\lambda^4 + (c_2 - 2)\lambda^2 - (c_2 - 1)(1 + 2c_2) = 0. \tag{9.34}$$

Constant c_2 is defined by

$$c_2 = \frac{1}{\gamma_L^3} \left[\mu + \frac{(1-\mu)\,\gamma_L^3}{(1 \mp \gamma_L)^3} \right] \qquad (9.35)$$

where the upper sign applies for orbits around L_1 and the lower sign for orbits around L_2. $\gamma_L = 1$ is defined by

$$\gamma_L = \left(\frac{G\,(m_{moon} + m_{earth})}{a^3} \right)^{\frac{1}{3}}. \qquad (9.36)$$

Higher-order approximations are available [9, 10]; however, this simple initial guess proved to be sufficient and demonstrated the robustness of the solution technique. For calculations of families of periodic orbits homotopy was employed, that is, the solution for a periodic orbit of a certain amplitude is used as an initial guess in the calculation of a periodic orbit with a similar amplitude and this process is continued.

9.4.1 Results

Periodic orbits of a given amplitude are obtained by specifying the initial position. For an orbit around L_1 with x amplitude A_x and z amplitude A_z, initial position is $r_1 = [L_1 + A_x, 0, A_z]$. Figure 9.6 shows a periodic orbit with $A_x = 1.0 \times 10^4$ km and $A_z = 5 \times 10^3$ km as well, as the initial guess from the first order Richardson approximation (Equation (9.33)) for a periodic orbit with amplitudes (A_x, A_z).

Periodic orbits of a given period are obtained by removing the constraints on the initial position and adding a constraint on the final time t_f, all other constraints

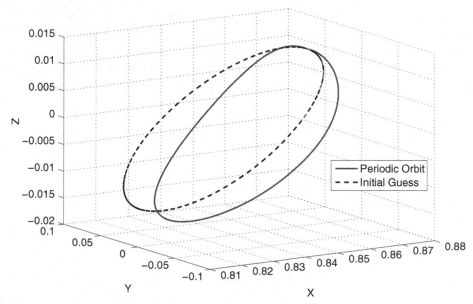

Figure 9.6. Single periodic orbit about Earth-Moon L_1.

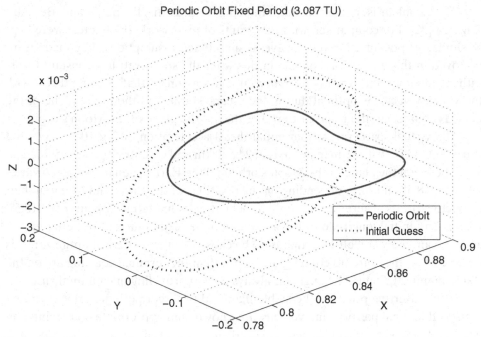

Figure 9.7. Single periodic orbit with period 3.087 TU about Earth-Moon Lagrange point L_1.

and the objective function remaining the same. The initial guess was a first order Richardson approximation for randomly selected amplitudes A_x and A_z. Figure 9.7 shows a periodic orbit with period 3.087 TU and the initial guess of the periodic orbit given to the numerical optimizer.

It is thus straightforward and efficient to use this method to create periodic orbits with desired amplitude or period. It would be possible to search for an orbit with amplitude and period in given ranges. The method could apply equally well to the generation of periodic orbits about the other Lagrange points ($L_2, \ldots, L_5$) and for different three body systems such as Earth-Sun-Spacecraft, Sun-Jupiter-Spacecraft and the like.

9.5 Optimal Earth Orbit to Lunar Orbit Transfer: Part 1—GTO to Periodic Orbit

The goal is to find an optimal, fuel-minimizing trajectory from a GTO about the Earth to a low lunar orbit (LLO) about the Moon via a periodic orbit about the L_1 Lagrange point. The first phase of the transfer is a thrust arc from GTO inserting into the stable manifold of the periodic orbit, followed by a coast along the stable manifold to the periodic orbit. The second phase is a coast along the unstable manifold toward the Moon followed by a thrust arc to insert into a low lunar orbit. In general there is no guarantee that the combination of two individually optimal trajectories is optimal. However, in this case the periodic orbit can be considered as a parking orbit that can be traversed with no cost, effectively decoupling the Earth-Halo transfer from the Halo-Moon transfer.

The problem is similar to that studied by Mingotti, Topputo, and Bernelli-Zazzera [22]. For comparison with the results of their work, parameters were kept as similar as possible. However, several simplifying assumptions they made were removed in this work. For example in this work the spacecraft has constant thrust rather than constant thrust *acceleration*; it is not assumed that aligning thrust with the velocity vector is optimal during the escape spiral, and the initial plane of the GTO is equatorial (the usual case) rather than in the plane of the Moon's orbit.

The virtual spacecraft has an initial thrust acceleration of $7 \times 10^{-4} m/s^{-2}$ and starts from GTO with semi major axis of 24,510 km and eccentricity 0.72345981. The angle between the plane of the Moon's orbit and the Earth's equatorial plane varies between 18.5° and 28.5° depending on the positions of the Earth and Moon in their respective orbits. Therefore, the initial inclination is limited to the interval (18.5°, 28.5°). The initial longitude of the ascending node (Ω), argument of periapse (ω), and true anomaly are left free, that is, to be chosen by the numerical optimizer. It is assumed that there are no coasting arcs other than the terminal coasting arc on the stable manifold, that is, the engine is always on and providing maximum thrust.

If the insertion point on the stable manifold of target periodic orbit is chosen *a priori* then final position and velocity are known, and appropriate constraints can be placed on the states at the final node and an optimal trajectory can be calculated readily. However, if the goal is simply to finish on the periodic orbit, this insertion point can almost certainly be improved upon. This constraint, that is, simply arriving on the periodic orbit, is problematic as there exists no analytical solution to the periodic orbit or stable manifold generation problems, hence there exists no terminal constraint of the form of Equation (9.21).

The solution developed in this work is to express the target periodic orbit in terms of parameters available to the optimizer. To achieve this, two additional parameters s and τ are introduced, hereafter referred to as the periodic orbit insertion point parameter, and the manifold insertion parameter respectively. The periodic insertion point parameter represents the normalized position around the periodic orbit from some arbitrary starting location (where $0 \leq s = t/T \leq 1$). The manifold insertion parameter represents the time taken to reach the periodic orbit from the manifold insertion point ($\tau \geq 0$). The states are then parameterized in terms of these new parameters.

A target periodic orbit with $A_z = 8,000$ km was generated by the procedure in Section 4. This procedure generates a discretized periodic orbit. For each state (x, y, z, v_x, v_y, v_z), a surface was fitted to create functions of the states in terms of the periodic orbit insertion parameter (s) and the manifold insertion parameter τ. A bicubic spline interpolant was fitted using the MATLAB curve fitting toolbox program *cftool*. These fitted surfaces are continuous and differentiable functions for the terminal states in terms of both insertion parameters (s, τ).

The quality of the initial guess significantly influences the ability of the NLP problem solver to obtain a convergent solution. It was chosen to have a feasible trajectory as an initial guess, that is, a trajectory that satisfies the initial and final constraints as well as the Runge-Kutta collocation constraints. This increases the likelihood that the optimizer will converge to a satisfactory solution. However, a valid trajectory that inserts into the target periodic orbit's stable manifold is not known *a priori* and must be computed. The initial guess trajectory was created using a three-step process.

(1) The initial orbit elements (ω, Ω) were chosen to be zero. The Equations of motion (9.3) were integrated forward with the assumption that the thrust is always aligned with the instantaneous velocity vector, that is, a thrust pointing angle of zero. The time of flight and the initial true anomaly (f) were selected so that at the end of the integration the spacecraft is in the vicinity of the target periodic orbit. The manifold insertion parameter (τ) was set to zero so the trajectory produced by this procedure targets the periodic orbit itself not the stable manifold.

(2) The next step was to create a valid trajectory that finished at a specified position on the periodic orbit but with the velocity unconstrained. A direct optimization was used with the Runge-Kutta transcription (Section 3.2). The initial conditions were the same as step 1, and there were no constraints applied to the final states. The objective function was the difference between the x, y, and z states at the final node (x_f, y_f, z_f) and the specified final position (x_s, y_s, z_s)

$$\phi(\tilde{x}) = \sqrt{(x_f - x_s)^2 + (y_f - y_s)^2 + (z_f - z_s)^2}. \tag{9.37}$$

The trajectory obtained with tangential thrust is used as the initial guess. A solution with an objective function value of zero, that is, reaching precisely the specified entry point, was obtained.

(3) The final step is to find a trajectory that reached the same position but that also had the velocity required to complete the periodic orbit. The procedure is similar to that of step 2 but with the x, y, and z states at the final node (x_f, y_f, z_f) now constrained to the specified final position (x_s, y_s, z_s). The objective then becomes minimization of the difference in the velocities at the final node $(v_{x_f}, v_{y_f}, v_{z_f})$ and the velocity specified at the end point $(v_{x_s}, v_{y_s}, v_{z_s})$. The objective function is then

$$\phi(\tilde{x}) = \sqrt{(v_{x_f} - v_{x_s})^2 + (v_{y_f} - v_{y_s})^2 + (v_{z_f} - v_{z_s})^2}. \tag{9.38}$$

The resulting trajectory is a feasible insertion trajectory that can be used as an initial guess for the problem of finding a fuel-minimizing insertion trajectory.

The numerical optimization procedure previously described was applied using the NLP parameters (Section 9.3) and Runge-Kutta constraints to generate optimal insertion trajectories for periodic orbits. The periodic orbit insertion parameter s, the manifold insertion parameters τ, and the initial moon position λ_m are now included as NLP parameters. The numerical method sought a trajectory that leaves Earth orbit and arrives at a point on the stable manifold of the periodic orbit, with the required velocity for insertion into the stable manifold, and coast to the periodic orbit such that the transfer is completed using minimum fuel.

The optimal trajectory obtained reaches the periodic orbit in 89.04856 days, the first 47.1023 days using maximum thrust and the final 41.9463 days coasting on the stable manifold. The mass fraction of propellant required is $m_p/m_0 = 0.09681$. The optimal trajectory is shown in Figure 9.8. The orbit elements at the point of departure

Table 9.1. *Orbit elements at moment of departure*
from GTO for the moon

semi-major axis (a)	24510*km*
eccentricity (e)	0.72346
inclination (i) (to lunar orbit plane)	18.5°
Longitude of ascending node (Ω)	−56.55057°
Argument of periapsis (ω)	159.11972°
True anomaly (f)	−65.74612°

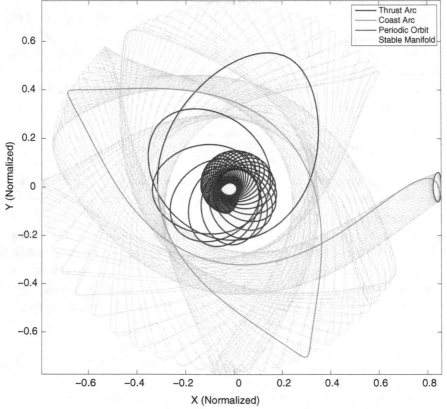

Figure 9.8. Optimal low-thrust trajectory to target orbit of amplitude $A_z = 8.0 \times 10^3$ km.

from the initial GTO are given in Table 9.1. Semi-major axis and eccentricity are
specified; the other elements are determined by the numerical optimiger.

9.6 Optimal Earth Orbit to Lunar Orbit Transfer: Part 2 – Periodic Orbit to Low-Lunar Orbit

The use of periodic orbits in an Earth-Moon low-thrust trajectory allows for the
possibility of long coast arcs on both the unstable and stable manifolds of the Halo
orbit, minimizing the time spent thrusting and hence minimizing fuel consumption.

For this second phase of the problem, the assumptions are the same as for the first phase, that is, the problem is the CR3BP; the Moon is in a circular orbit about the Earth and no other outside gravitational influences (Sun, Jupiter, and the like) or perturbations are considered. The spacecraft will start in GTO, that is an equatorial orbit, with an apogee at the radius of geostationary orbit 42,164 km. The initial orbit semi-major axis (a) and eccentricity (e) are the values for GTO (24,510 km and 0.72346, respectively). The angle between the equatorial plane of the Earth and the plane of the Moon's orbit varies between 18.5° and 28.5° depending on the relative positions of the Earth and Moon in their orbits, therefore the initial orbit inclination is restricted to the interval [18.5°, 28.5°]; the remaining orbit elements (Ω, ω, f) are free to be chosen by the optimizer, with the result as given in Table 9.1.

9.6.1 Halo Orbit-Moon Trajectory

The periodic orbit to Moon trajectory is qualitatively the opposite of the Earth-Halo trajectory. The spacecraft leaves the periodic orbit, coasts along the unstable manifold for a time, then starts its engine to complete the transfer to circular equatorial orbit about the Moon at an altitude of 100 km. Since the target orbit is a circular orbit in the reference plane, the longitude of the ascending node (Ω) and the argument of periapse (ω) are undefined. There is also a singularity in the orbit element variational Equations (9.5) due to zero inclination. To remove these singularities, equinoctial variables were used for this phase of the mission. For the Halo-Moon trajectory, the position on which the spacecraft leaves the unstable manifold is described by the same two parameters: the periodic orbit departure parameter (s) and manifold departure parameter (τ).

The final mass fraction from the Earth-Halo orbit trajectory $m_f/m_0 = 0.90319$ is now the initial value for the Halo-Moon trajectory. As for the Earth-Halo phase, a valid trajectory from the unstable manifold of the periodic orbit to low-lunar orbit is not known *a priori*, therefore it must be computed. It is computed using a three-step process similar to that used for the Earth-Halo phase.

(1) The Equations of motion (9.7) were integrated backward in time assuming tangential thrust opposite the direction of motion until the spacecraft had effectively escaped the Moon. The time of flight and final position within the circular orbit were chosen so that the spacecraft would be in the vicinity of the unstable manifold after the backward integration. The manifold departure parameters (s^*, τ^*) were chosen to minimize the Euclidian distance between the states at the initial node of the thrust arc ($x_0, y_0, z_0, v_{x_0}, v_{y_0}, v_{z_0}$) and the final states on the unstable manifold ($x_{s^*,\tau^*}, y_{s^*,\tau^*}, z_{s^*,\tau^*}, v_{x_{s^*,\tau^*}}, v_{y_{s^*,\tau^*}}, v_{z_{s^*,\tau^*}}$).

(2) The next step was to create a valid trajectory that began at the specified position on the unstable manifold of the periodic orbit but with the velocity unconstrained. A direct optimization was used with Runge-Kutta constraints and with the final states constrained to those used in part 1. There were no constraints applied to the initial states. The objective function was the difference between the x, y and z states at the initial node (x_0, y_0, z_0) and the specified initial position on the unstable

manifold $(x_{s*,\tau*}, y_{s*,\tau*}, z_{s*,\tau*})$.

$$\phi(\tilde{x}) = \sqrt{(x_0 - x_{s*,\tau*})^2 + (y_0 - y_{s*,\tau*})^2 + (z_0 - z_{s*,\tau*})^2}. \tag{9.39}$$

The trajectory obtained with tangential thrust is used as the initial guess. A solution with an objective function value of zero, that is, starting precisely at the specified exit point on the unstable manifold, was obtained.

(3) The final step is to find a trajectory that begins at the same position but that also starts with the velocity required to complete the periodic orbit. A similar procedure is used with the position states at the initial node (x_0, y_0, z_0) now constrained to the specified initial position $(x_{s*,\tau*}, y_{s*,\tau*}, z_{s*,\tau*})$. The objective then becomes the minimization of the difference in the velocities at the initial node $(v_{x_0}, v_{y_0}, v_{z_0})$ and the known velocity required at the specified starting point $(v_{x_{s*,\tau*}}, v_{y_{s*,\tau*}}, v_{z_{s*,\tau*}})$. The objective function is then

$$\phi(\tilde{x}) = \sqrt{(v_{x_0} - v_{x_{s*,\tau*}})^2 + (v_{y_0} - v_{y_{s*,\tau*}})^2 + (v_{z_0} - v_{z_{s*,\tau*}})^2}. \tag{9.40}$$

The resulting trajectory is a feasible trajectory that can be used as an initial guess for the problem of finding a fuel-minimizing trajectory from the periodic orbit about L_1 to low-lunar orbit.

The numerical optimization procedure previously described was applied using the NLP parameters (Section 9.3) and Runge-Kutta constraints to generate optimal departure trajectories from the periodic orbit to low-lunar orbit. The periodic orbit departure parameter s and the manifold departure parameter τ are now included as NLP parameters. The numerical method sought a trajectory that leaves the unstable manifold and arrives in a low-lunar orbit with an altitude of 100 km such that the transfer is completed using minimum fuel.

The optimal trajectory obtained reaches low lunar orbit in 34.2474 days, the first 15.2215 days coasting on the unstable manifold and the final 19.0260 days using maximum available thrust. The mass fraction of propellant required is $m_p/m_0 = 0.04229$. The trajectory is shown in Figures 9.9–9.11.

9.6.2 Combined Earth-Moon Trajectory

The two optimal trajectories can be combined to create a fuel-optimal trajectory from the Earth to the Moon via a periodic orbit about the interior (L_1) Earth-Moon Lagrange point. The characteristics of the combined trajectory are given in Table 9.2.

This trajectory has a shorter flight time than that computed by Mingotti et al. [22], 89.05 days versus 91.5 days, but uses slightly more propellant, that is, the propellant mass fraction is 0.09681 versus 0.0892. It was not expected that the results would be the same since in this work, as described in Section 9.5, several simplifying assumptions made in the previous study were removed.

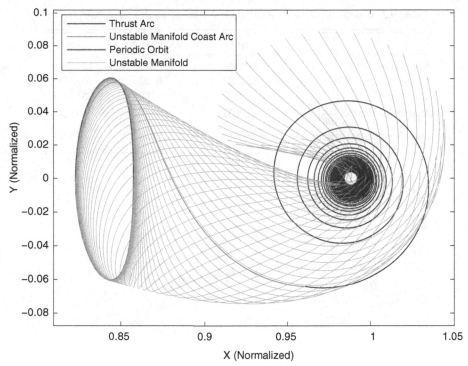

Figure 9.9. Optimal low-thrust trajectory to low-lunar orbit from a periodic orbit of amplitude $A_z = 8.0 \times 10^3$ km $X - Y$ plane.

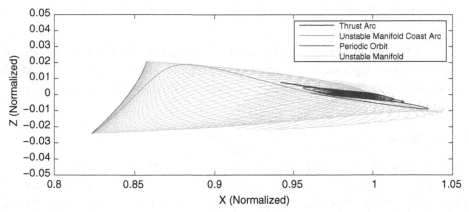

Figure 9.10. Optimal low-thrust trajectory to low-lunar orbit from a periodic orbit of amplitude $A_z = 8.0 \times 10^3$ km $X - Z$ plane.

9.7 Extension of the Work to Interplanetary Flight

The methods used here are not restricted to the problem of a Earth-to-Moon transfer. Any three-body system (for example, Earth-Sun, Sun-Jupiter, or Jupiter-Jovian Moon) has the necessary Lagrange points, periodic orbits, and stable and unstable manifolds to construct trajectories such as those shown in the previous

Table 9.2. *Characteristics of the complete Earth–Moon transfer trajectory.*

	First Leg	Second Leg	Total
t_{total} (days)	89.0486	34.2475	123.2960
$t_{powered}$ (days)	47.1023	19.0260	66.1282
t_{coast} (days)	41.9463	15.2215	57.1678
final mass fraction $\dfrac{m_f}{m_0}$	0.9032	0.9567	0.8641
mass fraction propellant $\dfrac{m_p}{m_0}$	0.0968	0.0433	0.1359

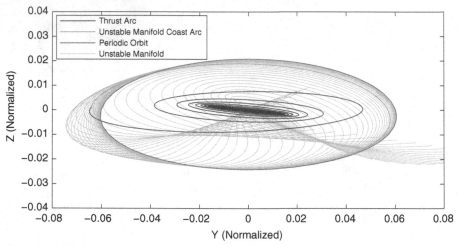

Figure 9.11. Optimal low-thrust trajectory to low-lunar orbit from a periodic orbit of amplitude $A_z = 8.0 \times 10^3$ km $Y-Z$ plane.

sections. Efficient interplanetary flight can be constructed by patching together individual optimal planet–stable manifold, unstable manifold–planet, or stable manifold–unstable manifold trajectories for various three-body systems.

9.8 Conclusions

This work demonstrates that periodic orbits about the Lagrange points can be generated by posing the problem as an optimization problem and using direct transcription to find the solution. Periodic orbits with specific properties (amplitude, period, energy, and so on) can be generated by altering the constraints on the state variables. This method is able to produce periodic orbits quickly (<30s on a 2GHz Intel Core 2 Duo), easily, and reliably even from poor initial guesses.

Many current and proposed missions take advantage of Lagrange point orbits [4, 5, 6] and their manifolds. These trajectories are the "on-ramp" to the Interplanetary Superhighway [2]. It is of great benefit to investigate how to reach these periodic orbits cheaply and it thus follows that very-efficient, low-thrust electric propulsion should be combined with the use of invariant manifolds to take advantage of these low-energy paths.

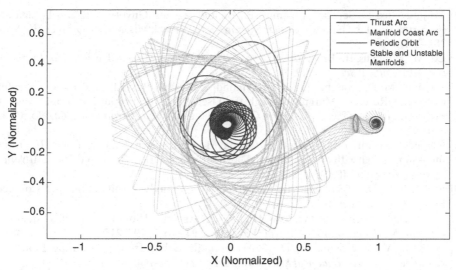

Figure 9.12. Optimal low-thrust trajectory to low-lunar orbit from GTO via a periodic orbit of amplitude $A_z = 8.0 \times 10^3$ km.

In this work, the same numerical optimization with direct transcription used for the generation of the periodic orbits was also used to optimize a complete planet-planet transfer via the interior Lagrange point (L_1). The specific example was a transfer from geostationary transfer orbit (GTO) to a low-lunar orbit. Optimal low-thrust trajectories were generated from the GTO to a periodic orbit via the stable manifold and also from the periodic orbit to low-lunar orbit. The optimization method minimizes the duration of the thrust arcs, thus minimizing the fuel consumed, by simultaneously determining all of the free parameters of the problem including the orientation of the departure orbit plane, the optimal point at which to enter the target orbit's stable manifold, and the thrust pointing angle time history. This is necessary as the insertion point chosen and the path to get there both influence the cost, that is, the cheapest (closest) point to reach is not necessarily the best point to insert into. Significant fuel savings can be realized by optimizing the insertion point simultaneously with optimizing the trajectory. The method demonstrated can be straightforwardly applied to other planet-planet transfers making use of the collinear Lagrange points.

REFERENCES

[1] Poincare, H. (1967) *New Methods of Celestial Mechanics. Volume II—Methods of Newcomb, Gylden, Lindstedt, and Bohlin*, NTIS, United States, Translation.

[2] Ross, S. (2004) *Cylindrical Manifolds and Tube Dynamics in the Restricted Three-Body Problem*, Ph.D. thesis, California Institute of Technology.

[3] Tripathi, R. K., Wilson, J., Cucinotta, F., Anderson, B., and Simonsen, L. (2003) Materials trade study for lunar/gateway missions, *Advances in Space Research*, **31**, No. 11, 2383–2388.

[4] Pilbratt, G. L. (2003) Herschel Space Observatory Mission Overview, SPIE, **4850**, 586–597.

[5] Triqueneaux, S., Sentis, L., Camus, P., Benoit, A., and Guyot, G. (2006) Design and Performance of the Dilution Cooler System for the Planck Mission, *Cryogenics*, **46**, No. 4, 288–297.

[6] Lo, M., Williams, B., Bollman, W., *et al.* (2001) Genesis Mission Design, *The Journal of the Astronautical Sciences*, **49**, No. 1, 169–184.

[7] Rayman, M. D., Fraschetti, T. C., Raymond, C. A., and Russell, C. T. (2007) Coupling of System Resource Margins through the Use of Electric Propulsion: Implications in Preparing for the Dawn Mission to Ceres and Vesta, *Acta Astronautica*, **60**, No. 10–11, 930–938.

[8] Russell, C., Barucci, M., Binzel, R., Capria, M., Christensen, U., *et al.* (2007) Exploring the Asteroid Belt with Ion Propulsion: Dawn Mission History, Status and Plans, *Advances in Space Research*, **40**, No. 2, 193–201.

[9] Richardson, D. (1980) Analytic Construction of Periodic Orbits about the Collinear Points, *Celestial Mechanics*, **22**, No. 3, 241–253.

[10] Gomez, G., and Marcote, M. (2006) High-Order Analytical Solutions of Hill's Equations, *Celestial Mechanics and Dynamical Astronomy*, **94**, No. 2, 197–211.

[11] Guibout, V., and Scheeres, D. (2004) Periodic Orbits from Generating Functions, *Advances in the Astronautical Sciences*, **116**, No. 2, 1029–1048.

[12] Gomez, G., Masdemont, J., and Simo, C. (1998) Quasihalo Orbits Associated with Libration Points, *Journal of the Astronautical Sciences*, **46**, No. 2, 135–176.

[13] Howell, K., and Pernicka, H. (1988) Numerical Determination of Lissajous Trajectories in the Restricted Three-Body Problem, *Celestial Mechanics*, **41**, No. 2, 107–124.

[14] Bryson, A. E. and Ho, Y-C. (1975) *Applied Optimal Control Revised Printing*, Hemisphere Publishing Corporation.

[15] Battin, R. H. (1987) An Introduction to the Mathematics and Methods of Astrodynamics. In *AIAA Education Series, AIAA*, Reston, VA.

[16] Prussing, J., and Conway, B. (1993) *Orbital Mechanics*. Oxford University Press, New York.

[17] Howell, K., Barden, B., and Lo, M. (1997) Application of Dynamical Systems Theory to Trajectory Design for a Libration Point Mission, *Journal of the Astronautical Sciences*, **45**, No. 2, 161–178.

[18] Herman, A. L., and Conway, B. A. (1996) Direct Optimization Using Collocation Based on High-Order Gauss-Lobatto Quadrature Rules, *Journal of Guidance, Control, and Dynamics*, **19**, No. 3, 592–599.

[19] Gill, P. E., Murray, W., and Saunders, M. A. (2005) SNOPT: An SQP Algorithm for Large-Scale Constrained Optimization, *SIAM Review*, **47**, No. 1, 99–131.

[20] Gill, P., Murray, W., Saunders, M. A., and Wright, M. H. (1993) *User's Guide for NZOPT 1.0: A Fortran Package For Nonlinear Programming*. McDonnell Douglas Aerospace, Huntington Beach, CA.

[21] Tarragó, P. (2007) *Study and Assessment of Low-Energy Earth-Moon Transfer Trajectories*. Master's Thesis, Université de Liége.

[22] Mingotti, G., Topputo, F., and Bernelli-Zazzera, F. (2007) Combined Optimal Low-Thrust and Stable-Manifold Trajectories to the Earth-Moon Halo Orbits. *New Trends in Astrodynamics and Applications III*, **886**, No. 1, 100–112.

10 Swarming Theory Applied to Space Trajectory Optimization

Mauro Pontani
Scuola di Ingegneria Aerospaziale, University of Rome "La Sapienza,"
Rome, Italy

Bruce A. Conway
Department of Aerospace Engineering, University of Illinois at
Urbana-Champaign, Urbana, IL

10.1 Introduction

The determination of optimal (either minimum-time or minimum-propellant-consumption) space trajectories has been pursued for decades with different numerical optimization methods. In general, numerical optimization methods can be classified as deterministic or stochastic methods. Deterministic gradient-based methods assume the continuity and differentiability of the objective function to be minimized. In addition, gradient-based methods are local in nature and require the identification of a suitable first-attempt "solution" in the region of convergence, which is unknown a priori and strongly problem dependent. These circumstances have motivated the development of effective stochastic methods in the last decades. These algorithms are also referred to as evolutionary algorithms and are inspired by natural phenomena. Evolutionary computation techniques exploit a population of individuals, representing possible solutions to the problem of interest. The optimal solution is sought through cooperation and competition among individuals. The most popular class of these techniques is represented by the genetic algorithms (GA) [1], which model the evolution of a species based on Darwin's principle of survival of the fittest. Differential evolution algorithms represent alternative stochastic approaches with some analogy with genetic algorithms, in the sense that new individuals are generated from old individuals and are eventually preserved after comparing them with their parents. Ant colony optimization [2] is another method, inspired by the behavior of ants, whereas the simulated annealing algorithm [2] mimics the equilibrium of large numbers of atoms during an annealing process.

The particle swarm optimization (PSO) technique, which is the methodology being addressed in this chapter, was first introduced by Eberhart and Kennedy [3, 4] in 1995 and belongs to the category of swarm intelligence methods [2, 5]. It mimics the unpredictable motion of bird flocks while searching for food, taking advantage of the mechanism of information sharing that affects the overall behavior of a swarm

[2–16]. The initial population that composes the swarm is randomly generated at the first iteration of the process. Each particle is associated with a position vector and a velocity vector at a given iteration. More specifically, the position vector includes the values of the unknown parameters of the problem, whereas the velocity vector determines the position update. Each particle represents a possible solution to the problem and corresponds to a specific value of the objective (or fitness) function. At the end of the process, the best particle (i.e., the best solution with reference to the objective function) is selected. Both the position and the velocity vector are updated in a single iteration. For each particle, the formula for velocity update includes three terms with stochastic weights; one of these terms is the so-called social component, related to the collective best position ever visited by a portion of the particles that form the swarm.

A number of options are available for implementing the PSO technique. First of all, different values for the stochastic weights have been proposed [10–12], and they seem to affect both the algorithm convergence and the capability to detect the global optimum. In addition, two different versions of the particle swarm exist [2, 5]: the *global* version, where the collective best position (associated with the social term in the velocity updating expression) is selected by considering the entire swarm, and the *local* version, where for each particle the collective best position is selected among the particles located in a proper neighborhood of the particle itself. Albeit less computationally efficient, in the scientific literature [2, 5, 7–11] the local version of the PSO algorithm, based on the definition of the neighbors of each particle, has been reported to be occasionally capable of avoiding local minima. Additional improvements based on the application of evolutionary operators to the particle swarm methodology are also reported by several researchers [17, 18].

The basic version of the particle swarm algorithm appears as very intuitive and is extremely easy to program. In addition, this kind of method is well suited for finding the globally optimal solution to an optimization problem and requires only the definition of the search space for the unknown parameters. Although computationally expensive with respect to gradient-based methods [13], in the scientific literature [19–21] the particle swarm technique is reported to be more efficient when compared to genetic algorithms, due to a reduced number of function evaluations. Despite its promising features and the vast number of papers devoted to this technique, most researchers concentrated on topological [22] and multimodal mathematical problems [15, 23], and only a limited number of applications appear of practical interest. In the scientific literature, several papers describe the use of the particle swarm methodology in the context of chemical processes [24, 25]. Fourie and Groenwold [26] employed the PSO technique for shape and size optimization in structural engineering. Most recently, Khurana et al. [27] applied the swarming theory to airfoil shape optimization. Bessette and Spencer [28, 29] successfully employed the particle swarm technique for space trajectory optimization, focusing on the optimization of time-limited orbital transfers [28] and impulsive interplanetary trajectories [29]. In both papers they claim that the PSO method outperforms the differential evolution algorithm (DE), with regard to convergence speed as well as to reliability. This statement is somewhat in contrast with what is affirmed by other researchers, such as Spaans and Mooji [30], who claim the superiority of differential evolution methods with respect to the particle swarm algorithm. Vasile et al. [31] investigated several algorithms for global optimization (including PSO and DE), employing different settings for each of them and using the success rate as the main index for the performance evaluation of each method. In their paper an improved algorithm based on differential evolution is proposed. Nevertheless, in the conclusions the authors remark on the importance of appropriately coupling the evolutionary technique with the problem structure, stating that the optimal choice of a method of solution is definitely problem dependent. Zhu et al. [32, 33] combined DE and PSO for satellite constellation design [32] and for determining the globally optimal low-thrust trajectory for asteroid exploration [33]. Lastly, Rosa Sentinella and Casalino [34] proposed a hybrid optimization procedure that runs three different optimizers – based on GA, DE, and PSO – in parallel. They applied this method to the optimization of multiple-impulse rendezvous trajectories and of Earth-to-Mars round-trip missions.

The work that follows adopts a basic, general-purpose global version of the PSO method, according to the general settings suggested in References 6–8. A simple MATLAB code has been implemented and applied to a variety of space trajectory optimization problems. In particular, the following applications have been considered:

(a) *Lyapunov periodic orbits around the collinear Lagrange points of the Earth-Moon system*. The problem consists of determining the initial position and velocity such that the motion around the collinear Lagrange points is indefinitely repeated in the context of a circular restricted three-body problem.
(b) *Lunar periodic orbits*. The problem is in determining the initial position and velocity such that the motion around the Moon is indefinitely repeated in the context of a circular restricted three-body problem.
(c) *Optimal four-impulse rendezvous*. The problem consists of determining the optimal locations, directions, and magnitudes of the four impulsive changes of velocity that allow performing a coplanar circle-to-circle rendezvous while minimizing propellant consumption (i.e., the characteristic velocity of the transfer trajectory).
(d) *Optimal low-thrust orbital transfer*. The problem consists of determining the thrust pointing angle time history that minimizes the time of flight while satisfying the terminal conditions for transferring a spacecraft from an initial circular orbit to a terminal circular orbit.

In this chapter the particle swarm technique, even in its simplest formulation and without interacting with other algorithms, is shown to be capable of effectively solving the previously mentioned space trajectory optimization problems with great numerical accuracy.

10.2 Description of the Method

In the context of space trajectories, the optimization problems of interest usually consist of minimizing a given objective function related with the time evolution of a dynamical system, which can be governed either by differential equations or by algebraic equations. The minimization is achieved by selecting the optimal values of the unknown parameters and time-varying variables. Several methodologies can be employed to translate the optimal control problems that involve continuous time-dependent control variables into parameter optimization problems. If the system dynamics are governed by a set of algebraic (nonlinear) equations the problem reduces to a nonlinear programming problem. Definitely, in both cases – in the presence of optimal control problems or nonlinear programming problems – the optimization process is aimed at finding the optimal values of a set of unknown parameters.

10.2.1 Unconstrained Optimization

Unconstrained parameter optimization problems can be stated as follows: determine the optimal values of the n unknown parameters $\{\chi_1, \ldots, \chi_n\}$ such that the objective function J is minimized. The time evolution of the dynamical system under consideration depends on $\{\chi_1, \ldots, \chi_n\}$, which are constrained to their respective ranges

$$a_k \leq \chi_k \leq b_k \quad (k = 1, \ldots, n). \tag{10.1}$$

As mentioned earlier in the chapter, the PSO technique is a population-based method, where the population is represented by a swarm of N particles. Each particle i $(i = 1, \ldots, N)$ is associated with a position vector $\boldsymbol{\chi}(i)$ and with a velocity vector $\boldsymbol{w}(i)$. The position vector includes the values of the n unknown parameters of the problem

$$\boldsymbol{\chi}(i) \triangleq [\chi_1(i) \quad \ldots \quad \chi_n(i)]^T \tag{10.2}$$

whereas the velocity vector, whose components are denoted with $w_k(i)$ $(k = 1, \ldots, n)$, determines the position update (both $\boldsymbol{\chi}$ and $\boldsymbol{w}$ are defined as n-dimensional column vectors). As the position components are bounded, the corresponding velocity components must be also constrained to suitable ranges

$$-(b_k - a_k) \leq w_k \leq (b_k - a_k) \Rightarrow -d_k \leq w_k \leq d_k \text{ if } d_k \triangleq b_k - a_k \ (k = 1, \ldots, n). \tag{10.3}$$

The limitations (10.3) are due to the fact that if $w_k > b_k - a_k$ or $w_k < a_k - b_k$ then, starting from any coordinate $\chi_k^{(j)}$ (at the iteration j), the updated coordinate $\chi_k^{(j+1)}$ $(=\chi_k^{(j)} + w_k)$ would violate the condition (10.1). If $\boldsymbol{a} \triangleq [a_1 \ \ldots \ a_n]^T, \boldsymbol{b} \triangleq [b_1 \ \ldots \ b_n]^T$, and $d \triangleq [d_1 \ \ldots \ d_n]^T$, the relationships (10.1) and (10.3) can be rewritten in compact form as

$$a \leq \chi \leq b \quad \text{and} \quad -d \leq w \leq d. \tag{10.4}$$

Each particle represents a possible solution to the problem and corresponds to a specific value of the fitness function. The expressions for position and velocity update determine the swarm evolution toward the location of the globally optimal position, which corresponds to the globally optimal solution to the problem under consideration. The initial population for the PSO algorithm is randomly generated by introducing N particles, whose positions and velocities are (stochastically) uniformly distributed in the respective search spaces, defined by Equation (10.4). The following steps compose the generic iteration j:

(a) for $i = 1, \ldots, N$:
   (i) evaluate the objective function associated with particle i, $J^{(j)}(i)$
   (ii) determine the best position ever visited by particle i up to the current iteration j, $\psi^{(j)}(i)$: $\psi^{(j)}(i) = \chi^{(l)}(i)$, where $l = \arg \min_{p=1,\ldots,j} J^{(p)}(i)$

(b) determine the global best position ever visited by the entire swarm, $Y^{(j)}$: $Y^{(j)} = \psi^{(j)}(q)$, where $q = \arg \min_{i=1,\ldots,N} \Im^{(j)}(i)$ and $\Im^{(j)}(i)$ $(i = 1, \ldots, N)$ represents the value of the objective function corresponding to the best position ever visited by particle i up to iteration j, that is, $\Im^{(j)}(i) = \min_{p=1,\ldots,j} J^{(p)}(i)$

(c) update the velocity vector. For each particle i and for each component $w_k(i)$ $(k = 1, \ldots, n; i = 1, \ldots, N)$:

$$w_k^{(j+1)}(i) = c_I w_k^{(j)}(i) + c_C[\psi_k^{(j)}(i) - x_k^{(j)}(i)] + c_S[Y_k^{(j)} - x_k^{(j)}(i)] \tag{10.5}$$

The *inertial, cognitive,* and *social* (stochastic) weights have the following expressions [6–8]

$$c_I = \frac{1 + r_1(0,1)}{2} \qquad c_C = 1.49445 r_2(0,1) \qquad c_S = 1.49445 r_3(0,1) \tag{10.6}$$

where $r_1(0,1)$, $r_2(0,1)$, and $r_3(0,1)$ represent three independent random numbers with uniform distribution between 0 and 1. Then:
   (i) if $w_k^{(j+1)}(i) < -d_k \Rightarrow w_k^{(j+1)}(i) = -d_k$
   (ii) if $w_k^{(j+1)}(i) > d_k \Rightarrow w_k^{(j+1)}(i) = d_k$

(d) update the position vector. For each particle i and for each component $\chi_k(i)$ $(k = 1, \ldots, n; i = 1, \ldots, N)$

$$\chi_k^{(j+1)}(i) = \chi_k^{(j)}(i) + w_k^{(j)}(i) \tag{10.7}$$

   (i) if $\chi_k^{(j+1)}(i) < a_k \Rightarrow \chi_k^{(j+1)}(i) = a_k$ and $w_k^{(j+1)}(i) = 0$
   (ii) if $\chi_k^{(j+1)}(i) > b_k \Rightarrow \chi_k^{(j+1)}(i) = b_k$ and $w_k^{(j+1)}(i) = 0$

The algorithm terminates when the maximum number of iterations N_{IT} is reached. The position vector of the global best particle, $\mathbf{Y}^{(N_{IT})}$, is expected to contain the optimal values of the unknown parameters, which correspond to the global minimum of J, denoted with $J_{opt}^{(N_{IT})}$. A fairly large number of iterations are set to ensure that the final solution is stable enough to be considered optimal. The central idea underlying the method is contained in the formula (10.5) for velocity updating. This formula includes three terms with stochastic weights: the first term is the so called *inertial* component and for each particle is proportional to its velocity in the preceding iteration; the second component is termed the *cognitive* component, directed toward the personal best position, that is, the best position experienced by the particle; and finally the third term is the *social* component, directed toward the global best position, that is, the best position yet located by any particle in the swarm. According to point (c), if a component w_k of the velocity vector violates Equation (10.3), then w_k is set to the minimum (maximum) value $-d_k$ (d_k). If a component χ_k of the position vector violates Equation (10.1), then χ_k is set to the minimum (maximum) value a_k (b_k), and the corresponding velocity component is set to 0. This ensures that in the successive iteration the update of the velocity component is not affected by the first term of Equation (10.5), which could lead the particle to again violating the constraint (10.1).

10.2.2 Constrained Optimization

Space trajectory optimization problems must be frequently modeled as constrained optimization problems – in other words, they involve equalities and/or inequalities regarding (directly or indirectly) the unknown parameters. The PSO algorithm described in the preceding subsection must be suitably adjusted to deal with constrained problems.

In general, evolutionary computation methods encounter difficulties in treating equality constraints [6–9, 14, 35], because they narrow considerably the search space where feasible solutions can be located. This is due to the fact that (nonredundant) equality constraints actually reduce the degree of freedom of the problem according to their number. In fact, m equality constraints reduce the degree of freedom by m. Therefore, in the presence of n unknown parameters, at most $m = n$ equality constraints are admissible ($m \leq n$)

$$d_r(\boldsymbol{\chi}) = 0 \qquad (r = 1, \ldots, m). \tag{10.8}$$

The most popular approach for dealing with these constraints consists in penalizing them by summing additional terms to the objective function

$$\tilde{J} = J + \sum_{r=1}^{m} \alpha_r \, |d_r(\boldsymbol{\chi})| \, . \tag{10.9}$$

This approach is employed also in this research. The values of the coefficients α_r ($r = 1, \ldots, m$) must be carefully chosen and are problem dependent. Small values can allow excessive constraint violations, whereas high values of α_r can render the problem ill-conditioned. The scientific literature includes a vast number of works addressing penalty methods [36, 37] (with special convergence properties) or alternative approaches [38]; however, in this work the coarse trial-and-error selection of suitable values of α_r proved to be a satisfactory approach.

Inequality constraints are less problematic because they reduce the search space of feasible solutions without decreasing the degree of freedom of the problem. For each particle, the simplest way of treating inequality constraints consists of assigning a fictitious infinite value to the fitness function if the particle violates at least one of them. In addition, the corresponding velocity is set to zero, so that the successive velocity update – according to Equation (10.5) – is affected only by the social term and by the cognitive term. This circumstance statistically leads the particle to a feasible region of the search space. Points (a)–(i) of the algorithm (in the preceding subsection) must be modified as follows:

(a) for $i = 1, \ldots, N$:
  (i) evaluate the inequality constraints. If (at least) one of them is violated *then* set $J^{(j)}(i) = \infty$ and $w^{(j+1)}(i) = 0$ and skip point (c), *else* evaluate the objective function associated with particle i, $J^{(j)}(i)$.

Further options exist for dealing with constraints (e.g., methods based on preserving feasibility of solutions, methods that distinguish between feasible and infeasible solutions, cf. References 14, 35–38). However, the two distinct, simple strategies described previously in this chapter for equality and inequality constraints will be shown to be quite effective for the problems considered in this chapter.

10.3 Lyapunov Periodic Orbits

Lyapunov periodic orbits represent a special class of planar periodic orbits around the collinear Lagrange points in the context of a circular restricted three-body problem. The problem investigated in this section is in determining the initial position and velocity of the third body such that its motion around the collinear Lagrange points of the Earth-Moon system is indefinitely repeated. The problem at hand is reformulated as an unconstrained optimization problem so that it can be solved by the PSO algorithm.

10.3.1 Problem Definition

The circular restricted three-body problem models the dynamics of three bodies with masses m_1, m_2, and m_3, under the assumption that $m_1 > m_2 \gg m_3 \approx 0$. This means that the mass of the third body, m_3, is considered negligible, whereas the remaining massive bodies, termed the primaries, describe counterclockwise circular orbits around their center of mass. This problem is conveniently described in synodic coordinates that represent a coordinate system rotating with the two primaries. The

x-axis connects the two primaries and is directed from body 1 to body 2, whereas the y-axis lies in the plane of their motion. This problem is analyzed by employing canonical units that represent a set of convenient normalized units. The distance unit (DU) is the (constant) distance between the two primaries (i.e., the Earth and the Moon, and therefore 1 DU = 384400 km), whereas the time unit (TU) is such that the two primaries complete a single orbit in a period equal to 2π TU (1 TU = 375190 sec). If μ_E and μ_M denote respectively the gravitational parameter of the Earth and of the Moon, these definitions of TU and DU imply that $\mu_E + \mu_M = 1$ DU^3/TU^2. After introducing the parameter $\mu \triangleq \mu_M/(\mu_E + \mu_M)$ (= 0.01215510), it is straightforward to rewrite the gravitational parameters of the two primaries as $\mu_E = 1 - \mu$ and $\mu_M = \mu$ (in DU^3/TU^2). Their respective positions along the x-axis are identified by $x_E = -\mu$ and $x_M = 1 - \mu$ (in DU), as shown in Figure 10.1.

With reference to the synodic frame illustrated in Figure 10.1, if (x, y) and (v_x, v_y) are the coordinates of position and velocity, the equations of motion of the planar circular restricted three-body problem are the following [39]

$$\dot{x} = v_x \qquad\qquad \dot{y} = v_y \tag{10.10}$$

$$\dot{v}_x = \frac{\partial \Omega}{\partial x} + 2v_y \qquad \dot{v}_y = \frac{\partial \Omega}{\partial y} - 2v_x \tag{10.11}$$

where

$$\Omega \triangleq \frac{x^2 + y^2}{2} + \frac{1 - \mu}{\sqrt{(x + \mu)^2 + y^2}} + \frac{\mu}{\sqrt{(x + \mu - 1)^2 + y^2}}. \tag{10.12}$$

It is relatively straightforward to demonstrate that an integral exists for this dynamical system [39]: the *Jacobi integral*, whose value is referred to as the *Jacobi constant* and denoted with C

$$C = 2\Omega - (v_x^2 + v_y^2). \tag{10.13}$$

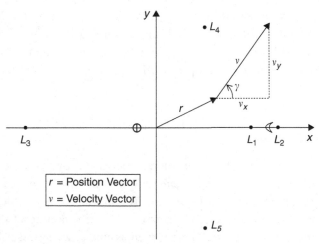

Figure 10.1. Synodic reference frame.

The value of C remains unchanged and is associated with the energy of the dynamical system. Zero-velocity curves are the geometrical loci where $v_x = v_y = 0$, and constitute the boundary of the region where the third-body motion is allowed. It is worth reporting the values of C_1, C_2, and C_3, which represent the values of the Jacobi constant corresponding to the situations when the zero-velocity curves contain the Lagrange points L_1, L_2, or L_3, respectively:

$$C_1 = 3.188383 \qquad C_2 = 3.172196 \qquad C_3 = 3.012152 \qquad \text{(in DU}^2/\text{TU}^2\text{)}. \quad (10.14)$$

These values define four intervals for the Jacobi constant, corresponding to different geometries of the region where the third-body motion is allowed [39].

With regard to the equations of motion (10.10)–(10.11), after introducing the angle γ that the velocity vector forms with the x-axis, due to (10.13), the two components v_x and v_y can be rewritten as

$$v_x = \sqrt{2\Omega - C} \cos \gamma \qquad \text{and} \qquad v_y = \sqrt{2\Omega - C} \sin \gamma. \quad (10.15)$$

As a result

$$\tan \gamma = \frac{v_y}{v_x}. \quad (10.16)$$

The time derivative of Equation (10.16) is

$$\frac{\dot\gamma}{\cos^2 \gamma} = \frac{\dot v_y v_x - \dot v_x v_y}{v_x^2} \Rightarrow \dot\gamma = \frac{\dot v_y v_x - \dot v_x v_y}{v_x^2 + v_y^2} = \frac{(\Omega_y - 2v_x)v_x - (\Omega_x + 2v_y)v_y}{2\Omega - C} \quad (10.17)$$

where $\Omega_x \triangleq \partial\Omega/\partial x$ and $\Omega_y \triangleq \partial\Omega/\partial y$. After replacing v_x and v_y with their respective expressions depending on γ, Equation (10.17) becomes

$$\dot\gamma = \frac{\Omega_y \cos \gamma - \Omega_x \sin \gamma}{\sqrt{2\Omega - C}} - 2. \quad (10.18)$$

Therefore, the existence of the Jacobi integral allows replacing Equations (10.11) with Equation (10.18). As a result, the set of equations employed to determine periodic orbits is composed of (10.10) and (10.18), also referred to as the *Birkhoff's equations* [39]. They represent the *state equations* for the dynamical system under consideration.

In the context of the circular restricted three-body problem, for a given Jacobi constant, in the synodic coordinate system an orbit is termed *periodic* if it is indefinitely repeated. This means that after the period T the three variables $\{x, y, \gamma\}$ assume their respective initial values

$$x(T) = x(0) \qquad y(T) = y(0) \qquad \gamma(T) = \gamma(0) + 2\rho\pi \quad (\rho \in \mathbb{Z}). \quad (10.19)$$

Lyapunov orbits are periodic clockwise trajectories around the Lagrange points L_1, L_2, and L_3. These orbits are also symmetrical with respect to the x-axis. Thus there exists an initial position, on a Lyapunov orbit, along the x-axis with initial velocity aligned with the y-axis. Hence, for a given Jacobi constant, in the synodic coordinate system a Lyapunov periodic orbit is such that after the period T the three variables $\{x, y, \gamma\}$ assume their respective initial values

$$x(T) = x(0) \qquad y(T) = y(0) = 0 \qquad \gamma(T) = \gamma(0) = \frac{\pi}{2} + 2\rho\pi \quad (\rho \in \mathbb{Z}). \quad (10.20)$$

The problem can be easily translated into an optimization problem if the following objective function is introduced

$$J = |x(T) - x(0)| + |y(T)|$$
$$+ \min\left\{ \mathrm{mod}\left[\left| \gamma(T) - \frac{\pi}{2} \right|, 2\pi \right], \mathrm{mod}\left[-\left| \gamma(T) - \frac{\pi}{2} \right|, 2\pi \right] \right\}. \quad (10.21)$$

The third term of J is a piecewise linear function of $\gamma(T)$, yields a maximum value equal to π if $\gamma(T) = \pi/2 + \pi + 2\rho\pi$ ($\forall \rho \in \mathbb{Z}$), and vanishes if $\gamma(T) = \pi/2 + 2\rho\pi$ ($\forall \rho \in \mathbb{Z}$). The objective function J has a global minimum equal to 0, provided that at least one periodic orbit exists. Hence, the problem consists of finding the initial condition $x(0)$ and the period T corresponding to the global minimum of J.

10.3.2 Numerical Results

For this example, the PSO algorithm uses a population of 30 particles ($N = 30$) and is run for 500 iterations ($N_{IT} = 500$). Each particle is associated with the values of the two unknown parameters of the problem: $\chi = [x(0) \quad T]^T$. The optimization process is repeated for three different cases (labeled with [a], [b], and [c]), corresponding to the collinear Lagrange points L_1, L_2, and L_3, whose coordinates in canonical units are

$$x_{L1} = 0.836893 \qquad x_{L2} = 1.155700 \qquad x_{L3} = -1.005065. \quad (10.22)$$

For each case, the optimal values of the unknown parameters are sought in the following ranges

$$x_{Li} - 0.15 \, \mathrm{DU} \le x(0) \le x_{Li} \quad (i = 1, 2, 3) \qquad \text{and} \qquad 3 \, \mathrm{TU} \le T \le 8 \, \mathrm{TU} \quad (10.23)$$

and C is set to 3.01 $\mathrm{DU}^2/\mathrm{TU}^2$. Table 10.1 reports the results obtained with the PSO algorithm after 500 iterations, that is, the globally optimal values of the unknown parameters as well as the related value of the fitness function $J_{opt}^{(N_{IT})}$. Figure 10.2 shows the objective evolution as a function of the iteration index j, whereas Figure 10.3 portrays the corresponding Lyapunov orbits in the synodic reference frame. In two cases $J_{opt}^{(j)} < 10^{-10}$ after 300 iterations, and this circumstance testifies to the effectiveness and numerical accuracy of the PSO method. In addition, the fact that χ includes

Table 10.1. *Results related with the Lyapunov*
orbits (C = 3.01 DU²/TU²)

	$x(0)$ (DU)	T (TU)	$J_{opt}^{(N_{IT})}$
L_1	0.775284	4.142957	$4.357 \cdot 10^{-12}$
L_2	1.027601	4.369107	$6.043 \cdot 10^{-6}$
L_3	−1.050020	6.218442	$4.811 \cdot 10^{-11}$

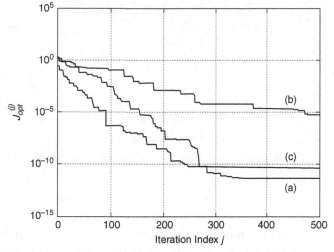

Figure 10.2. Lyapunov orbits: objective evolution as a function of the iteration index.

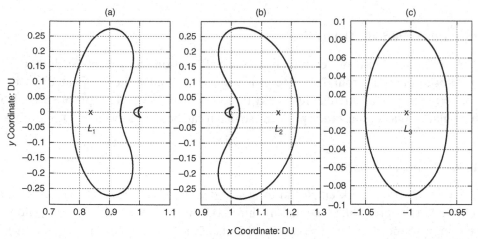

Figure 10.3. Lyapunov orbits about the collinear Lagrange points (C = 3.01 DU²/TU²).

only two parameters has the favorable consequence that the objective function J can be represented as a spatial surface in the neighborhood of the global optimal solution $Y^{(N_{IT})}$. Figure 10.4 illustrates this surface (for the Lyapunov orbit about the Lagrange point L_2), showing that the convergence region is extremely reduced in size and irregular in shape. As an immediate consequence, gradient-based methods

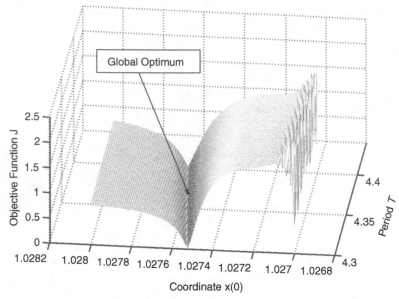

Figure 10.4. Lyapunov orbit around L_2: surface associated with $J(x(0), T)$.

are likely to be unsuccessful in finding a Lyapunov orbit, unless a sufficiently accurate guess is provided. This is what most researchers did for decades, for instance by employing analytical expansions near the Lagrange points as first attempt solutions [40]. In contrast, the PSO algorithm does not need any guess and successfully finds Lyapunov orbits with great accuracy.

10.4 Lunar Periodic Orbits

The problem consists of determining the initial position and velocity such that the third-body motion around the Moon is stable (i.e., indefinitely repeated) in the context of a planar circular restricted three-body problem. This problem is also formulated as an optimization problem, so that it can be solved by the PSO algorithm.

10.4.1 Problem Definition

The general framework employed to investigate the problem at hand is the same adopted in the preceding section. The third-body motion is described in the synodic reference frame, and the same canonical units of the previous section are employed (therefore, 1 DU = 384400 km and 1 TU = 375190 sec). The position of the Earth and of the Moon are identified by $x_E = -\mu$ and $x_M = 1 - \mu$ (in DU) (respectively), and the equations of the third-body motion (also termed *state equations*) are represented by Equations (10.10) and (10.18).

A variety of lunar periodic orbits has been known since the 1960s [41]. However, in this study the orbits with an initial position along the x-axis and with initial velocity aligned with the y-axis are sought. This means that $y(0) = 0$ and $\gamma(0) = \pi/2$. For a given Jacobi constant C, an orbit is periodic if after the period T the three variables $\{x, y, \gamma\}$ assume their respective values at the initial time. The problem can be easily translated into an optimization problem if an objective function identical to the one employed for the previous example is introduced

$$J = |x(T) - x(0)| + |y(T)|$$
$$+ \min\left\{\mathrm{mod}\left[\left|\gamma(T) - \frac{\pi}{2}\right|, 2\pi\right], \mathrm{mod}\left[-\left|\gamma(T) - \frac{\pi}{2}\right|, 2\pi\right]\right\}. \tag{10.24}$$

The global minimum of J equals 0, provided that at least one periodic orbit exists.

A single additional condition is needed to ensure that the periodic orbit is actually an orbit around the Moon. Let β denote the angle that the position vector relative to the Moon forms with the x-axis. Then

$$\tan\beta = \frac{y}{x - x_M}. \tag{10.25}$$

Equation (10.25) implies that the following equation holds for β

$$\dot{\beta} = \frac{\sqrt{2\Omega - C}}{(x - x_M)^2 + y^2}[(x - x_M)\sin\gamma - y\cos\gamma] \tag{10.26}$$

with $\beta(0) = \pi$ (if $x(0) < x_M$ and $y(0) = 0$). This equation is integrated together with the state equations (10.10) and (10.18). The third body completes at least a single loop around the Moon if

$$|\beta(T) - \beta(0)| \geq 2\pi. \tag{10.27}$$

10.4.2 Numerical Results

The PSO algorithm uses a population of 30 particles ($N = 30$) and is run for 500 iterations ($N_{IT} = 500$). Each particle is associated with the values of the two unknown parameters of the problem: $\chi = [x(0) \quad T]^T$. The PSO technique assigns an infinite value of the objective function to the particles that violate the inequality constraint (10.27).

With the intent of finding periodic orbits with different features, the optimization process is repeated for six different cases, corresponding to two distinct time intervals for T and three distinct intervals over the x-axis for $x(0)$

$$1\,\mathrm{TU} \leq T \leq 3\,\mathrm{TU} \quad \text{or} \quad 3\,\mathrm{TU} \leq T \leq 8\,\mathrm{TU} \tag{10.28}$$

$$x_{L1} \leq x(0) \leq 0.87\,\mathrm{DU} \quad \text{or} \quad 0.87\,\mathrm{DU} \leq x(0) \leq 0.9\,\mathrm{DU} \quad \text{or} \quad 0.9\,\mathrm{DU} \leq x(0) \leq x_M. \tag{10.29}$$

The Jacobi constant C is set to $3.01\,\mathrm{DU}^2/\mathrm{TU}^2$. For three out of the six cases (labeled with [a], [b], and [c]), the PSO algorithm was able to find periodic orbits around the Moon. The related results are reported in Table 10.2 (in which the number of loops

Table 10.2. *Results related to the lunar periodic orbits* ($C = 3.01\ DU^2/TU^2$)

$x(0)$ (DU)	T (TU)	$J_{opt}^{(N_{IT})}$	number of loops	class
0.891403	1.446013	$1.794 \cdot 10^{-12}$	1	C
0.945452	5.719834	$2.757 \cdot 10^{-6}$	2	H_2
0.853238	6.036339	$8.054 \cdot 10^{-10}$	4	unknown

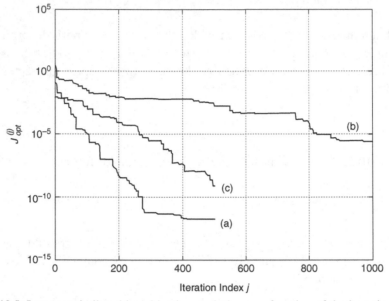

Figure 10.5. Lunar periodic orbits: objective evolution as a function of the iteration index.

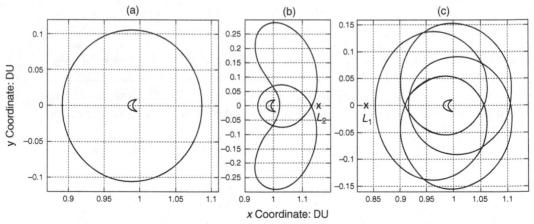

Figure 10.6. Periodic orbits around the Moon ($C = 3.01\ DU^2/TU^2$).

refers to the loops in the period T; not necessarily all the loops are around the Moon). Figure 10.5 illustrates the objective evolution as a function of the iteration index for each case, whereas Figure 10.6 portrays the corresponding lunar periodic orbits. For the two-loop orbit the number of iterations was increased to 1,000, because the

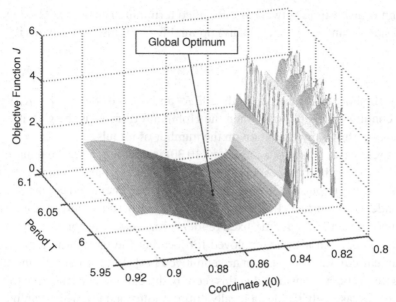

Figure 10.7. Four-loop lunar periodic orbit: surface associated with $J(x(0), T)$.

objective function was not sufficiently close to 0 after 500 iterations. However, even 1,000 iterations are not prohibitive, requiring only 150 minutes of CPU time (running MATLAB) using a 1.6 GHz AMD Sempron processor on an Acer Aspire 3102WLMi laptop computer, assuming relative and absolute error tolerance for the numerical integration of the equations of motion of 10^{-10} and 10^{-12}, respectively. According to the classification given by Broucke [41], the single-loop orbit is a C-type orbit, whereas the two-loop orbit can be recognized to belong to the H_2 class of periodic orbits. The fact that only two unknown parameters are involved in the problem of interest has the favorable consequence that the objective function (depending on $x(0)$ and T) can be associated with a spatial surface. Figure 10.7 illustrates this surface for the case corresponding to the four-loop orbit and points out that the convergence region is extremely reduced in size and irregular in shape. This circumstance implies again that local algorithms are unlikely to converge unless a very accurate guess solution is provided. In contrast, for this problem as well, the PSO method does not need any guess and successfully finds the desired solution with great accuracy.

10.5 Optimal Four-Impulse Orbital Rendezvous

The orbital rendezvous of a chaser spacecraft with a target vehicle (respectively denoted with "C" and "T" henceforward) in a specified time can be achieved by two (or more) impulses. If the orbits of the two spacecraft are coplanar and the time is sufficiently large, the globally optimal (two-impulse) solution is the Hohmann transfer. More precisely, the Hohmann transfer represents the optimal rendezvous if the specified transfer time, denoted with Δt_T, is not less than the sum of two terms: (1) the time needed for a correct phasing between C and T, Δt_P; (2) the

Hohmann transfer time between the specified terminal orbits, Δt_H. This means that the Hohmann transfer is the (globally) optimal two-impulse rendezvous if

$$\Delta t_T \geq \Delta t_P + \Delta t_H. \qquad (10.30)$$

If the terminal orbits are circular with radii R_1 and R_2, if $0.08376 \leq R_2/R_1 \leq 11.939$, and the condition (10.30) holds, then the Hohmann transfer is the globally optimal rendezvous without any limitation on the number of impulses.

For those cases where the condition (10.30) is violated, additional impulses can allow performing the rendezvous with substantial propellant savings with respect to the use of two impulses. Analytical results exist for linearized coplanar circle-to-circle rendezvous. In fact, Prussing [42, 43] proved that as many as four impulses are required for an optimal time-fixed rendezvous. Analytical solutions arising from the linear theory can also be employed for locating the optimal rendezvous when the nonlinear equations of motion are employed [44]. Yet for the four-impulse rendezvous that is being considered in this section, the extension of the linear theory to the nonlinear case only yielded a locally optimal solution [44], which can be outperformed by a completely different four-impulse rendezvous trajectory, as shown by Colasurdo and Pastrone [45].

10.5.1 Problem Definition

The problem consists of determining the optimal locations, directions, and magnitudes of the four impulses that allow performing the orbital rendezvous between the chaser spacecraft C and the target vehicle T in a specified time. Both spacecraft are placed in the same circular orbit of radius R_1, and their angular displacement at the initial time is $\Delta \xi = \pi$. In this case the time needed for a correct phasing for a Hohmann transfer tends to infinity ($\Delta t_P \to \infty$), and therefore the Hohmann transfer cannot represent the optimal rendezvous trajectory, regardless of the transfer time Δt_T. As both spacecraft are placed in the same (initial) circular orbit, the first impulse can be assumed to occur at the initial time without any loss of generality. If ξ denotes the angular displacement from the axis corresponding to the initial position of C, then the initial conditions for the two spacecraft are the following

$$v_r^{(C)}(t_0^-) = 0 \qquad v_\theta^{(C)}(t_0^-) = \sqrt{\frac{\mu}{R_1}} \qquad r^{(C)}(t_0^-) = R_1 \qquad \xi^{(C)}(t_0^-) = 0 \qquad (10.31)$$

$$v_r^{(T)}(t_0^-) = 0 \qquad v_\theta^{(T)}(t_0^-) = \sqrt{\frac{\mu}{R_1}} \qquad r^{(T)}(t_0^-) = R_1 \qquad \xi^{(T)}(t_0^-) = \pi \qquad (10.32)$$

where μ represents the gravitational parameter of the attracting body, $v_r^{(C/T)}$ and $v_\theta^{(C/T)}$ denote the radial and the horizontal components of the velocity (respectively), and $r^{(C/T)}$ is the radius. The initial time t_0 is set to 0 (the superscripts "$-$" and "$+$" refer to the instant immediately before and after the application of an impulse). At the terminal time t_f, both the position and the velocity of the two spacecraft must coincide, and for C this circumstance leads to the following terminal conditions:

$$v_r^{(C)}\left(t_f^+\right) = 0 \quad v_\theta^{(C)}\left(t_f^+\right) = \sqrt{\frac{\mu}{R_1}} \quad r^{(C)}\left(t_f^+\right) = R_1 \quad \xi^{(C)}\left(t_f^+\right) = \pi + t_f\sqrt{\frac{\mu}{R_1^3}}.$$

$$(10.33)$$

What will be demonstrated is that the following six parameters define the rendezvous trajectory

$$\{\Delta v_1, \delta_1, \Delta v_2, \delta_2, \Delta E_1, \Delta E_2\} \qquad (10.34)$$

where Δv_1 and Δv_2 represent the magnitudes of the first two impulsive changes of velocity, δ_1 and δ_2 their respective directions (relative to the local horizontal), and finally ΔE_1 and ΔE_2 denote the eccentric anomaly variations during the first two Keplerian arcs of the trajectory. In fact, after the first impulse, the velocity components v_r and v_θ change to

$$v_r(t_0^+) = v_r(t_0^-) + \Delta v_1 \sin \delta_1 \quad \text{and} \quad v_\theta(t_0^+) = v_\theta(t_0^-) + \Delta v_1 \cos \delta_1 \qquad (10.35)$$

whereas $r(t_0^+) = r(t_0^-) = R_1$ (the superscript "C" is omitted henceforth). The second impulse occurs at the time t_1. In the time interval $[t_0^+, t_1^-]$, the trajectory is Keplerian, and therefore the following relationships hold

$$a_1 = \frac{\mu r(t_0^+)}{2\mu - r(t_0^+)[v_r^2(t_0^+) + v_\theta^2(t_0^+)]} \qquad (10.36)$$

$$e_1 = \sqrt{1 - \frac{r^2(t_0^+)v_\theta^2(t_0^+)}{\mu a_1}} \qquad (10.37)$$

$$\cos[f(t_0^+)] = \frac{v_\theta(t_0^+)}{e_1}\sqrt{\frac{a_1(1 - e_1^2)}{\mu}} - \frac{1}{e_1} \quad \text{and} \quad \sin[f(t_0^+)] = \frac{v_r(t_0^+)}{e_1}\sqrt{\frac{a_1(1 - e_1^2)}{\mu}}$$

$$(10.38)$$

where a_1 and e_1 respectively represent the semi-major axis and the eccentricity of the first Keplerian arc of the trajectory, and $f(t_0^+)$ denotes the true anomaly at t_0^+. The eccentric anomaly at t_0^+, $E(t_0^+)$, is given by the well known formulas

$$\sin[E(t_0^+)] = \frac{\sin[f(t_0^+)]\sqrt{1 - e_1^2}}{1 + e_1 \cos[f(t_0^+)]} \quad \text{and} \quad \cos[E(t_0^+)] = \frac{\cos[f(t_0^+)] + e_1}{1 + e_1 \cos[f(t_0^+)]}. \qquad (10.39)$$

Immediately before the second impulsive change of velocity, the eccentric anomaly $E(t_1^-)$ is $E(t_1^-) = E(t_0^+) + \Delta E_1$. The corresponding true anomaly, $f(t_1^-)$, can be derived by employing the counterparts of Equations (10.39)

$$\sin[f(t_1^-)] = \frac{\sin[E(t_1^-)]\sqrt{1 - e_1^2}}{1 - e_1 \cos[E(t_1^-)]} \quad \text{and} \quad \cos[f(t_1^-)] = \frac{\cos[E(t_1^-)] - e_1}{1 - e_1 \cos[E(t_1^-)]}. \qquad (10.40)$$

Hence, the radius and the velocity components before the second impulse are given by

$$r(t_1^-) = a_1\{1 - e_1 \cos[E(t_1^-)]\} \tag{10.41}$$

$$v_r(t_1^-) = \sqrt{\frac{\mu}{a_1(1 - e_1^2)}} e_1 \sin[f(t_1^-)] \qquad v_\theta(t_1^-) = \sqrt{\frac{\mu}{a_1(1 - e_1^2)}}\{1 + e_1 \cos[f(t_1^-)]\}. \tag{10.42}$$

The second impulse changes the velocity components v_r and v_θ analogously to what occurs at the application of the first impulse (cf. Equations [10.35]). In the time interval $[t_1^+, t_2^-]$ the trajectory is Keplerian, and therefore the same steps that lead to determining a_1, e_1, $f(t_0^+)$, and $E(t_0^+)$ (for the first Keplerian arc) can be repeated to calculate a_2, e_2, $f(t_1^+)$, and $E(t_1^+)$ (for the second Keplerian arc). Immediately before the third impulsive change of velocity, the eccentric anomaly is $E(t_2^-) = E(t_1^+) + \Delta E_2$. The corresponding true anomaly, $f(t_2^-)$, can be derived in the same way in which $f(t_1^-)$ was obtained from $E(t_1^-)$. The radius and the velocity components before the third impulse are given by

$$r(t_2^-) = a_2\{1 - e_2 \cos[E(t_2^-)]\} \tag{10.43}$$

$$v_r(t_2^-) = \sqrt{\frac{\mu}{a_2(1 - e_2^2)}} e_2 \sin[f(t_2^-)] \qquad v_\theta(t_2^-) = \sqrt{\frac{\mu}{a_2(1 - e_2^2)}}\{1 + e_2 \cos[f(t_2^-)]\}. \tag{10.44}$$

The times at which the second and third impulse occur can be easily calculated through Kepler's law

$$t_1 = \sqrt{\frac{a_1^3}{\mu}}\{\Delta E_1 - e_1\{\sin[E(t_1^-)] - \sin[E(t_0^+)]\}\} \tag{10.45}$$

$$t_2 = t_1 + \sqrt{\frac{a_2^3}{\mu}}\{\Delta E_2 - e_2\{\sin[E(t_2^-)] - \sin[E(t_1^+)]\}\}. \tag{10.46}$$

The third and fourth impulsive changes of velocity can be determined (in their respective magnitudes and directions) through the use of Lambert's theorem. In fact, at the time t_2, both the angular separation from the desired final position and the time available to complete the rendezvous are known, as well as the radius at t_2 (given by Equation [10.43]) and the terminal radius (which is R_1). In particular, the angular separation between the position vector at t_2 and the final position is given by

$$\Delta \xi_L = \xi(t_f) - [f(t_1^-) - f(t_0^+)] - [f(t_2^-) - f(t_1^+)] \tag{10.47}$$

whereas the time available to complete the rendezvous is simply

$$\Delta t_L = t_f - t_2. \tag{10.48}$$

Hence, given $r(t_2)$, $r(t_f)$ $(= R_1)$, $\Delta\xi_L$, and Δt_L, a standard Lambert solver, which follows the steps described in [46], yields the value of the semi-major axis a_3, as well as the eccentricity e_3 of the third Keplerian arc of trajectory, which finally connects C with T in the specified time. It is worth mentioning that multiple-revolution solutions are not considered for the third Keplerian arc. Once a_3 and e_3 are known, the true anomalies $f(t_2^+)$ and $f(t_f^-)$ can be obtained through the following three relations:

$$r(t_2) = \frac{a_3(1 - e_3^2)}{1 + e_3 \cos[f(t_2^+)]} \tag{10.49}$$

$$R_1 = \frac{a_3(1 - e_3^2)}{1 + e_3 \cos[f(t_f^-)]} \tag{10.50}$$

$$f(t_f^-) = f(t_2^+) + \Delta\xi_L. \tag{10.51}$$

After inserting Equation (10.51) into Equation (10.50), the two relationships (10.49) and (10.50) yield $f(t_2^+)$

$$\cos[f(t_2^+)] = \frac{p_3 - r(t_2)}{e_3 r(t_2)} \tag{10.52}$$

$$\sin[f(t_2^+)] = \frac{r(t_2)R_1 e_3 + R_1 p_3 e_3 \cos \Delta\xi_L - r(t_2)p_3 e_3 - r(t_2)R_1 e_3 \cos \Delta\xi_L}{r(t_2)R_1 e_3^2 \sin \Delta\xi_L} \tag{10.53}$$

where $p_3 \triangleq a_3(1 - e_3^2)$. The values of $f(t_2^+)$ and $f(t_f^-)$ (given by Equation [10.51]), allow obtaining both the components of the velocity at the corresponding times, $(v_r(t_2^+), v_\theta(t_2^+))$ and $(v_r(t_f^-), v_\theta(t_f^-))$. At the times t_2 and t_f, the velocity impulsively changes, and relationships formally identical to Equations (10.35) hold. As a result, it is straightforward to derive the magnitudes and directions of the last two impulsive changes of velocity

$$\Delta v_3 = \sqrt{[v_r(t_2^+) - v_r(t_2^-)]^2 + [v_\theta(t_2^+) - v_\theta(t_2^-)]^2} \tag{10.54}$$

$$\sin \delta_3 = \frac{v_r(t_2^+) - v_r(t_2^-)}{\Delta v_3} \quad \text{and} \quad \cos \delta_3 = \frac{v_\theta(t_2^+) - v_\theta(t_2^-)}{\Delta v_3} \tag{10.55}$$

$$\Delta v_4 = \sqrt{[v_r(t_f^+) - v_r(t_f^-)]^2 + [v_\theta(t_f^+) - v_\theta(t_f^-)]^2} = \sqrt{v_r^2(t_f^-) + \left[\sqrt{\frac{\mu}{R_1}} - v_\theta(t_f^-)\right]^2} \tag{10.56}$$

$$\sin \delta_4 = \frac{v_r(t_f^+) - v_r(t_f^-)}{\Delta v_4} = \frac{-v_r(t_f^-)}{\Delta v_4} \quad \text{and} \quad \cos \delta_4 = \frac{v_\theta(t_f^+) - v_\theta(t_f^-)}{\Delta v_4} = \frac{\sqrt{\frac{\mu}{R_1}} - v_\theta(t_f^-)}{\Delta v_4}. \tag{10.57}$$

The globally optimal rendezvous, which is sought by the PSO algorithm, minimizes the characteristic velocity of the overall orbital maneuver. This means that the objective function for the problem at hand is

$$J = \Delta v_1 + \Delta v_2 + \Delta v_3 + \Delta v_4. \tag{10.58}$$

The Keplerian arcs that compose the overall rendezvous trajectory are assumed to be of elliptic type. This means that the following constraints must hold

$$a_1 > 0 \qquad a_2 > 0 \tag{10.59}$$

in conjunction with the following time constraint [46]

$$\Delta t_L > \frac{1}{3}\sqrt{\frac{2}{\mu}}[s^{3/2} - \text{sign}(\sin \Delta \xi_L)(s - c)^{3/2}] \tag{10.60}$$

where

$$c = \sqrt{[r(t_2)]^2 + R_1^2 - 2r(t_2)R_1 \cos \Delta \xi_L} \quad \text{and} \quad s = \frac{r(t_2) + R_1 + c}{2}. \tag{10.61}$$

Due to Lambert's theorem (cf. Reference [46]), this last constraint ensures that also the third Keplerian arc is of elliptic type (i.e., $a_3 > 0$).

10.5.2 Numerical Results

For the problem at hand, the PSO algorithm employs 50 particles ($N = 50$) and is run for 1,000 iterations ($N_{IT} = 1,000$). Each particle includes the values of the six unknown parameters

$$\chi = \begin{bmatrix} \Delta v_1 & \delta_1 & \Delta v_2 & \delta_2 & \Delta E_1 & \Delta E_2 \end{bmatrix}^T. \tag{10.62}$$

The problem is solved by employing a normalized set of units: the radius of the initial orbit represents the distance unit (DU), whereas the time unit (TU) is such that $\mu = 1 \, \text{DU}^3/\text{TU}^2$. The transfer time Δt_T is set to 4.6π TU ($t_f = \Delta t_T = 4.6\pi$ TU). The search space is defined by the following inequalities

$$0 \, \frac{\text{DU}}{\text{TU}} \leq \Delta v_i \leq 0.2 \, \frac{\text{DU}}{\text{TU}} \quad 0 \leq \delta_i \leq 2\pi \quad 0 \leq \Delta E_i \leq 4\pi \quad (i = 1, 2). \tag{10.63}$$

The PSO algorithm assigns an infinite value of the objective function to the particles that violate at least one of the inequality constraints (10.59) and (10.60).

Tables 10.3 and 10.4 collect the results of the optimization process, that is, the optimal locations, directions, and magnitudes of the four impulses associated with the (globally) optimal rendezvous trajectory, as well the values of the semi-major axis and eccentricity of the three Keplerian arcs. Figure 10.8 illustrates the objective evolution as a function of the iteration index and points out that the PSO algorithm needs only 300 iterations to locate the globally optimal solution. The remaining

Table 10.3. *Results related with the optimal four-impulse rendezvous*

Impulse	Δv (DU/TU)	δ (deg)	r (DU)	ξ (deg)	t (TU)
1	0.0511846	192.776	1	0	0
2	0.0307293	182.507	0.832237	203.928	2.996029
3	0.0307293	357.493	0.832237	84.072	11.455296
4	0.0511846	12.776	1	288	14.451326

Table 10.4. *Four-impulse rendezvous: results for the Keplerian arcs of the trajectory*

Keplerian arc	1	2	3
Semi-major axis a (DU)	0.911399	0.857262	0.911399
Eccentricity e	0.0979352	0.0540120	0.0979351

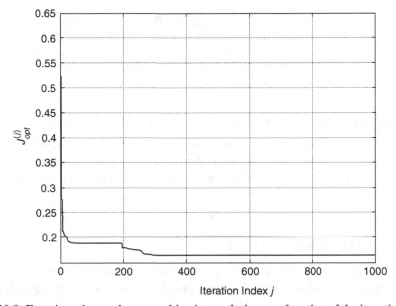

Figure 10.8. Four-impulse rendezvous: objective evolution as a function of the iteration index.

iterations are employed for refinement. Figure 10.9 portrays the optimal rendezvous trajectory, corresponding to a minimum value of J equal to 0.163828 DU/TU. The solution obtained with the PSO method is virtually indistinguishable from that found by Colasurdo and Pastrone [45], and exhibits a high degree of symmetry. In fact, from Tables 10.3 and 10.4, it is apparent that

$$\Delta v_1 \simeq \Delta v_4 \quad \Delta v_2 \simeq \Delta v_3 \quad \mathrm{mod}(\delta_1 - \delta_4, 2\pi) = \pi \quad \mathrm{mod}(\delta_2 + \delta_3, 2\pi) = \pi$$

$$r(t_1) \simeq r(t_2) \quad \xi(t_1) - \xi(t_0) \simeq \xi(t_f) - \xi(t_2) \quad t_1 - t_0 \simeq t_f - t_2 \quad a_1 \simeq a_3 \quad e_1 \simeq e_3.$$

$$\tag{10.64}$$

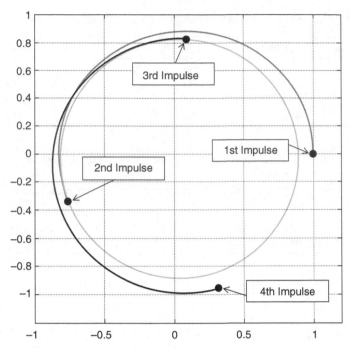

Figure 10.9. Optimal four-impulse rendezvous trajectory.

For the problem at hand it is relatively straightforward to demonstrate that an optimal solution must be symmetric (i.e., that all the relations (10.64) must hold precisely). Hence, the fact that the relationships (10.64) are satisfied to a great accuracy (up to six digits) actually represents an additional proof of optimality of the solution found with the PSO methodology.

10.6 Optimal Low-Thrust Orbital Transfers

In this section the optimization of low-thrust orbital transfers between two coplanar circular orbits is considered. The problem consists of determining the thrust pointing angle time history that minimizes the time of flight while satisfying the terminal conditions for injection into the terminal orbit. For the problem at hand, the necessary conditions for optimality are employed to express the control variable (i.e., the thrust pointing angle) as a function of the costate.

10.6.1 Problem Definition

At the initial time t_0 ($= 0$), the spacecraft is placed in a circular orbit of radius R_1. The terminal circular orbit has radius R_2. Therefore, if v_r, v_θ, and r represent, respectively, the radial and the horizontal components of velocity and the radius, the initial conditions (at t_0) and the final conditions (at t_f) are given by

$$v_r(t_0) = 0 \quad v_\theta(t_0) = \sqrt{\frac{\mu}{R_1}} \quad r(t_0) = R_1 \tag{10.65}$$

$$v_r(t_f) = 0 \quad v_\theta(t_f) = \sqrt{\frac{\mu}{R_2}} \quad r(t_f) = R_2 \qquad (10.66)$$

where μ denotes the gravitational parameter of the attracting body. The spacecraft is assumed to employ a constant (low) thrust during the entire time of flight, and therefore the thrust-to-mass ratio (T/m) has the following expression

$$\frac{T}{m} = \frac{T}{m_0 - \frac{T}{c}(t - t_0)} = \frac{cn_0}{c - n_0(t - t_0)}. \qquad (10.67)$$

In Equation (10.67), m_0 is the initial spacecraft mass, and therefore n_0 represents the thrust-to-mass ratio at t_0, whereas c is the effective exhaust velocity of the propulsive system. Note that Equation (10.67) implies that minimizing the time of flight will also maximize the remaining mass, which means it will also minimize the propellant required.

The spacecraft equations of motion (also referred to as the *state equations*) involve the two components of velocity and the radius r

$$\dot{v}_r = -\frac{\mu - rv_\theta^2}{r^2} + \frac{T}{m}\sin\delta \quad \dot{v}_\theta = -\frac{v_r v_\theta}{r} + \frac{T}{m}\cos\delta \quad \dot{r} = v_r. \qquad (10.68)$$

The angle δ is the thrust pointing angle (relative to the local horizontal). The state vector is $x \triangleq [x_1 \ x_2 \ x_3]^T = [v_r \ v_\theta \ r]^T$, whereas the control vector includes δ only: $u \triangleq \delta$. If t_0 is set to 0, the objective function to be minimized is given by

$$J = t_f. \qquad (10.69)$$

To state the necessary conditions for optimality, a Hamiltonian and a function of terminal conditions are introduced as

$$H = \lambda_1 \left[-\frac{\mu_E - x_3 x_2^2}{x_3^2} + \frac{T}{m}\sin u \right] + \lambda_2 \left[-\frac{x_1 x_2}{x_3} + \frac{T}{m}\cos u \right] + \lambda_3 x_1 \qquad (10.70)$$

$$\Phi = t_f + \upsilon_1 x_{1f} + \upsilon_2 \left[x_{2f} - \sqrt{\frac{\mu}{R_2}} \right] + \upsilon_3 [x_{3f} - R_2] \quad (x_{kf} \triangleq x_k(t_f), k = 1, 2, 3) \qquad (10.71)$$

where $\lambda \ (\triangleq [\lambda_1 \ \lambda_2 \ \lambda_3]^T)$ is the (time-varying) adjoint (or costate) variable conjugate to the state Equations (10.68), and $\upsilon \ (\triangleq [\upsilon_1 \ \upsilon_2 \ \upsilon_3]^T)$ is the time-independent adjoint variable conjugate to the boundary conditions (10.66). The necessary conditions for optimality include the following set of adjoint equations for λ^* (in this section the superscript "*" will denote the optimal value of the respective variable henceforth)

$$\dot{\lambda}_1^* = -\lambda_3^* + \frac{x_2^* \lambda_2^*}{x_3^*} \quad \dot{\lambda}_2^* = \frac{-2x_2^* \lambda_1^* + x_1^* \lambda_2^*}{x_3^*} \quad \dot{\lambda}_3^* = \frac{(x_2^*)^2 \lambda_1^* - x_1^* x_2^* \lambda_2^*}{(x_3^*)^2} - \frac{2\mu \lambda_1^*}{(x_3^*)^3}. \qquad (10.72)$$

In addition, the optimal control δ^* can be expressed as a function of the costate through the Pontryagin minimum principle

$$\delta^* = \arg\min_{\delta} H \Rightarrow \cos\delta^* = -\frac{\lambda_2^*}{\sqrt{(\lambda_1^*)^2 + (\lambda_2^*)^2}} \quad \text{and} \quad \sin\delta^* = -\frac{\lambda_1^*}{\sqrt{(\lambda_1^*)^2 + (\lambda_2^*)^2}}.$$

(10.73)

Lastly, as the final time is unspecified, the following transversality condition must hold:

$$H(t_f) + \frac{\partial\Phi}{\partial t_f} = 0 \Rightarrow \frac{cn_0}{c - n_0 t_f^*}\sqrt{[\lambda_1^*(t_f^*)]^2 + [\lambda_2^*(t_f^*)]^2} - 1 = 0.$$

(10.74)

The necessary conditions for optimality allow translating the optimal control problem into a two-point boundary-value problem involving Equations (10.65), (10.66), (10.68) in conjunction with Equations (10.72)–(10.74). The unknowns are represented by the state x, the control u, the adjoint variables λ and v, and the time of flight t_f.

The method of solution employed for this problem is based on the following points:

(a) the control is expressed as a function of the costate through Equation (10.73)
(b) the adjoint equations (10.72) are numerically integrated together with the state equations (10.68) after picking the time of flight t_f and the values of the components of λ at the initial time
(c) the final conditions (10.66) and the transversality condition (10.74) are checked

The problem reduces to the determination of four unknown parameters $(\lambda_1(0), \lambda_2(0), \lambda_3(0), t_f)$ that lead the dynamical system to satisfying the four boundary conditions (10.66) and (10.74).

10.6.2 Constraint Reduction

As mentioned in Section 10.2, equality constraints reduce the search space where feasible solutions can be located. However, what will be demonstrated is that for the problem at hand the transversality condition can be neglected by the PSO algorithm.

To do this, one has first to recognize the special structure of the costate equations (10.72), which are homogeneous in λ. This circumstance implies that if an optimization algorithm is capable of finding some initial value of λ such that

$$\lambda_1(0) = a\lambda_1^*(0) \quad \lambda_2(0) = a\lambda_2^*(0) \quad \lambda_3(0) = a\lambda_3^*(0) \quad (a > 0)$$

(10.75)

then the same proportionality (10.75) holds between λ and the optimal λ^* at any t, due to homogeneity of Equations (10.72). Moreover, the control u can be written as a function of λ through Equation (10.73), and one can recognize that $\boldsymbol{u}$ coincides with the optimal control $\boldsymbol{u}^*$

$$\cos\delta = -\frac{\lambda_2}{\sqrt{\lambda_1^2 + \lambda_2^2}} = -\frac{a\lambda_2^*}{a\sqrt{(\lambda_1^*)^2 + (\lambda_2^*)^2}} = -\frac{\lambda_2^*}{\sqrt{(\lambda_1^*)^2 + (\lambda_2^*)^2}} \equiv \cos\delta^* \quad (10.76)$$

$$\sin\delta = -\frac{\lambda_1}{\sqrt{\lambda_1^2 + \lambda_2^2}} = -\frac{a\lambda_1^*}{a\sqrt{(\lambda_1^*)^2 + (\lambda_2^*)^2}} = -\frac{\lambda_1^*}{\sqrt{(\lambda_1^*)^2 + (\lambda_2^*)^2}} \equiv \sin\delta^*. \quad (10.77)$$

This circumstance implies that if the conditions (10.75) hold then the final conditions (10.66) are fulfilled at the minimum final time t_f^*. In contrast, the transversality condition is violated, because the value of $H(t_f^*)$, due to Equation (10.74), turns out to be

$$H(t_f^*) = -\frac{cn_0}{c - n_0 t_f^*}\sqrt{\lambda_1^2(t_f^*) + \lambda_2^2(t_f^*)}$$

$$= -a\frac{cn_0}{c - n_0 t_f^*}\sqrt{[\lambda_1^*(t_f^*)]^2 + [\lambda_2^*(t_f^*)]^2} = -a \neq -1. \quad (10.78)$$

Therefore, provided that the proportionality condition holds, the optimal control $\boldsymbol{u}^*$ can be determined without considering the transversality condition, which in fact is ignorable in this context. Thus this condition is discarded in order to reduce the number of equality constraints considered by the PSO algorithm, with the intent of improving its performance. Once a costate λ proportional to the optimal costate λ^* has been determined, it can be suitably scaled. In fact, from Equation (10.78), it is apparent that the proportionality coefficient is simply given by $a = -H(t_f^*)$. In short, the PSO algorithm can follow three steps:

(a) consider only the three conditions (10.66) as the equality constraints of the problem
(b) once proper values of $\{\lambda_1(0), \lambda_2(0), \lambda_3(0), t_f\}$ (such that Equations [10.66] are satisfied to the desired accuracy) have been determined, calculate $a(= -H(t_f^*))$
(c) scale λ by the coefficient a to obtain λ^*, which fulfills also the transversality condition (10.74)

10.6.3 Numerical Solution

Each particle that forms the swarm includes the four unknown parameters

$$\chi = \begin{bmatrix} \lambda_1(0) & \lambda_2(0) & \lambda_3(0) & t_f \end{bmatrix}^T. \quad (10.79)$$

The problem is solved by employing a normalized set of units: the radius of the initial orbit represents the distance unit (DU), whereas the time unit (TU) is such that $\mu = 1\ \mathrm{DU}^3/\mathrm{TU}^2$. Two orbital transfers are considered as illustrative examples, corresponding to two different radii of the terminal orbit: (a) $R_2 = 2$ DU and (b) $R_2 = 5$ DU. The optimal values of the four unknown parameters are sought in the following ranges

$$1 \text{ TU} \le t_f \le 100 \text{ TU} \quad \text{and} \quad -1 \le \lambda_k(0) \le 1 \quad (k = 1, 2, 3). \qquad (10.80)$$

For both cases, the following values of c and n_0 are employed: $c = 1.5$ DU/TU and $n_0 = 0.01$ DU/TU^2.

Three equality constraints (related with the terminal conditions for orbit injection) are involved in the problem at hand. The control is expressed as a function of the costate through the necessary conditions for optimality (10.73), and therefore the satisfaction of the terminal conditions for orbit injection suffices to ensure the optimality of the solution. Hence, the following fitness function can be considered by the PSO algorithm

$$\tilde{J} = \sum_{k=1}^{3} \alpha_k |d_k| \quad \text{with} \quad d_1 = v_r(t_f) \quad d_2 = v_\theta(t_f) - \sqrt{\frac{\mu}{R_2}} \quad d_3 = r(t_f) - R_2. \quad (10.81)$$

For this example, the PSO algorithm uses a population of 50 particles ($N = 50$) and is run for 500 iterations ($N_{IT} = 500$); α_k is set to 1 (for $k = 1, 2, 3$). Figure 10.10 illustrates the objective evolution as a function of the iteration index and points out that the PSO algorithm produces very accurate results also after 300 iterations in both cases. Figure 10.11 illustrates the optimal trajectories and the corresponding

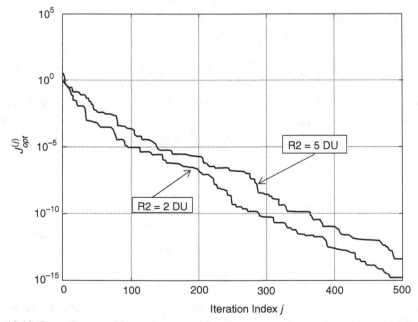

Figure 10.10. Low-thrust orbital transfers: objective evolution as a function of the iteration index.

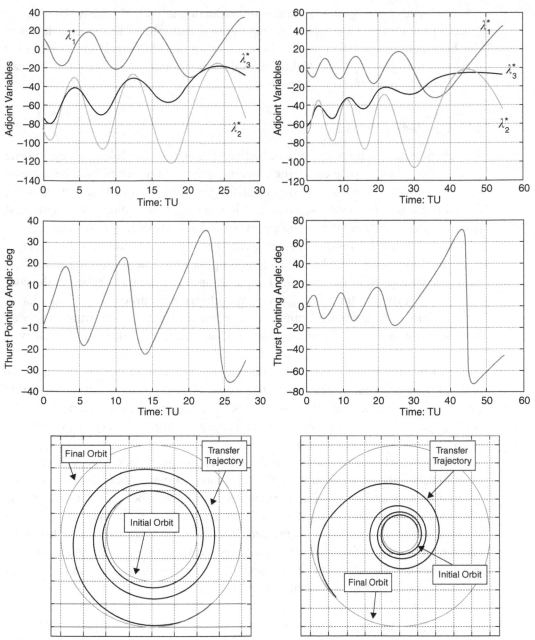

Figure 10.11. Optimal low-thrust transfers: adjoint variables, optimal control time histories and related trajectories for two cases: column (a) $R_1 = 2$ DU; column (b) $R_2 = 5$ DU.

optimal control law time histories and adjoint variables for the two cases taken into account. The two transfers of interest are completed in the following minimum times:

$$(a) \; R_2 = 2 \text{ DU}: \quad t_f^* = 27.970 \text{ TU}$$

$$(b) \; R_2 = 5 \text{ DU}: \quad t_f^* = 54.544 \text{ TU}$$

It is worth noticing that the constraint reduction allows arbitrarily defining the search space for the initial values of the Lagrange multipliers. This means that they can be sought in the interval $-1 \leq \lambda_k(0) \leq 1$ by the PSO algorithm, and only *a posteriori* their correct values (fulfilling also the transversality condition (10.74)) can be recovered, as discussed in Section 10.6.2.

10.7 Concluding Remarks

This chapter describes the use of an effective stochastic methodology for optimizing space trajectories, namely the particle swarm optimization technique. This method does not require continuity and differentiability of the objective function, does not need any guess to generate a solution, and is well suited for finding global optima. It is also based on very intuitive concepts, which makes it very easy to program.

A variety of space trajectory optimization problems are considered and a simple implementation of the PSO methodology is applied to their solution. The method proves to be capable of determining Lyapunov and periodic lunar orbits in the context of the restricted three-body problem with great accuracy, after formulating the problem as a parameter optimization problem. Then the algorithm has been successfully applied to the optimization of a four-impulse rendezvous trajectory, a well-known problem admitting multiple local minima. With regard to this problem, the PSO method proves extremely effective and accurate in locating the globally optimal solution. Finally, the last section considers the optimization of low-thrust orbital transfers, where the time of flight is to be minimized. For this problem the necessary conditions for optimality are employed to express the control variable as a function of the adjoint variables.

This research demonstrates that the PSO technique indeed represents an efficient, reliable, and accurate method of determining optimal space trajectories despite its intuitiveness and simplicity. However, as alternative stochastic algorithms, the PSO methodology can occasionally encounter difficulties when dealing with constrained problems (especially in the presence of equality constraints). This study successfully employs two distinct approaches for treating equality and inequality constraints and shows the way of reducing the number of equality constraints, albeit with regard to a particular type of optimization problem.

The number of particles to employ in each application depends on the problem complexity. A greater number of unknown parameters require a greater number of particles. Of course, an increased density of particles allows a more effective exploration of the search space, and this circumstance implies an enhanced probability of detecting the globally optimal solution, particularly with regard to problems with multiple (local) minima. For all of the optimization problems considered in this chapter, the number of particles (as well as the number of iterations) was adjusted up to obtaining the desired result at the first attempt.

In general, in the presence of unsatisfactory results, several options are available to a potential user for improving the performance attainable by a PSO algorithm. First, the number of particles and iterations can be increased to achieve more satisfactory results. Secondly, the fine-tuning of the weighting coefficients in the formula for velocity update could yield improved results, at least with regard to specific cases. However, it should be noted that the choice of weighting coefficients (10.6) gave satisfactory results for the several qualitatively different orbital optimization problems of this chapter. In addition, two versions of the PSO algorithm exist (i.e., the local version and the global version), with their respective advantages and drawbacks. The local version can be employed if the global version is suspected to have yielded a local minimum. However, swarming theory is a fast-developing discipline, and one can conclude that the proper choice of PSO algorithm is problem-dependent.

REFERENCES

[1] Goldberg, D. E. (1989) *Genetic Algorithms in Search, Optimization, and Machine Learning*, Addison Wesley, Boston, MA.

[2] Engelbrecht, A. P. (2007) *Computational Intelligence. An Introduction*, 2nd ed., Wiley, Chichester, UK.

[3] Eberhart, R., and Kennedy, J. (1995) *A New Optimizer Using Particle Swarm Theory*, Proceedings of the Sixth International Symposium on Micromachine and Human Science, Nagoya, Japan.

[4] Kennedy, J., and Eberhart, R. (1995) *Particle swarm optimization*, Proceedings of the IEEE International Conference on Neural Networks, Piscataway, NJ.

[5] Kennedy, J., and Eberhart, R. C. (2001) *Swarm Intelligence*, Academic Press, 1st ed., San Diego, CA, 2001.

[6] Hu, X., and Eberhart, R. (2002) *Solving Constrained Nonlinear Optimization Problems with Particle Swarm Optimization*, Proceedings of the Sixth World Multiconference on Systemics, Cybernetics and Informatics (SCI 2002), Orlando, FL.

[7] Hu, X., Shi, Y., and Eberhart, R. (2004) *Recent Advances in Particle Swarm*, Proceedings of the IEEE Congress on Evolutionary Computation (CEC 2004), Portland, OR.

[8] Hu, X., Eberhart, R. C., and Shi, Y. (2003) *Engineering Optimization with Particle Swarm*, Proceedings of the IEEE Swarm Intelligence Symposium (SIS 2003), Indianapolis, IN.

[9] Eberhart, R. C., and Shi, Y. (2001) *Particle Swarm Optimization: Developments, Applications, and Resources*, Proceedings of the IEEE Congress on Evolutionary Computation (CEC 2001), Seoul, Republic of Korea.

[10] Eberhart, R. C., and Shi, Y. (2000) *Comparing Inertia Weights and Constriction Factors in Particle Swarm Optimization*, Proceedings of the IEEE Congress on Evolutionary Computation (CEC 2000), La Jolla, CA.

[11] Carlisle, A., and Dozier, G. (2001) *An Off-The-Shelf PSO*, Proceedings of the Workshop on Particle Swarm Optimization, Indianapolis, IN.

[12] Clerc, M. (1999) *The Swarm and the Queen: Towards a Deterministic and Adaptive Particle Swarm Optimization*, Proceedings of the IEEE Congress on Evolutionary Computation (CEC 1999), Washington, DC.

[13] Venter, G., and Sobieszczanski-Sobieski, J. (2003) Particle Swarm Optimization, *AIAA Journal*, **41**, No. 8, 1583–1589.

[14] Parsopoulos, K. E., and Vrahatis, M. N. (2002) Particle Swarm Optimization Method for Constrained Optimization Problems, *Intelligent Technologies – Theory and Applications: New Trends in Intelligent Technologies, Frontiers in Artificial Intelligence and Applications series*, **76**, edited by P. Sincak, J. Vascak, V. Kvasnicka, J. Pospichal, 214–220.

[15] Mendes, R., Kennedy, J., and Neves, J. (2004) The Fully Informed Particle Swarm: Simpler, Maybe Better, *IEEE Transactions on Evolutionary Computation*, **8**, No. 3, 204–210.

[16] Kitayama, S., Yamazaki, K., and Arakawa, M. (2006) *Adaptive Range Particle Swarm Optimization*, AIAA 2006-6912 Proceedings of the 11th AIAA/ISSMO Multidisciplinary Analysis and Optimization Conference, Portsmouth, VA.

[17] Higashi, N., and Iba, H. (2003) *Particle Swarm Optimization with Gaussian Mutation*, Proceedings of the IEEE Swarm Intelligence Symposium (SIS 2003), Indianapolis, IN.

[18] Kalivarapu, V., and Winer, E. (2008) *Implementation of Digital Pheromones in Particle Swarm Optimization for Constrained Optimization Problems*, AIAA 2008-1974 Proceedings of the 49th AIAA/ASME/ASCE/AHS/ASC Structures, Structural Dynamics and Materials Conference, Schaumburg, IL.

[19] Angeline, P. J. (1998) Evolutionary Optimization Versus Particle Swarm Optimization: Philosophy and Performance Differences, *Evolutionary programming VII, Lecture Notes in Computer Science*, **1447**, Springer, 601–610.

[20] Eberhart, R. C., and Shi, Y. (1998) Comparison Between Genetic Algorithms and Particle Swarm Optimization, *Evolutionary Programming VII, Lecture Notes in Computer Science*, **1447**, Springer, 611–616.

[21] Hassan, R., Cohanim, B., and de Weck, O. (2005) *A Comparison of Particle Swarm Optimization and the Genetic Algorithm*, AIAA 2005-1897 Proceedings of the 46th AIAA/ASME/ASCE/AHS/ASC Structures, Structural Dynamics and Materials Conference, Austin, TX.

[22] Fourie, P. C., and Groenwold, A. A. (2001) Particle Swarms in Topology Optimization, *Proceedings of the Fourth World Congress of Structural and Multidisciplinary Optimization*, Liaoning Electronic Press, 1771–1776.

[23] Parsopoulos, K. E., and Vrahatis, M. N. (2004) On the Computation of All Global Minimizers Through Particle Swarm Optimization, *IEEE Transactions on Evolutionary Computation*, **8**, No. 3, 211–224.

[24] Cockshott, A. R., and Hartman, B. E. (2001) Improving the Fermentation Medium for Echinocandin B Production Part II: Particle Swarm Optimization, *Process Biochemistry*, **36**, No. 7, 661–669.

[25] Ourique, C. O., Biscaia, E. C., and Pinto, J. C. (2002) The Use of Particle Swarm Optimization for Dynamical Analysis in Chemical Processes, *Computers & Chemical Engineering*, **26**, No. 12, 1783–1793.

[26] Fourie, P. C., and Groenwold, A. A. (2002) The Particle Swarm Optimization Algorithm in Size and Shape Optimization, *Structural and Multidisciplinary Optimization*, **23**, No. 4, 259–267.

[27] Khurana, M. S., Winarto, H., and Sinha, A. K. (2008) *Application of Swarm Approach and Artificial Neural Networks for Airfoil Shape Optimization*, AIAA 2008-5954 Proceedings of the 12th AIAA/ISSMO Multidisciplinary Analysis and Optimization Conference, Victoria, Canada.

[28] Bessette, C. R., and Spencer, D. B. (2006) Optimal Space Trajectory Design: A Heuristic-Based Approach, *Advances in the Astronautical Sciences*, Univelt Inc., San Diego, CA, **124**, 1611–1628; AAS paper 06-197.

[29] Bessette, C. R., and Spencer, D. B. (2006) *Identifying Optimal Interplanetary Trajectories through a Genetic Approach*, AIAA 2006-6306 AIAA/AAS Astrodynamics Specialist Conference and Exhibit, Keystone, CO.

[30] Spaans, C. J., and Mooij, E. (2009) *Performance Evaluation of Global Trajectory Optimization Methods for a Solar Polar Sail Mission*, AIAA 2009-5666 AIAA Guidance, Navigation, and Control Conference, Chicago, IL.

[31] Vasile, M., Minisci, E., and Locatelli, M. (2008) *On Testing Global Optimization Algorithms for Space Trajectory Design*, AIAA 2008-6277 AIAA/AAS Astrodynamics Specialist Conference and Exhibit, Honolulu, HI.

[32] Zhu, K., Li, J., and Baoyin, H. (2010) Satellite scheduling considering maximum observation coverage time and minimum orbital transfer fuel cost, *Acta Astronautica*, **66**, 220–229.

[33] Zhu, K., Jiang, F., Li, J., and Baoyin, H. (2009) Trajectory Optimization of Multi-Asteroids Exploration with Low Thrust, *Transactions of the Japan Society for Aeronautical and Space Sciences*, **52**, No. 175, 47–54.

[34] Rosa Sentinella, M., and Casalino, L. (2009) Cooperative Evolutionary Algorithm for Space Trajectory Optimization, *Celestial Mechanics and Dynamical Astronomy*, **105**, No. 1–3, 211–227.

[35] Michalewicz, Z., and Schoenauer, M. (1996) Evolutionary Algorithms for Constrained Parameter Optimization Problems, *Evolutionary Computation*, **4**, No. 1, 1–32.

[36] Prasad, B. (1981) A Class of Generalized Variable Penalty Methods for Nonlinear Programming, *Journal of Optimization Theory and Applications*, **35**, No. 2, 159–182.

[37] Fletcher, R. (1987) *Practical methods of optimization*, Wiley, Chichester, UK.

[38] Koziel, S., and Michalewicz, Z. (1999) Evolutionary Algorithms, Homorphous Mappings, and Constrained Parameter Optimization, *Evolutionary Computation*, **7**, No. 1, 19–44.

[39] Szebeheli, V. (1967) *Theory of Orbits*, Academic Press, New York, NY.

[40] Richardson, D. L. (1980) Analytical Construction of Periodic Orbits About the Collinear Points, *Celestial Mechanics and Dynamical Astronomy*, **22**, No. 3, 241–253.

[41] Broucke, R. A. (1968) *Periodic Orbits in the Restricted Three-Body Problem With Earth-Moon Masses*, JPL Technical Report 32-1168, Pasadena, CA.

[42] Prussing, J. E. (1969) Optimal Four-Impulse Fixed-Time Rendezvous in the Vicinity of a Circular Orbit, *AIAA Journal*, **7**, No. 5, 928–935.

[43] Prussing, J. E. (1970) Optimal Two- and Three-Impulse Fixed-Time Rendezvous in the Vicinity of a Circular Orbit, *AIAA Journal*, **8**, No. 7, 1221–1228.

[44] Prussing, J. E., and Chiu, J.-H. (1986) Optimal Multiple-Impulse Time-Fixed Rendezvous Between Circular Orbits, *Journal of Guidance, Control, and Dynamics*, **9**, No. 1, 17–22.

[45] Colasurdo, G., Pastrone, D. (1994) *Indirect Optimization Method for Impulsive Transfer*, AIAA 94-3762 Proceedings of the AIAA/AAS Astrodynamics Conference, Scottsdale, AZ.

[46] Prussing, J. E., and Conway, B. A. (1993) *Orbital Mechanics*, Oxford University Press, New York, NY.

Index

Adjoint:
  variable, 3, 8, 9, 72, 74–76, 285, 290
  equations, 18, 159, 164, 285
Ant colony optimization, 12, 263
Arc
  coast, 9, 72–75, 81, 256–261
  constrained, 7, 38, 139, 146–152
  finite burn, 80, 81, 85
  intermediate thrust, 20
  Keplerian, 279–282
  maximum thrust, 20, 72–75
  no-thrust, 24
  unconstrained, 139
Argument of perigee, 57
Automatic node placement, 70, 71, 78

Bang-bang, 19
Basin–hopping algorithm, 193
Bellman's principle, 7, 8, 210
Betts, J., 3, 13
Birkhoff's equations, 271
Bolza form, 4
Boundary condition, 11
Branch and bound, 203

Calculus of variations, 3, 4, 16, 18, 35, 112, 177, 239
Canonical units, 21, 59, 62, 73, 270
Canonical form, 49
Cardinal node, 43, 50
Cassini1 problem, 187
Central difference approximation, 91
Cluster pruning, 194, 195
Coast:
  arc, 37, 72–75, 89, 108, 259–261
  initial, 24, 26, 28, 35, 73, 99
  terminal, 26, 31, 34
Collocation, 7, 9, 10–15, 38–40, 50, 75–78, 114, 248, 251, 262
  points, 249

Coordinates:
  Cartesian, 4, 52, 53, 67, 179, 180, 207
  cylindrical, 52
  Earth-centered inertial, 56
  polar, 52
  synodic, 269
Coordinate transformation, 37, 56, 59
Copernicus, 79
Corner, 146, 149
Costate
  equations, 124
  variable, 7, 38, 108, 112, 116, 120, 124, 130
  vector, 87–90, 108, 124, 137

Deep Space, 1 2, 239
Defect, 41–46, 58, 63, 66 72, 73, 75
  scaled, 66
Delaunay variables, 4, 40, 52
Differential evolution, 12, 178, 192, 203, 215
Direct collocation, 13, 251
Direct transcription, 8, 39, 47, 52, 56, 58, 60, 62, 72, 261
Direct method, 3, 9, 10, 12, 16, 47, 65, 113, 239

Edelbaum equation, 161
Electric propulsion, 2, 37, 112, 138, 262
Elliptic elements, 4, 40, 49, 52, 53
Ephemeris, 1, 57, 85
Ephemerides, 187, 204, 208
Equilibrium point, 238, 239, 245
Epoch, 80, 94, 97, 186–188, 197–199, 208
  reference, 80, 99
Equinoctial elements, 4, 49, 52–59, 242–244, 257
Euler-Lagrange equations, 8, 37, 74, 76, 125, 142, 143, 146, 147

Evolutionary algorithm, 3, 9, 13, 64, 203,
   215, 236, 237, 263, 293
Exhaust velocity, 17, 60, 62, 73, 82, 85, 88,
   112, 285

Forward difference approximation, 91
Frame
   base, 84, 95
   function output, 84
   impulsive maneuver, 84
   propagation, 84
   state input, 84
   steering, 86
Free-return, 80, 93, 94, 96

Galileo spacecraft, 2, 60
Gauss–Lobatto, 9, 13, 43, 71, 75, 76, 249–251
Genetic algorithm, 3, 10, 64, 178, 190, 192,
   203, 263, 264
Global
   optimization, 178, 190, 195, 202, 214, 265
   convergence, 215, 217
Gravity:
   assist, 1, 60, 178, 194, 204–207, 221–223
   constant, 60, 142, 153, 170
   field, 17, 23, 64
   gradient, 18
   loss, 74, 105
   universal constant of, 85
Grid refinement, 40, 67–71
GTOC1 problem, 187

Hamiltonian, 5, 6, 9, 18–20, 23, 35, 76, 88,
   116, 119, 142, 155, 171, 285
   augmented, 145
   constancy of, 6, 172
Hamilton–Jacobi equation, 3, 7
Hermite–Simpson, 9, 12, 70, 73, 75
Hohmann transfer, 3, 4, 28, 35, 36, 223, 229,
   277–278

Impulse, 6, 16, 81
Impulsive maneuver, 2, 81, 181
Indirect method, 3, 6, 8, 12, 16, 38, 112,
   124, 239
Initial guess, 12, 62–65, 73, 101, 124–126,
   129, 164, 171, 239, 250–255, 258
Integration
   explicit, 51, 67
   implicit, 8, 9, 12, 39, 41, 66, 67, 70, 250
   numerical, 8, 63, 89, 115
Interplanetary superhighway, 238, 260
Interpolation error, 68, 70
Invariant manifold, 2, 260
Island model, 193

Jacobi constant, 271, 272, 275
Jacobian, 48, 91, 92
J_2, 141, 158, 169
Jump condition, 146

Keplerian, 2, 4, 279–283
Kepler's
   equation, 53, 54, 280
   problem, 184
Knot, 58–60
Kuhn-Tucker multipliers, 9, 72

Lagrange
   coefficients, 184, 185
   multiplier, 3, 38, 87, 136, 140, 141, 145,
      171, 290
   points, 238, 245, 246, 265, 269,
      271, 272
Lagrange's planetary equations, 114, 140
Lambert's problem, 10, 25, 30, 182, 185, 205,
   208, 233, 280–282
Lawden spiral, 12
Legendre–Gauss–Lobatto point, 43
Longitude:
   eccentric, 54
   mean, 53–55
   of ascending node, 57, 115, 137, 254–257
   true, 54, 59
Low-energy transfer, 2, 260, 262
Lyapunov orbit, 265, 269, 272–274, 290

Manifold, 2, 13, 239, 246, 247, 253–262
Mass
   discontinuity, 81
   flow rate, 17, 107
Mayer form, 87
Mean motion, 52, 53, 137, 142, 240
Mini-max problem, 92, 110
Minimum:
   global, 10, 11, 13, 65, 209, 215, 217, 251,
      272, 275
   local, 6, 10, 13, 40, 72, 92, 211, 217, 291
Monodromy matrix, 246, 247
Monotonic basin hopping, 190, 203, 214
Multi-agent collaborative search, 211
Multi-start technique, 203, 214, 217, 218,
   226, 235

NEAR spacecraft, 2
Necessary conditions, 3–6, 8, 9, 11, 12, 16–21,
   31, 37–39, 72, 85, 87, 113, 116, 136,
   140, 239, 285–288, 290
Node, 40–51, 80–83, 129–134, 248–258
Noncontemporaneous, 27

Nonlinear programming problem, 3, 8, 42–45, 72, 92, 207, 239, 266
Nonsingular elements, 52
NSTAR engine, 62

Objective function, 60–62, 72, 80, 90–97, 103, 107–109, 183, 187–190, 207–212, 224, 233, 247, 255, 263–269, 285, 290
Oblateness, 55, 63, 114–116, 137
Optimal steering law, 117, 118
Optimality condition, 89, 117, 136, 145, 164, 171
Orbital
    elements, 52, 53, 73, 84, 93, 95, 115, 116, 119, 120, 124, 137, 242
    averaging, 113, 115–119, 124, 136
Osculating, 84, 114, 115, 120, 125
OTIS, 9, 67

Parameter space, 11
Particle swarm optimization, 11, 178, 190, 192, 200, 263–265, 290–292
Penalty function, 11, 251
Periodic orbit, 79, 238–240, 245–247, 250–262, 269, 271, 272, 275
Perturbation, 1–3, 6, 31, 33, 37, 53, 55, 59, 116, 141, 158, 159, 242, 244, 245, 247
Poincare', H., 238, 245
Pontryagin minimum principle, 5, 8, 37, 286
Population, 10, 11, 64, 192, 211, 212, 215, 264, 266, 267
Power-limited, 22, 61
Primer:
    vector, 3, 6, 9, 16, 19, 21, 23, 26, 30, 32, 33, 72, 88, 108
    vector equation, 19, 23
Propellant:
    consumption, 17, 79, 265
    flow rate, 17, 49
    mass fraction, 255, 258, 260
    mimimizing, 1, 16, 17, 60, 113, 132–136,
Pseudospectral method, 9, 12, 39, 45–47, 50, 51

Quasi-circular, 112, 118, 119, 124–126

Rendezvous, 7, 18, 25, 28, 53, 64, 73, 180, 265, 277–284, 290
Resonant:
    swing-by, 206, 221
    orbits, 229, 230, 233
Robustness, 8, 39, 52, 53, 59, 110, 235, 252
Rocket: 1, 2, 20, 37, 60, 97
    equation, 22, 60, 82, 104, 105, 126
Rosetta problem, 188

Runge–Kutta,
    integration, 3, 39, 51, 175
    parallel shooting, 47–49, 75, 250, 254, 255, 257, 258

Scaling: 59, 65–67, 68, 108
    factor, 67, 172
Segment: 40–51, 65–71, 80–90, 93–104, 202, 248–251
    length, 43, 44, 69, 71
Semimajor axis, 53, 54, 59, 142, 154–158
Semiminor axis, 55
SEPSPOT, 112, 124–126
Shape-based trajectory, 3, 10, 64
Shooting method, 8, 12, 38, 48, 112, 124, 246
Simulated annealing, 178, 192, 195, 199
Singular arc, 20
Solar radiation pressure, 2, 55, 85
Solar sail spacecraft, 4, 62
Space pruning, 178, 186, 194
Specific impulse: 17, 60, 62, 97, 112, 113, 117, 134, 136, 137, 179, 238
    constant, 17, 97
    variable, 22, 23, 134
Sphere of influence, 40, 56, 59, 60, 180
State variable, 3, 6, 7, 17, 18, 38, 39, 47, 48, 52, 56, 58, 66, 83, 140, 142, 145, 146, 251
State vector, 4, 17, 48, 73, 80, 82–90, 93, 114, 115, 137, 179, 206, 207, 250
Static/Dynamic control, 7
Steering angle, 116, 117, 119, 124, 137
Steering program: 120
    anti-tangent, 132
    pitch, 119–123
    thrust, 124
    tangent, 120, 126
    yaw, 123, 124
Switching function, 6, 19, 72, 74, 89, 108
Sufficient conditions, 16, 21, 23, 36

Tangent steering, 120, 126
Three body problem, 2, 4, 52, 59, 79, 239, 240, 241, 250, 251, 259, 260, 269–271, 274, 290
Third-body, 55, 242, 269, 271, 275
Thrust:
    acceleration, 5, 17–19, 55, 61, 63, 116, 117, 123, 136, 141, 143, 170, 242, 254
    arc, 8, 10, 20, 37, 39, 108, 256–261
    continuous, 1, 2, 61, 72, 112, 113, 118, 123
    direction, 5, 6, 16, 18, 21, 63, 85–87, 89, 106, 116, 119, 120, 123, 137
    low, 2–4, 6, 7, 10, 12–15, 37, 48, 49, 52, 53, 60–64, 80, 87, 97, 112–115, 130, 132, 136, 181, 194, 238–244, 260, 261, 284, 290

maximum, 20, 72, 97, 254, 255
tangential, 6, 255, 257, 258
velocity-pointing, 255
vector, 23, 49, 86, 87, 90, 98, 107, 108, 116,
  124, 140, 172
Tisserand graph, 203
Transition matrix, 24, 26, 36, 246
Transversality condition, 87, 90, 147, 160,
  286, 287, 290
Trapezoid rule, 9, 12, 70, 125

Tsiolkovsky equation (cf. rocket
  equation), 60
Two-point-boundary-value problem, 5, 38,
  116, 136, 182, 286

Variation matrix, 246, 247

Zero-velocity curve, 271

Printed in the United States
By Bookmasters